Stimuli-responsive Drug Delivery Systems

Biomaterials Science Series

Editor-in-chief:
Roger Narayan, *University of North Carolina and NC State University, USA*

Series editors:
Pankaj Vadgama, *Queen Mary University of London, UK*
Nan Huang, *Southwest Jiaotong University, China*

Titles in the Series:
1: Stimuli-responsive Drug Delivery Systems

How to obtain future titles on publication:
A standing order plan is available for this series. A standing order will bring delivery of each new volume immediately on publication.

For further information please contact:
Book Sales Department, Royal Society of Chemistry, Thomas Graham House, Science Park, Milton Road, Cambridge, CB4 0WF, UK
Telephone: +44 (0)1223 420066, Fax: +44 (0)1223 420247,
Email: booksales@rsc.org
Visit our website at www.rsc.org/books

Stimuli-responsive Drug Delivery Systems

Edited by

Amit Singh
AllExcel Inc., USA
Email: nanobioamit@gmail.com

and

Mansoor M. Amiji
Northeastern University, USA
Email: m.amiji@northeastern.edu

ROYAL SOCIETY OF CHEMISTRY

Biomaterials Science Series No. 1

Print ISBN: 978-1-78801-113-6
PDF ISBN: 978-1-78801-353-6
EPUB ISBN: 978-1-78801-466-3
Print ISSN: 2397-1401
Electronic ISSN: 2397-141X

A catalogue record for this book is available from the British Library

The Royal Society of Chemistry is a charity, registered in England and Wales, Number 207890, and a company incorporated in England by Royal Charter (Registered No. RC000524), registered office: Burlington House, Piccadilly, London W1J 0BA, UK, Telephone: +44 (0) 207 4378 6556.

For further information see our web site at www.rsc.org

Printed in the United Kingdom by CPI Group (UK) Ltd, Croydon, CR0 4YY, UK

Preface

Advances in material science and nanotechnology have revolutionized several fields, but the single and most significant impact has been in the field of medicine and diagnostics. With a deeper understanding of complex medical conditions, such as cancer, there is a paradigm shift in treatment approaches and the current development in the field focuses on patient-centric "*personalized therapy*" instead of the prior concept of a "*magic bullet*". Cancer cells and the tumor microenvironment exhibits diverse cellular, molecular and physiological features that vary not only with the type of cancer, but also from patient to patient within the same type of cancer. As the diseases continue to evolve and adapt to the existing therapeutic options, so have the efforts to devise novel therapeutic strategies through scientific innovations to circumvent these challenges. Even though the molecular fingerprint of diseases such as cancer remains unique from patient to patient, the physiological microenvironment of the disease in general exhibit similarities in properties. Therefore, the past decade has seen efforts to exploit the subtle differences in the physiological profile of the diseases and develop materials that are sensitive to these triggers. This book specifically focuses on the development of such smart drug delivery options, based on the sound knowledge of disease profile and precision in engineering of materials to achieve the desired physico-chemical properties.

In the book *Stimuli-Responsive Drug Delivery Systems*, the first chapter lays down the ground rules for the need for smart stimuli-responsive materials for drug and gene delivery and introduces the fundamental concepts involved in the design of such materials. The chapter is aims to educate the reader about the basic physiological differences at a disease site, as well as the essential concepts involved in designing vectors for delivery of chemical drugs, as well as complex biological therapeutic molecules such as DNA, RNA and protein.

Biomaterials Science Series No. 1
Stimuli-responsive Drug Delivery Systems
Edited by Amit Singh and Mansoor M. Amiji

Published by the Royal Society of Chemistry, www.rsc.org

This chapter is followed by a detailed account of the chemical manipulations that are popularly used for designing smart materials. The second chapter discusses the choice of materials available for drug delivery and the chemical modifications that impart the triggered release property to the materials.

Building on the backbone of the first two chapters, chapters 3 to 12 discuss different classes of stimuli-responsive materials that have been developed for drug delivery applications. These chapters provide an in-depth assessment of the state-of-the-art materials that have been studied for triggered release of material in different categories with adequate detailed examples from preclinical or clinical settings. The focus of these chapters has been to provide a comprehensive knowledge about the respective fields through the developments made in the past decade. Readers will therefore find an updated information pool on the subject matter consolidated by the experts in the field, which is otherwise scattered into numerous individual scientific reports. Chapter 13 of the book is dedicated specifically for the discussion of devices that show stimuli-responsiveness and have potential application in drug delivery.

The final chapter of the book introduces the regulatory perspective on the challenges in development of these materials. Any material designed for therapeutic use in humans has to undergo very stringent selection criteria for safety and efficacy and regulatory agencies across the globe have guidelines that have to be met before an experiment material becomes a "*drug*". This chapter outlines some major impediments that are encountered along the way of drug development involving smart materials. This chapter is unique to this book and is an aspect of drug development that is seldom discussed in other books on similar themes. This chapter is an attempt to make the readers aware of the problems and pitfalls that often leads to an unsuccessful story of an otherwise promising "*drug candidate*".

We hope that the book *Stimuli-Responsive Drug Delivery Systems* will provide the reader with a flavor of the subject matter and incite a deeper interest in the multidisciplinary and highly active research field of nanomedicine, nanotechnology and advanced drug delivery systems. This book is a product of the shared enthusiasm of the publishers, editors and the authors that there is a need for a consolidated material on the topic of stimuli-responsive materials and we hope that the book will serve as a valuable resource to the interested scientific community.

Amit Singh
West Haven, Connecticut, USA

Mansoor M. Amiji
Boston, Massachusetts, USA

Dedication

I dedicate this book to my family - my parents, sisters and brother. I would also dedicate this book to my friends and colleagues for their support and encouragement through the years.

Amit Singh

I dedicate this book to my family - my lovely wife and our three wonderful daughters. I also dedicate this book to my past and present postdoctoral associates and graduate students, who have made tremendous contributions to the research success of my group.

Mansoor M. Amiji

Biomaterials Science Series No. 1
Stimuli-responsive Drug Delivery Systems
Edited by Amit Singh and Mansoor M. Amiji

Published by the Royal Society of Chemistry, www.rsc.org

Contents

Biomaterials Science Series No. 1
Stimuli-responsive Drug Delivery Systems
Edited by Amit Singh and Mansoor M. Amiji

Published by the Royal Society of Chemistry, www.rsc.org

CHAPTER 1

Fundamentals of Stimuli-responsive Drug and Gene Delivery Systems

VLADIMIR P. TORCHILIN

Northeastern University, Department of Pharmaceutical Sciences, Center for Pharmaceutical Biotechnology and Nanomedicine, 360 Huntington Ave, 02115, Boston, 02151, USA
*E-mail: v.torchilin@northeastern.edu

1.1 Introduction

The use of drug-loaded nanoparticulate pharmaceutical carriers (drug delivery systems, DDS) can overcome many of the limitations characteristic of free therapeutic entities, such as low stability, fast inactivation or degradation *in vivo*, non-specific toxicity, poor solubility, unfavorable pharmacokinetics, and poor biodistribution. Nanocarriers can be designed/engineered that may entrap both hydrophilic and hydrophobic drugs, increase their stability and longevity *in vivo*, provide controllable drug release, change drug pharmacokinetics in any required way (in other words, drug's pharmacokinetics can be replaced with the specifically designed pharmacokinetics of the pharmaceutical carrier/delivery system), minimize undesirable side effects, and allow targeting of drugs to pathological organs and tissues of interest, or even to individual cells and intracellular organelles.[1–3] Currently, there

Biomaterials Science Series No. 1
Stimuli-responsive Drug Delivery Systems
Edited by Amit Singh and Mansoor M. Amiji

Published by the Royal Society of Chemistry, www.rsc.org

exists a broad variety of nanoparticulate carriers meeting different practical requirements, which includes liposomes, polymeric micelles, polymeric nanoparticles, niosomes, solid lipid nanoparticles, dendrimers, nanoemulsions, inorganic nanoparticles, nanoshells, nanotubes, and many others.[1,4–7]

If required, nanoparticulate DDS can be specifically engineered to selectively change certain parameters/functions (for example, enhance drug release or intracellular uptake) in response to certain intrinsic stimuli characteristics of pathological tissues or external stimuli applied from the outside of the body.[4,8–11] The intrinsic/internal stimuli that are characteristic for the pathological areas (tumors, infarcts, athrosclerotic lesions, sites of infection or areas of transplant rejection) make those areas different from normal tissues, include local changes of different intensity in pH,[12,13] temperature,[14–16] redox conditions,[17,18] and the expression of certain biologically/enzymatically active molecules.[19,20] External stimuli include magnetic field, heat, light (including laser beams), and ultrasound, see Figure 1.1.[8,11]

A better understanding of local changes in the microsurroundings of pathological areas, including tumors, has allowed for the development of novel materials and engineering principles in order to build DDS capable of specifically reacting to certain "abnormal" conditions of affected tissues.[21] Thus, for example, many lipids and polymers can undergo certain changes in their physical–chemical properties as a response to even insignificant

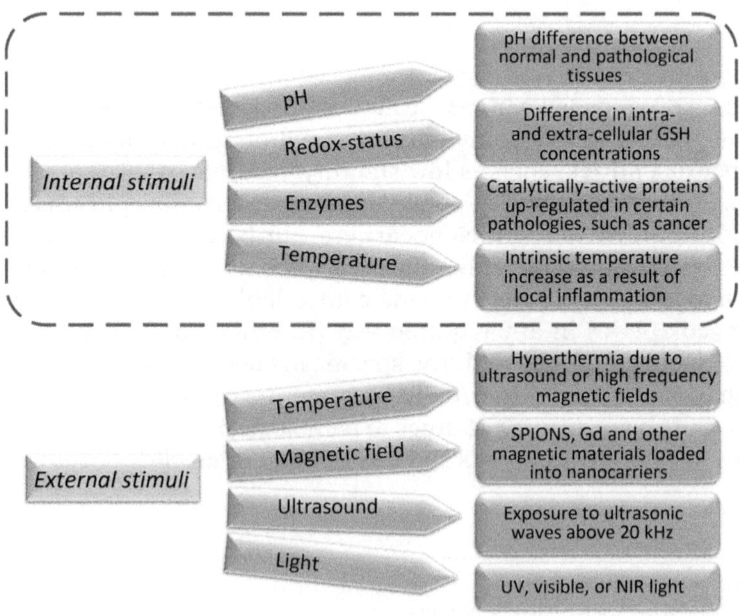

Figure 1.1 Types of possible stimuli and their origin.

changes in pH or temperature, which can be used to engineer smart DDS.[22] Upon exposure to intrinsic or external stimuli, such stimuli-sensitive DDS switch on certain functions controlling drug release or the efficacy of the intracellular uptake/penetration.[4] Drug release can be controlled by the structural changes in a microcontainer type DDS loaded with the drug (such as carrier degradation or an increase in its permeability) or by the cleavage of chemical bonds used to attach the drug to the carrier. The specific biochemistry of various affected areas helps to engineer stimuli-sensitive DDS responding to these conditions, which can allow for precise temporal and spatial drug delivery improving its effectiveness and minimizing off-target effects.[4,23,24]

We will present here an overview of stimuli-sensitive DDS. Redox potential, pH, hypoxia, and overexpression of certain catalytically-active molecules represent a group of intrinsic factors/stimuli characteristics of affected/pathological areas/tissues in the body. Magnetic field, light, and ultrasound belong to a group of external stimuli. Temperature can serve as both an intrinsic stimuli (local temperature increase characteristic of the inflamed areas) and as an external stimuli if applied from outside the body, or generated inside a certain area in the body under the action of external stimuli (such as high frequency magnetic field).

1.2 pH-Sensitive DDS

Low pH is a well-established property of tumors and inflamed and hypoxic areas. In 1930, Warburg and co-workers showed that tumor cells preferentially convert glucose and other substrates to lactic acid (pK 3.7), even under aerobic conditions, and it was thought that lactic acid was the main source of tumor acidity. However, it is now known that the decreased pH is also related to the high levels of CO_2 and the increased expression and activity of vascuolar type proton pumps.[25–27] Deficiencies in tumor perfusion, hypoxia, and metabolic abnormalities associated with uncontrolled cell growth and proliferation (tumorigenesis) also contribute to the low pH of tumors.

A pH-sensitive DDS has to protect the loaded drug when in the blood (pH 7.3–7.4) and release the drug or expose some of the special functions (such as cell penetrating ability) when in the interstitial space of tumors (pH 6.8–7.2) or in the intracellular compartments, such as cytoplasm (pH slightly above 7.0), endosomes (pH 5–6), and lysosomes (pH 4.5–5.5).[28–30] This allows for the massive drug release, which is especially important in tumor treatment, as if high concentrations of free cytotoxic drug appear intracellularly, it is possible to exceed the efflux capacity of drug transporters (such as P-gp) and kill the tumor cells efficiently. Drug release can also occur in slightly acidic extracellular fluids of tumor tissue after tumor accumulation *via* the enhanced permeability and retention (EPR) effect. In endosomes or lysosomes, which have an even lower pH value than extracellular fluids, pH-triggered drug release proceeds with high efficiency when assisted by hydrolysis (with enzymes such as catepsin B).[31,32]

DDS for pH-triggered drug release usually contain protonizable components (amino, imidazolyl, and carboxyl groups), which at acidic pH values can induce the DDS destabilization and drug release as a result of protonation. Thus, adriamycin-loaded poly(L-histidine)-*b*-PEG-containing mixed micelles are non-protonated and stable at pH 7.4, at pH 6.6 they undergo protonation of the histidine residues, which destabilizes the micelle structure and results in burst release of a loaded anticancer drug.[33] pH-sensitive polymers (containing acidic–carboxylic or sulfonic groups or basic ammonium salts groups) undergo pH-sensitive conformational changes resulting in dissociation, destabilization (such as collapsing or swelling), or changes in the partition coefficient drug/vehicle.[34] DDS with acid-labile chemical bonds are also used for pH-triggered drug release. Chemical bonds, such as the ortho ester bond, the acetal/ketal bond, and the hydrazone bond can break down in acidic conditions. When a drug is bound to the DDS *via* such bonds, it will be released upon acidification.[34,35] Doxorubicin conjugated onto hydrophobic segments of the folate-modified amphiphilic hyperbranched block-copolymer, H40-P(LA–DOX)-*b*-PEG–OH/FA, *via* the acid-labile hydrazone linkage were released rapidly at acidic pH typical of the tumor microenvironment and endocomes of tumor cells.[35] CO_2-generating pH-responsive DDS represents an interesting strategy enabling the intracellular drug release in lysosomes. In a study by Liu and coworkers, the bicarbonate ion was encapsulated in the liposomes composed of HSPC, Chol, and $mPEG_{2000}$–DSPE. In an acidic medium (pH 5.0), the liposomes released CO_2 gas with the formation of carbonic acid and corresponding hydronium ions, which led to fast drug release from the liposomes. This allowed the DDS to overcome the resistance of breast cancer cells (MCF-7R) to doxorubicin.[36]

An interesting approach to pH-mediated drug delivery suggests the use of charge-conversional or charge-reversal DDS,[37,38] which are negatively charged at neutral pH and convert their surface charge into a positive one in response to the lowered pH. Hybrid calcium phosphate (CaP)/charge-conversional polymer nanopreparations were loaded with siRNA against VEGF and tested against pancreatic cancer showing high gene silencing efficiency in a BxPC3 tumor model.[38] These DDS contained a block copolymer of PEG and the endosomolytic polyanionic poly(*N'*-{*N'*-[(*N*-*cis*-aconityl)-2-aminoethyl]-2-aminoethyl}aspartamide) (PAsp), which is stable at pH 7.4, but degrades at pH 5.5 inside endosomes or lysosomes due to cleavage of the *cis*-aconitic amide bonds, reverting back to the parent polycation.

A pH-sensitive micellar DDS for the delivery of plasmid DNA was engineered by the conjugation of phospholipids, such as phosphatidyl ethanolamine (PE), to low molecular weight polyethylene imine (PEI), and such PEI–PE conjugates effectively complexed DNA.[39] The PEI–PE/DNA complexes were mixed with low pH-degradable PEG–hydrazone–PE co-polymer and

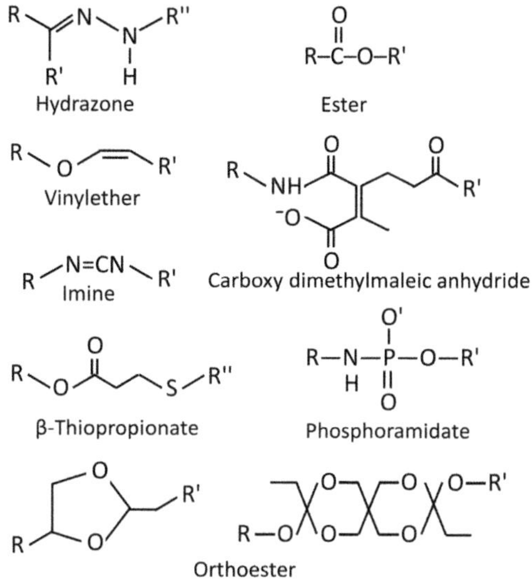

Figure 1.2 Some examples of pH-sensitive bonds/groups.

formed DDS with high transfection efficacy at acidic pH values due to PEG detachment and efficient uptake of remaining positively charged DNA complexes by cells.

Another strategy is the use of detachable functions or coatings of DDS, such as DDS with a protective polyethylene glycol (PEG) cover. Detachment of the PEG coat upon exposure to acidified pH could be very important, since the PEG coat may interfere with DDS interaction with the cell and/or retard the release of the encapsulated drug/DNA. pH-sensitive links (diorto esters, vinyl esters, cystein-cleavable lipopolymers, double esters, and hydrazones) can be used to attach PEG to the surface of DDS, which degrade fast at lowered pH eliminating the PEG cover and allowing for efficient intracellular delivery.[40]

See some examples of pH-sensitive bonds/linkers are shown in Figure 1.2.

1.3 Redox Potential-sensitive DDS

Glutathione (GSH) is a crucial tripeptide generated in the cell cytoplasm, which acts as the main reducing agent in cells facilitating the thiol-disulfide exchange reaction.[41,42] The intracellular concentration (2–10 mM) is especially prominent in certain regions, such as the cytosol, mitochondrion, and nucleus and is dramatically higher than in the blood (about 2–20 μM), where the disulfide exchange reactions are minimal.[43,44] The GSH/glutathione disulfide (GSSG) redox couple imparts the cells with their oxidative property and is kept under a reductive condition by NADPH and glutathione reductase. Other redox couples also contribute towards the intracellular levels of

GSH, namely $NADH/NAD^+$, $NADPH/NADP^+$ and thioredoxinred/thioredoxinox. In addition to the cytosolic reducing conditions the endosome and the lysosome also have a redox potential that can be used to promote endosomal and subsequent lysosomal escape.[45] The redox potential in the endosome is modulated by a specific reducing enzyme, gamma interferon-inducible lysosomal thiol reductase (GILT) in the co-presence of reducing agents such as cysteine and not GSH, while the lysosome's reductive environment is controlled by high amounts of cysteine-like thiols.[45,46] Interestingly, in a tumor mass, the GSH concentration is much higher than in normal tissues, and this difference makes GSH a promising candidate stimulus for redox-mediated drug delivery.

The reduction of the disulfide bonds in redox-sensitive DDS results in a rapid disruption of the nanocarrier, as well as an increase in the cytotoxic activity of therapeutics, first of all, in cancer.[42,46–50] The rate and efficacy of the reduction depends on many factors including the exact chemical structure of the bond. Poly(L-lysine)-based DNA-loaded DDS were developed by conjugating poly(L-lysine) through an N-terminal cysteine to a PEG chain by the disulfide bond or thioether bond. The nanoparticles containing the hindered disulfides (thioester) were not cleaved in the extra-cellular space of the two different cancer cell lines (HeLa and HuH7), while the nanoparticles with the unhindered bonds (disulfide) underwent the cleavage. The rate of dePEGylation was shown to be cell line dependent, suggesting different amounts of thiols are secreted from different cell types.[51]

Redox-responsive polyphosphate nanosized assemblies based on amphiphilic hyperbranched multiarm copolyphosphates (HPHSEP-*star*-PEPx) with redox responsive backbone were prepared, and the GSH-mediated intracellular drug delivery was investigated in the HeLa cells. The results suggested that DOX-loaded HPHSEP-*star*-PEPx micelles show higher cell proliferation inhibition against glutathione monoester pretreated HeLa cells compared to the nonpretreated ones.[52] A redox-responsive micellar nanopreparation assembled from the single disulfide bond-bridged block polymer of poly(ε-caprolactone) and poly(ethyl ethylene phosphate) (PCL–SS–PEEP) achieved more drug accumulation and retention in MCF-7/ADR breast cancer cells. The system rapidly released its DOX payload in response to the intracellular reductive environment and also significantly enhanced the cytotoxicity of doxorubicin to MCF-7/ADR cancer cells.[53]

Cho and coworkers prepared redox-sensitive polymeric nanoparticles from a monomer containing TMBQ (trimethyl-locked benzoquinone) as the redox-sensitive group. TMBQ is a substance that forms a lactone *via* intramolecular cyclization by a two-electron reduction, either chemically or enzymatically. Paclitaxel was loaded into these polymeric nanoparticles and a triggered release of paclitaxel was observed by sodium dithionite-mediated reduction.[54]

Redox-responsive DDS have been used widely for intracellular delivery of siRNA and genes. Conjugation of anti-GFP siRNA to the cationic lipid DOPE *via* a disulfide bond resulted in a product, which could be incorporate into PEG_{2000}–PE micelles. The resultant nanosized micelles effectively protected siRNA against nuclease-mediated degradation and released siRNA in the presence of 10 mM GSH; a 30-fold higher GFP downregulation was obtained compared to free siRNA.[55] Furthermore, survivin siRNA was conjugated to phosphothioethanol (PE) *via* a disulfide bond and the resulting siRNA–S–S–PE conjugate was incorporated into PEG_{2000}–PE micelles. A significant decrease of cell viability and a down-regulation of survivin protein levels were achieved after the treatment with survivin siRNA micelles in several cancer cell lines. The down-regulation of survivin provided a significant sensitization of the cells to paclitaxel in both sensitive and resistant cancer cell lines.[56]

Bioreducible cationic lipids or polymers containing disulfide bonds, such as cholesterol disulfide cationic lipids,[57] branched poly(disulfide amine),[58] hyperbranched poly(amido amine) (PAMAM),[59,60] and disulfide based polyethyleneimine (SS–PEI),[61] have been synthesized. Redox-sensitive micelles containing the above or similar component have been successfully used for the delivery of genes, anti-sense oligonucleotides, and siRNA.[62–64] Bioreducible, non-viral carriers were suggested for the intracellular delivery of siRNA against human telomerase reverse transcriptase.[61] The authors used a biodegradable PEI of ~800 Da, containing multiple disulfide bonds (SS–PEI), which effectively binds siRNA to form nano-sized positively charged complexes that easily destabilize and release siRNA in a reducing environment both *in vitro* and *in vivo*.

In addition to systems that respond to a reductive environment for the delivery of cargo, nanopreparations that are sensitive to hypoxic conditions (such as in certain tumor areas and in infarct zones) are also being developed. Perche *et al.* incorporated azobenzene as a hypoxia-responsive bioreductive linker between PEG and PEI in lipid–PEI–PEG copolymer micelles loaded with siRNA for hypoxia-specific detachment of PEG and effective uptake of remaining positively charged siRNA complexes with PEI by cells.[65]

Some examples of redox-sensitive bonds/linkers are shown in Figure 1.3.

1.4 Enzyme-sensitive DDS

Many efforts have been made to understand the molecular events occurring during carcinogenesis and the signaling pathways participating in cancer progression. The interactions between the tumor cells and the tumor microenvironment play a pivotal role in this process. For accelerated growth and proliferation, cancer cells overexpress certain molecules, including proteins with enzymatic activity, which can be considered as cancer markers or local stimuli making a tumor microenvironment different from a normal one.

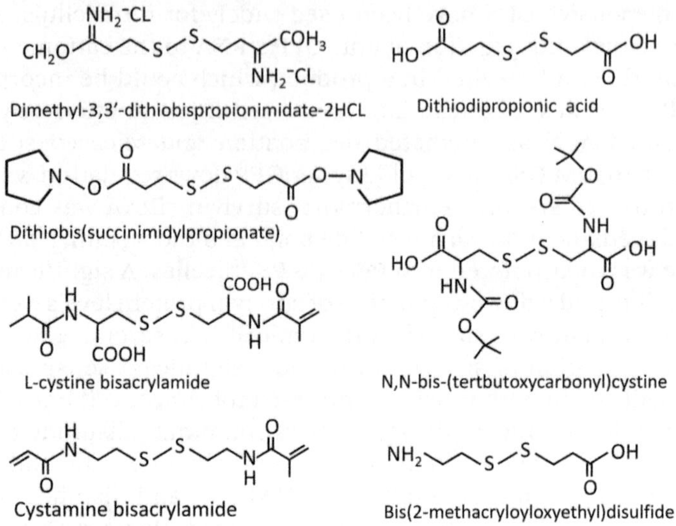

Figure 1.3 Some examples of redox-sensitive bonds/groups.

Thus, DDDS is sensitive to the cation of locally overexpressed enzymes typical for many cancers and may represent a promising strategy for drug delivery to tumors.[66]

Matrix metalloproteinases (MMPs), are the main mediators of the changes in the microenvironment which exist during cancer progression. MMP activity, especially MMP-2 and MMP-9, contributes to tumor growth and differentiation, the multistep processes of tumour angiogenesis, invasion and metastasis including proteolytic degradation of the extracellular matrix (ECM), alteration of the cell–cell and cell–ECM interaction, migration and angiogenesis.[67–70] The alterations in certain local enzyme expression and composition (such a MMPs) can be not only considered as biomarkers for cancer diagnosis and prognosis, but also provide an opportunity to design DDS that are capable of releasing their drug loads *via* enzyme-triggered mechanisms at the tumor site. Thus, paclitaxel was conjugated to the octapeptide Gly-Pro-Leu-Gly-Ile-Ala-Gly-Gln known to be selectively and efficiently hydrolyzed by the MMP2 yielding the PTX–AcG conjugate. As a result, PTX concentrations needed to arrest the cell cycle in the G2/M phase were 100-fold lower when PTX–AcG was used compated to the free drug. *In vivo*, tumor-bearing mice treated with PTX–AcG had more and larger areas of necrosis and fewer proliferating cells in tumor sections than mice treated with free PTX, confirming the activation of the conjugate by MMP2 at the tumor sites.[71] In another study, a self-assembling, MMP2-sensitive, TATp-modified micellar nanopreparation composed of the paclitaxel prodrug (PEG_{2000}–MMP2 clevable peptide–PTX) and two other polymers, TATp–PEG_{1000}–phosphoethanolamine (PE) (a cell-penetrating enhancer) and PEG_{1000}–PE (a nanocarrier building block) was formulated in an aqueous environment with high PTX loading. PTX was located in the hydrophobic "core" of the mixed polymeric micelle, and covered and protected by a hydrophilic PEG shell providing prolonged circulation and good tumor accumulation *via* the EPR effect. When inside the MM2-overexpressing tumor, the peptide link between the drug and PEG was cleaved, the cell-penetrating function become exposed and effectively brought the drug inside cancer cells resulting in high anticancer activity and low side effects.[72]

MMP-cleavable substrates (typically, cleavable peptides) have been used for the delivery of multiple drugs and imaging agents.[73–75] A functionalized gold nanoparticle (AuNP) for tumor imaging and drug delivery was recently reported, where DOX was conjugated to the AuNPs by a thiol–Au bond with a MMP-2-cleavable peptide substrate.[74] The nanoparticles were modified with PEG for prolonged circulation. The presence of a thiol–Au bond led to accelerated drug release once the particles were internalized *via* endocytosis. Injection of the AuNP into SCC-7 tumor-bearing mice caused rapid release of the drug, which resulted in efficient tumor growth inhibition.

Huang *et al.* engineered the MMP2-sensitive DDS combining the siRNA against VEGF and DOX in nanoparticles built from a dendrigraft poly-L-lysine with a cell-penetrating peptide (CPP) conjugated to PEG.[76] The CPP was masked using a covalently attached pH-sensitive peptide, with a MMP2-sensitive linker. In circulation, the masking peptide shielded the CPP and exposed it only upon exposure to lower pH conditions and elevated MMP2 levels to activate the CPP. These nanopreparations were tested on glioma models both *in vitro* and *in vivo* and promoted apoptosis and anti-angiogenesis simultaneously.

In another approach, the tetra-peptide sequence, GFLG (Gly-Phe-Leu-Gly), sensitive to cathepsin B, an enzyme known to be overexpressed by tumor cells, was inserted as a cleavable linker between DOX and PEGylated dendrimer (mPEGylated dendrimer–GFLG–DOX) for enzyme-triggered doxorubicin release in tumors. *In vivo* tumor growth inhibition was improved 2 fold with reduced side-effects compared to a standard DOX formulation at an equal dose.[77]

1.5 Thermo-sensitive DDS

Local hyperthermia is characteristic of cancer and inflammation, and the temperature difference between normal and pathological tissues may potentially serve as a trigger for the design of temperature-sensitive DDS. In addition, external heat sources can also be used locally to control the tissue temperature.[78] Different strategies can be applied for the external heating of the tissue, such as: dielectric heating by microwave irradiation, ohmic heating by electrode-applied high frequency currents, optical laser heating with fiberoptics, ultrasound application, interstitial laser photocoagulation, and even water bath heating.[79,80] When the pathology (tumor) site is heated, an increase in the endothelial pore size and blood flow takes place, resulting an increased extravasation of DDS.[24] Hyperthermia is also responsible for decreased DNA synthesis, induction of heat shock proteins, altered protein synthesis, disruption of microtubule organizing centers, changes in the expression of receptors and binding of growth factors, and ultimately, changes in the cellular morphology.[81]

To achieve a temperature-sensitive response, liposomes could be prepared with lipids, which have a specific gel-to-liquid phase transition temperature providing the maximal liposome membrane destabilization and drug release at temperatures of 40–42 °C within a 30–60 min period in clinical hyperthermia treatment protocols. Liposomal composition including DPPC/MSPC/DSPE-PEG$_{2000}$ (90 : 10 : 4 mole ratio) is a good example of such a system. Myristoyl stearoyl phosphatidyl choline (MSPC) is a lysolipid which facilitates the rapid release of a drug at 40–42 °C while the PEG-lipid enhances liposomal circulation lifetimes and promotes their accumulation within tumors *via* the EPR effect.[82–84] One of the most commonly used and studied lipids for use in temperature-sensitive DDS is dipalmitoylphosphatidylcholine (DPPC), which has an ideal gel-to-liquid crystalline transition temperature (T_g) of 41 °C.[78,85–88]

Thermoresponsive polymeric materials have been used widely in formulating temperature-sensitive DDS, which have a sharp transition temperature, where they become either soluble or insoluble (the lower critical solution temperature, *LCST*). When the thermoresponsive component is combined to/with a hydrophilic polymer, the thermoresponsive block forms the hydrophobic core, and when the thermoresponsive component is combined with a hydrophobic polymer, this results in the formation of micelles with a thermoresponsive shell.[89] Poly(*N*-isopropylacrylamide) (PNIPAAm) derivatives are the most common thermosensitive polymers used in DDS, they transforms to a hydrophilic structure below LCST and a compact hydrophobic structure above LCST.[90] The thermo-sensitive amphiphilic block copolymer, P-(*N*,*N*-isopropylacrylamide-*co*-*N*-hydroxymethylacrylamide)-*b*-caprolactone [P-(NIPAAm-*co*-NHMAAm)-*b*-PCL] with a LCST of about 38 °C was used to prepare DOX-loaded micelles, which effectively inhibited tumor growth in nude mice.[91]

Thermo-sensitive comb-like copolymers of methoxy PEG (mPEG) blocks and hydrophobic polyacrylate (PA) backbones with thermo-sensitive PNIPAM graft chains (mPEG-*b*-PA-*g*-PNIPAM) have been used to prepare micelles loaded with camptothecin, which released the drug at temperatures above 40 °C.[92] Thermo-sensitive polyelectrolyte complex micelles suggested for delivery of 5-fluorouracil (5-FU) were composed of two biocompatible graft copolymers, chitosan-*g*-poly(*N*-isopropylacrylamide) (CS-*g*-PNIPAM) and carboxymethyl cellulose-*g*-poly(*N*-isopropylacrylamide) (CMC-*g*-PNIPAM).[93]

Elastin-like polypeptides (ELPs) investigated as vehicles for thermoresponsive delivery[94–96] are genetically encoded biopolymers which show a phase transition temperature similar to the LCST observed with some polymers, wherein they are soluble at low temperatures, but can phase-separate into a gel-like phase above a critical transition temperature.[94]

1.6 Magnetically-sensitive DDS

Magnetically-sensitive nanopreparations have been used widely in biomedical applications for enhancement of magnetic resonance imaging contrast, magnetic field-assisted radionuclide therapy, hyperthermia, and tissue-specific release of therapeutic agents. The most popular agent, iron oxide, is used as either maghemite (γ-Fe_2O_3) or magnetite (Fe_3O_4), which are converted in the required materials by precipitation of magnetite in a solution containing stabilizing agents (biocompatible polymers, such as dextran, PEG, PEI, polyvinyl alcohol) or by adsorption of polymers on the magnetic particle surface after their synthesis. Upon administration, the localization of such materials in the body for different purposes is controlled by an external magnetic field.[24,97–100]

When the size of iron oxide nanoparticles is reduced to less than 20 nm, they gain a single domain and become superparamagnetic. Superparamagnetic iron oxide nanoparticles (SPIONs) have iron oxides in the core and are coated with biocompatible polymers and can be functionalized with drugs, proteins or plasmids.[101] Following the removal of the magnetic field, SPIONs lose their magnetization and become highly dispersed, which prevents their aggregation and recognition by the MPS. The are biocompatable and convenient as drug delivery and imaging systems.[102,103] DOX was incorporated into the polymeric coat of thermally cross-linked SPIONs (TCL-SPIONs) bearing carboxylic groups through electrostatic interactions between the positively charged DOX and the negatively charged carboxyls. The preparation controlled by the external magnetic field provided good tumor growth inhibition, and allowed effective tumor imaging in mice bearing lung carcinoma.[104] Anti-HER2/neu (HER, herceptin) antibody-modified, pH-sensitive drug-delivering magnetic nanoparticles (HER-DMNPs) loaded with DOX provided rapid drug release and allowed for the use of *in vivo* MRI in real time.[105]

Vinblastin-loaded magnetic cationic liposomes (MCLs) significantly improved vascular uptake of MCLs in a murine melanoma model under the action of an external magnet, reducing tumor nodules in metastatic sites compared to the control group without the magnet.[106] Fe_3O_4/Au nanoparticles with an iron oxide core coated with a layer of gold were modified with PEG and loaded with DOX. The preparation provided DOX concentration in the liver exposed to the magnetic field much higher than in controls with no magnetic field.[107] Complexes of cationic lipids with plasmid DNA associated with SPIONs can be concentrated onto target cells using magnetic fields.[108] Transferrin-conjugated (for better targeting of glioma cells) magnetic silica poly(D,L-lactic-*co*-glycolic acid) (PLGA) nanoparticles co-loaded with DOX and PTX demonstrated an effective delivery across the blood–brain barrier.[109]

MCLs loaded with SPIONs could be guided to the tumor site using an external magnet and their accumulation and biodistribution could be monitored using MRI, as well as biodistribution analysis. In melanoma-bearing mice two-fold greater accumulation of the administered MCLs in tumor masses was achieved by magnetic targeting.[110]

An interesting application of magnetic nanoparticles is generating magnetic hyperthermia. The action of the alternating magnetic field (AMF) of sufficient strength onto magnetic particles localized in the desired area (tumor) causes the magnetic moment inside these particles to oscillate, which is converted to heat through hysteresis losses or Néel relaxation and transferred to the surrounding environment with the efficacy depending on the strength and duration of exposure to the AMF.[111] At a temperature of 41–43 °C, the damage to normal cells is reversible, while tumor cells are killed.[112] Cationic albumin-conjugated magnetite nanoparticles have also been suggested as a novel method for hyperthermia in cancer therapy.[113]

1.7 Ultrasound-sensitive DDS

Ultrasound is a well-established clinical imaging modality[114] with microbubbles used as the ultrasound contrast agents to enhance imaging resolution. In addition, ultrasound can be safely used for triggered drug delivery and release, by the localized destruction of DDS. With this in mind, ultrasound-responsive DDS accumulated in desired areas are made leaky by locally applied ultrasound to release incorporated drugs. Such release depends on the time of ultrasound application, its frequency and power density, and the type of ultrasound; pulsed or continuous wave ultrasound, as well as on drug lipophilicity.[115,116] Low frequency ultrasound (20–100 kHz) releases drug from DDS more effectively and can penetrate deeper into the tissue than high-frequency ultrasound (1–3 MHz),[117,118] although the latter does not allow for sharp focusing.[117] For *in vivo* applications, low-frequency ultrasound is more appropriate for larger and deep-seated pathological areas (tumors), while high-frequency ultrasound is used for small and superficial lesions (tumors).[117,118] Ultrasound-triggered drug release was studied with Pluronic-based polymeric micelles.[119–123] Folate-modified P105 micelles loaded with DOX have been subjected to the action of low frequency ultrasound (70 kHz) resulting in drug release, which increased with the increase in the ultrasound power intensity of ultrasound.[123] High intensity focused ultrasound was used to trigger the drug release from poly(ethylene glycol)-*b*-poly[*N*-(2-hydroxypropyl) methacrylamide-lactate] (mPEG-*b*-p(HPMAm-Lac$_n$) micelles.[124]

Ultrasound also enhances cell membrane permeabilization when applied in a high frequency pulse regime and allows for better penetration of DDS into tissues and cells, producing higher drug concentration, for example, in tumors.[125] The cavitation of microbubbles under the action of a ultrasonic field can cause transient, reversible cell membrane permeabilization.[126] Microbubbles may serve as drug carriers, as they perform the dual functions of target-specific delivery by responding to ultrasound and can deliver their payload intracellularly because of membrane permeabilization under the action of ultrasound.[127–131] Since larger microbubbles (2–10 μm) have short circulation times (minutes),[126,130] the use of nanobubbles was suggested,[130,132,133] or bubble combinations with another nano-DDS, such as liposomes.[134,135] Thus, antitumoral activity of liposome-encapsulated DOX (Caelyx®) was enhanced by ultrasound application in a mouse tumor model.[136] DOX-loaded liposomes were conjugated to the surface of microbubbles *via* a biotin–avidin linkage. Such complexes provided rapid cellular uptake, enhancement of DOX nuclear accumulation, and less drug efflux in the resistant cells treated by these complexes under the action of ultrasound, compared to just DOX-loaded liposomes with and without addition of verapamil or liposomes complexes with bubbles without applied ultrasound. The authors concluded that their DDS in combination with ultrasound was as an effective system to sensitize cells and overcome MDR.[137] PTX and siRNA co-loaded ultrasound-responsive nanobubbles have also been developed to treat hepatocellular carcinoma (HCC) by hetero-assembly of polymeric micelles and liposomes with nanobubbles. Such complexes have been given intravenously to nude mice bearing human HepG2 xerografts and an external low-frequency ultrasound was applied at the tumor site providing effective codelivery of the active component to the tumor. Antiapoptotic response in HepG2 cells was effectively suppressed by the co-delivered siRNA targeting an antiapoptosis gene (BCL-2 siRNA) during PTX chemotherapy.[138]

DDS termed eLiposomes have also been described, which represent liposome nanodroplets encapsulating an emulsion (usually, pefluorinated hydrocarbons) capable of phase change, as well as a therapeutic agent.[139,140] Ultrasound causes the emulsion droplet to change from a liquid to a gas, which increases the volume/pressure within the liposome, rupturing it and provoking the drug release.

1.8 Light-sensitive DDS

Photoactivation is an another option that can be used to trigger drug release from DDS. Light provides a very broad range of adjustable parameters (*i.e.* wavelength, duration, and intensity) and the light source can be controlled both spatially and temporaly to regulate drug release.[141] Wavelengths in the range of 700–1000 nm, near-infrared light (NIR), are reported to be more suitable for biomedical applications than UV or visible light, because NIR deeply penetrates tissues with high spatial precision and shows less risk of damage to normal cells.[142,143] The photo-responsive groups or chromophores are usually encapsulated within, or conjugated to the surface of the DDS.[144] A variety of mechanisms can induce light-triggered release of cargoes from the DDS, including photo-isomerization, photo-cross-linking, photosensitization-induced oxidation, light-triggered reversible or irreversible switch in polarity, photo-decross-linking, or photo-degradation of the polymer backbone.[145,146] Azobenzene and spiropyran have been used as reversible photo-responsive groups in photoresponsive DDS. The azobenzene group undergoes light-induced, extremely rapid and fully reversible, *E* (*trans*)-*Z* (*cis*) isomerization,[147] while a reversible isomerization occurs between spiropyran and merocyanine, following UV and visible light exposure. By a different mechanism, NIR irradiation causes irreversible photocleavage of chromophores like 2-diazo-1,2-napthoquinone (DNQ), altering the polarity of the nanocarrier, as well as destabilizing its structure, followed by the release of the incorporated drugs.[142] Photopolymerization or photocrosslinking involve a polymerizable double bond that can be irradiated. After the photopolymerization of double bonds in the hydrophobic portion of the bilayer, the bilayer is disrupted by pore creation and allows the release of the encapsulated drug.[145,148,149]

Polymeric micelles and liposomes have been the two most widely utilized nanocarriers for the development of light-responsive vehicles.[150–152] However, other nanocarriers like dendrimers,[153] silica nanoparticles,[154] gold nanoparticles,[155] and nanogels[156] have also been employed. Liposomes composed of dioleoylphosphatidylethanolamine and 2-(hexadecyloxy)-cinnamic acid were reported, where photo-dimerization of the cinnamic acid residue destabilized the liposomal membrane and enabled the release of the loaded cargo.[157] Photo-sensitive DOX-loaded liposomes have been described, made of diplamitoyl phosphatidylcholine (DPPC), a photopolymerizable diacetylene phospholipid (1,2 bis(tricosa-10,12-diynoyl)-*sn-glycero*-3-phosphocholine), and DSPE–PEG_{2000},[148] the drug release from which was initiated by visible light. NIR light-sensitive polymeric micelles have been used for the delivery of doxorubicin.[158]

Gold nanostructures have a high surface plasmon resonance that allows them to absorb NIR radiation and convert the photons into heat, which is either used to trigger the release of chemotherapeutics from NIR responsive DDS or for photothermal ablation (PTA), by increasing the temperature of the surrounding tissues.[145] Thus, DOX-loaded NIR light-sensitive liposomes have been prepared with hollow gold nanospheres attached to the liposome membrane.[159] Drug release was observed from the liposomes upon NIR irradiation resulting in increased cytotoxicity against cancer cells both *in vitro* and *in vivo*.

Photodynamic therapy (PDT) is a promising strategy for treating various types of cancer. In PDT, a photosensitizer (or a precursor) under the action of light reacts with oxygen in the tissue and generates reactive oxygen species (ROS) including singlet oxygen and free radicals that destroy cancer cells/damage DNA. This photophysical mechanism results in cellular toxicity and necrosis, apoptosis or autophagy.[160,161] The most commonly used photosensitizers are porphyrin derivatives, dyes such as phthalocyanins and napthalocyanins and chlorophyll-based derivatives. Many drawbacks of conventional photosensitizers (such as non-specific toxicity) have been overcome with the use of liposomal DDS,[162–164] as seen in the review by Lim *et al.*[165]

Table 1.1 presents some examples of stimuli-sensitive DDS, which can destabilize and release their cargoes to provide better cell penetration under the action of various stimuli as described above.

Table 1.1 Some examples of stimuli-sensitive DDS for monotherapy.

DDS	Sensitive components	Stimulus	Active substance	Outcome	Reference
Nanoparticles	Bioreducible poly (*b*-amino esters) (PAEs), poly[bis (2-hydroxylethyl)-disulfide-diacrylate tetraethylene-pentamine] (PAP)	Redox potential	siRNA against P-gp and survivin	Low IC50 of doxorubicin in MDR cells, down-regulation of P-gp and survivin, decreased tumor volumes *in vivo*	166
Micelle	MMP2 sensitive GPLGIAGQ peptide	Enzyme MMP2 enzyme	Paclitaxel	Enhanced antitumor activity	72
Copper sulphide nanaparticles	Gelatine	Enzyme gelatinase	Doxorubicin	Enhanced cytotoxicity	167
Liposome	Lipid C6/phospholipase A(2)	Enzyme phospholipase A(2)	Retinoid	Enhanced cytotoxicity	168
Hydrogel	Phe-Pip-Arg-Ser/thrombin	Enzyme thrombin	Heparin	Autoregulation of heparin release	169
Hydrogel nanocapsules	KNRVK/plasmin	Enzyme protease	VEGF	Blood vessel formation induced by enzyme-responsive release of VEGF	170
Nanogel	Fibrinogen-*graft*-PNIPAAm	Temperature	5-Floururacil	Triggering of apoptosis and enhanced toxicity *in vitro*	171
PEGylated fullerene/iron oxide nanocomposites	Magnetically sensitive nanoparticles	Magnetic field	Iron oxide	T2-weighted magnetic resonance imaging and photodynamic therapy, targeted drug delivery	172
Silica nanoparticles	Magnetically sensitive nanoparticles	Magnetic field	Mn(2+) ions	T1-weighted magnetic resonance imaging	173
Gadolinium oxide (Gd_2O_3) nanoplates	Magnetically sensitive nanoparticles	Magnetic field	Gd_2O_3	Multimodal imaging in biomedical applications	174

(*continued*)

Table 1.1 (*continued*)

DDS	Sensitive components	Stimulus	Active substance	Outcome	Reference
Long circulating magnetoliposomes	Magnetically sensitive nanoparticles	Magnetic field	Superparamagnetic iron oxide nanoparticles	T2-weighted magnetic resonance imaging	175
Mesoporous silica nanoparticles	Theo-nitrobenzyl	Light 980 nm NIR irradiation	Doxorubicin	Controllable drug delivery and fluorescence imaging	176
Mesoporous silica nanoparticles	Nitroveratryl carbamate-protected aminopropyl-functionality	Light, 350 nm UV irradiation	Doxorubicin	Light- and pH-responsive release of doxorubicin	177
Dendritic micellar nanocarrier	Diazonaphthoquinone	Light 808 nm NIR irradiation; 365 nm UV irradiation	Doxorubicin	Light-sensitive release of doxorubicin	178
Micelle	*Meta*-tetra (hydroxyphenyl) chlorine	Light emitting diodes with the peak intensity at about 660 nm	*Meta*-tetra (hydroxyphenyl) chlorine	Photodynamic therapy	179
Nanoemulsions	Perfluoropentane or perfluoro-15-crown-5-ether	Ultrasound	Paclitaxel	Ultrasonic tumor imaging and targeted chemotherapy (1 MHz or 5 MHz)	130, 180–182
Liposome	Perfluoropropane gas	Ultrasound	pDNA	Ultrasound imaging and gene delivery (frequency: 1 MHz, duty: 50%, burst rate: 2.0 Hz, intensity: 1.0 W cm^{-2}, time: 2 min)	183

Hetero-assembly of siRNA-loaded micelles and gas-cored liposomes	Octafluoropropane	Ultrasound	siRNA	Tumor gene therapy (frequency of 1 MHz, pulse repetition frequency of 1 kHz with 50% a duty cycle, intensity of 1.0 W cm^{-2} and exposure time for 1 min)	184
Liposome-microbubble complexes	Perfluoropropane	Ultrasound	Paclitaxel	Ultrasound-triggered tumor-targeted chemotherapy (10 ms burst length, 1% duty cycle, 1 Hz pulse repetition frequency and 10 min sonication duration)	185 and 186
Micelle	Perfluoropentane or perfluoro-15-crown-5-ether	Ultrasound	Doxorubicin	Ultrasound-mediated intracellular and nuclear trafficking (MHz continuous wave or pulsed ultrasound with 33% duty cycle at 3.4 W cm^{-2} nominal power density)	187

1.9 Stimuli-sensitive DDS for Combination Therapy: Case of Cancer

Combination therapy, for example, the co-administration of two or more therapeutic agents acting through different mechanisms to achieve a therapeutic response better than with each of these agent individually, is widely used for the treatment of many diseases including cancer.[188] Such therapy has also been used to overcome multi-drug resistance in cancer, by combining the cancer cell anti-apoptotic defense mechanism of a modulator with chemotherapeutic agents.[189] A number of siRNA-based treatments that target genes involved in cancer cell survival mechanisms have also been evaluated in combination with conventional anti-cancer drugs.

As already mentioned, DDS have several favorable features and can play a significant role in the delivery of agents for combination therapy. Thus, for example, for chemically different drugs with different PK, DDS can impart a consistent and unified PK profile, biodistribution and stability, and deliver them simultaneously to the target site.[190] An important feature of DDS as described above, is the ability to provide a triggered/controlled release of payloads in response to various stimuli. Using such smart preparation, multiple payloads can be co-released/co-delivered within the same required site. Many of such combination preparations can include biological macromolecules such as siRNA, DNA, or antibodies, which can work effectively in combination with conventional anti-cancer agents for better cancer therapy.

The general principles of the engineering of stimuli-sensitive combination DDS are the same as those described and discussed above, so there is no need to repeat this. The following several tables (Tables 1.2–1.6) present some interesting and promising combination preparations based on stimuli-sensitive DDS.

1.10 Concluding Remarks

Stimuli-responsive DDS demonstrate a better control over the temporal and spatial release of drugs compared to traditional DDS. Still, a careful understanding of stimuli-sensitivity of the designed DDS is required to predict/minimize off-target effects. Combining targeting moieties with the stimuli-sensitivity can further improve control over the distribution and localization of such systems and minimize side-effects. The future challenge will be to develop DDS responding to biomarkers present in very low concentration ranges, and engineer DDS with certain signal amplification features. Another trend is incorporating multiple functions into DDS allowing for simultaneous detection, diagnosis, and therapy of a disease using a single nanoparticle. Such DDS are garnering a great deal of interest from the scientific community, but are still largely the subject of academic research. Their clinical translation may be difficult, but definitely not impossible.

Table 1.2 pH-sensitive DDS for combination therapy.

System	Composition	Payloads for co-delivery		Outcome	References
		Payload 1	Payload 2		
Amphiphilic linear-dendritic prodrugs	Methoxypoly-(ethylene glycol) (MPEG)-*b*-poly(amidoamine) (PAMAM)	DOX conjugated *via* hydrazone bond	10-Hydroxy-camptothecin (HCPT)	pH-dependent release of both drugs resulted in effective uptake and inhibited growth	182 and 191
Hollow mesoporous silica nanoparticle (HMSNP)	Mesoporous silica nanoparticles coated with folic acid conjugated with polyethyleneimine (PEI–FA)	DOX	siRNA against Bcl-2	pH-responsive intracellular drug/siRNA release minimized pre-release, reduced off-target effects, suppressed expression of anti-apoptotic protein Bcl-2, and increased apoptosis	192
Cationic micellar nanoparticles (micelleplexes)	Dual pH-responsive poly (2-(dimethylamino)ethyl methacrylate)-*block*-poly (2-(diisopropylamino)-ethyl methacrylate) (PDMA-*b*-PDPA) di-block copolymers complexed with siRNA	PTX	siRNA against Bcl-2l	Down-regulation of anti-apoptotic Bcl-2, increased apoptosis, promising for overcoming MDR	193
Polymeric micelles	DOX conjugated with poly (styrene-*co*-maleic anhydride) (SMA) derivative with adipic dihydrazide (ADH) *via* an acid-cleavable hydrazone bond	DOX	Disulfiram (DSF)	Enhanced cytotoxicity, strong apoptotic response effective inhibitory effect on the growth in resistant breast cancer models	194
Prodrug nanoparticles	Poly (lactic acid) conjugated – cisplatin prodrug modified-D-alpha-tocopheryl polyethylene glycol 1000 succinate(TPGS), carboxy group modified co-polymer (TCPNPs)	Cisplatin	Docetaxel	pH-sensitive release of drugs; TPGS-cisplatin prodrug NPs (TCPNPs) showed better therapeutic efficacy than individual drugs, lowered IC50s observed in HER-2 overexpressing cells	195

Table 1.3 Redox-sensitive DDS for combination therapy.

		Payloads for co-delivery			
System	Composition	Payload 1	Payload 2	Outcome	References
Micelles	Biodegradable polymer methoxyl-poly-(ethylene glycol)-*block*-poly(lactide-*co*-2-methyl-2-carboxyl-propylene carbonate) with pendant carboxyl groups	Hydrophilic oxaliplatin pro-drug (Oxa(IV)–COOH)	Hydrophobic daunorubicin (DRB)	Reduced systematic toxicity and greater synergistic effect than combination of free drugs both *in vitro* and *in vivo*	196
Nanoparticles	Bioreducible poly (*b*-amino esters) (PAEs), poly[bis(2-hydroxylethyl)-disulfide-diacrylate tetraethylenepentamine] (PAP)	iMdr-1-shRNA	iSurvivin-shRNA	Low IC50 of DOX in MDR cells, down-regulation of P-gp and survivin, decreased tumor size *in vivo*	197
Micelles	Arginine-grafted poly (cystamine-bisacrylamide-diaminohexane) (ABP)	PTX	DNA	Increased cellular uptake, low cytotoxicity in non-reductive conditions	198
Mesoporous silica nanoparticle (MSNP)	Amino-terminated alkyl chains with disulfide bonds functionalized on nanoparticle surface by the reaction of *S*-(2 aminoethylthio)-2 thiopyridine hydrochloride (SATH) and thiol-modified nanoparticles (MSNPSH)	Negatively charged ssDNA	DOX	Enhanced cellular internalization and significant apoptosis induction	199
Polyplex	Supramolecular self-assembled inclusion complex prepared from PTX, star-shaped cationic polymer containing *g* cyclodextrin (*g*-CD), multiple oligoethylenimine (OEI) arms with folic acid (FA) conjugated *via* a disulfide linker	PTX	Plasmid DNA (pDNA) encoding luciferase or p53 gene	Enhanced transfection, significant apoptosis	200

Table 1.4 Enzyme-sensitive DDS for combination therapy.

System	Composition	Payloads for co-delivery		Outcome	References
		Payload 1	Payload 2		
Theranostic nanohybrids	Graphene oxide (GO) grafted with poly(ethylenimine)-*co*-poly (ethylene glycol) (PEI–PEG) *via* MMP2 cleavable PLGLAG peptide linkage	DOX	DNA	Enhanced drug efficacies against tumor cells, more efficient transfection comparable to that with PEI_{25k}	201
Dendrimers	Dendritic thiolated hyperbranched polyglycerol	Maleimide-bearing DOX prodrug	Maleimide-bearing methotrexate prodrug	Drug release in the presence of cathepsin B, high cytotoxicity against human tumor cell lines	202
Polymeric micelles	MMP-2 sensitive (PEG–pp–PEI–PE) co-polymer	PTX	siRNA (anti-GFP or anti-survivin)	Improved tumor targeting, tumor cell internalization and synergistic anti-tumor activity of co-loaded PTX and siRNA	203

Table 1.5 Temperature-sensitive DDS for combination therapy.

System	Composition	Payloads for co-delivery		Outcome	References
		Payload 1	Payload 2		
Hydrogel	Linoleic acid-coupled poloxamer thermo-sensitive hydrogel	PTX	Akt1 shRNA	Synergistic anti-cancer effects, inhibition of tumor growth *in vivo* by inhibition of Akt1 signaling and induction of apoptosis	204
Nanogel	Fibrinogen-*graft*-poly (*N*-isopropyl-acrylamide) PNIPAAm	5-Floururacil	Megestrol acetate	Triggering of apoptosis and enhanced toxicity *in vitro*	171
Thermo-responsive barrier gel	Injectable thermo-gelling poly(di-ethylaminoethyl methacrylate) (PDEAEM)-Pluronic F127 (PL)-PDEAEM pentablock copolymer (PB)	PTX	DNA	Enhanced transfection efficiency and anti-cancer effects of paclitaxel *in vitro*	205

Table 1.6 Magnetic-field sensitive DDS for combination therapy.[a]

System	Composition	Payloads for co-delivery		Outcome	References
		Payload 1	Payload 2		
Nanoparticles	Magnetic mesoporous silica nanoparticles (MMSNs)	DOX	PTX or rapamycin	Better internalization in A549 cells, enhanced apoptosis and tumor growth inhibition compared to single-drug MMSNs	206
Nanoparticles	PLGA- magnetic NPs conjugated to herceptin	PTX	Carboplatin or rapamycin	Increased cellular uptake and enhanced synergistic effect with herceptin	207
Double emulsion (W/O/W) nano-capsules (DEC)	PVA (MW 16k) shell, iron oxide (IO) NP localized in the PVA shell, IVO24, a peptide targeting cancer cells, attached to DEC	PTX	DOX	Magnetic field (50 kHz) induced controllable drug release, enhanced dual magneto-chemotherapy and magneto-hyperthermia *in vitro* and *in vivo*	208

[a]PVA-Poly(vinyl alcohol).

References

1. V. P. Torchilin, *Adv. Drug Delivery Rev.*, 2006, **58**, 1532.
2. D. Peer, J. M. Karp, S. Hong, O. C. Farokhzad, R. Margalit and R. Langer, *Nat. Nanotechnol.*, 2007, **2**, 751.
3. O. C. Farokhzad and R. Langer, *ACS Nano*, 2009, **3**, 16.
4. E. Fleige, M. A. Quadir and R. Haag, *Adv. Drug Delivery Rev.*, 2012, **64**, 866.
5. A. Z. Wang, R. Langer and O. C. Farokhzad, *Annu. Rev. Med.*, 2012, **63**, 185.
6. M. Ferrari, *Nat. Rev. Cancer*, 2005, **5**, 161.
7. V. P. Torchilin, *Nat. Rev. Drug Discovery*, 2005, **4**, 145.
8. V. Torchilin, *Eur. J. Pharm. Biopharm.*, 2009, **71**, 431.
9. M. S. Shim and Y. J. Kwon, *Adv. Drug Delivery Rev.*, 2012, **64**, 1046.
10. Y. L. Colson and M. W. Grinstaff, *Adv. Mater.*, 2012, **24**, 3878.
11. S. Mura, J. Nicolas and P. Couvreur, *Nat. Mater.*, 2013, **12**, 991.
12. L. E. Gerweck and K. Seetharaman, *Cancer Res.*, 1996, **56**, 1194.
13. I. F. Tannock and D. Rotin, *Cancer Res.*, 1989, **49**, 4373.
14. F. Colotta, P. Allavena, A. Sica, C. Garlanda and A. Mantovani, *Carcinogenesis*, 2009, **30**, 1073.
15. S. H. Gromkowski, J. Yagi and C. A. Janeway Jr, *Eur. J. Immunol.*, 1989, **19**, 1709.
16. C. Stefanadis, C. Chrysochoou, D. Markou, K. Petraki, D. B. Panagiotakos, C. Fasoulakis, A. Kyriakidis, C. Papadimitriou and P. K. Toutouzas, *J. Clin. Oncol.*, 2001, **19**, 676.
17. A. Russo, W. DeGraff, N. Friedman and J. B. Mitchell, *Cancer Res.*, 1986, **46**, 2845.
18. G. Saito, J. A. Swanson and K. D. Lee, *Adv. Drug Delivery Rev.*, 2003, **55**, 199.
19. R. de la Rica, D. Aili and M. M. Stevens, *Adv. Drug Delivery Rev.*, 2012, **64**, 967.
20. J. E. Ghadiali and M. M. Stevens, *Adv. Mater.*, 2008, **20**, 4359.
21. N. Kamaly, Z. Xiao, P. M. Valencia, A. F. Radovic-Moreno and O. C. Farokhzad, *Chem. Soc. Rev.*, 2012, **41**, 2971.
22. J. Zhuang, M. R. Gordon, J. Ventura, L. Li and S. Thayumanavan, *Chem. Soc. Rev.*, 2013, **42**, 7421.
23. R. Cheng, F. Meng, C. Deng, H. A. Klok and Z. Zhong, *Biomaterials*, 2013, **34**, 3647.
24. V. P. Torchilin, *Nat. Rev. Drug Discovery*, 2014, **13**, 813.
25. J. R. Griffiths, *Br. J. Cancer*, 1991, **64**, 425.
26. G. Helmlinger, A. Sckell, M. Dellian, N. S. Forbes and R. K. Jain, *Clin. Cancer Res.*, 2002, **8**, 1284.
27. S. Fais, A. De Milito, H. You and W. Qin, *Cancer Res.*, 2007, **67**, 10627.
28. L. E. Gerweck and K. Seetharaman, *Cancer Res.*, 1996, **56**, 1194.
29. X. L. Shi, G. J. Mao, X. B. Zhang, H. W. Liu, Y. J. Gong, Y. X. Wu, L. Y. Zhou, J. Zhang and W. Tan, *Talanta*, 2014, **130**, 356.

30. S. Boeckle, J. Fahrmeir, W. Roedl, M. Ogris and E. Wagner, *J. Controlled Release*, 2006, **112**, 240.
31. D. Schmaljohann, *Adv. Drug Delivery Rev.*, 2006, **58**, 1655.
32. Y. Liu, W. Wang, J. Yang, C. Zhou and J. Sun, *Asian J. Pharm. Sci.*, 2013, **8**, 159.
33. E. S. Lee, H. J. Shin, K. Na and Y. H. Bae, *J. Controlled Release*, 2003, **90**, 363.
34. J. Liu, Y. Huang, A. Kumar, A. Tan, S. Jin, A. Mozhi and X. J. Liang, *Biotechnol. Adv.*, 2014, **32**, 693.
35. M. Prabaharan, J. J. Grailer, S. Pilla, D. A. Steeber and S. Gong, *Biomaterials*, 2009, **30**, 5757.
36. J. Liu, H. Ma, T. Wei and X. J. Liang, *Chem. Commun.*, 2012, **48**, 4869.
37. Y. Lee, T. Ishii, H. Cabral, H. J. Kim, J. H. Seo, N. Nishiyama, H. Oshima, K. Osada and K. Kataoka, *Angew. Chem., Int. Ed. Engl.*, 2009, **48**, 5309.
38. F. Pittella, K. Miyata, Y. Maeda, T. Suma, S. Watanabe, Q. Chen, R. J. Christie, K. Osada, N. Nishiyama and K. Kataoka, *J. Controlled Release*, 2012, **161**, 868.
39. R. R. Sawant, S. K. Sriraman, G. Navarro, S. Biswas, R. A. Dalvi and V. P. Torchilin, *Biomaterials*, 2012, **33**, 3942.
40. S. Ganta, H. Devalapally, A. Shahiwala and M. Amiji, *J. Controlled Release*, 2008, **126**, 187.
41. F. Q. Schafer and G. R. Buettner, *Free Radical Biol. Med.*, 2001, **30**, 1191.
42. Y. Li, K. Xiao, J. Luo, W. Xiao, J. S. Lee, A. M. Gonik, J. Kato, T. A. Dong and K. S. Lam, *Biomaterials*, 2011, **32**, 6633.
43. S. E. Moriarty-Craige and D. P. Jones, *Annu. Rev. Nutr.*, 2004, **24**, 481.
44. D. Giustarini, A. Milzani, I. Dalle-Donne and R. Rossi, *Blood Cells, Mol., Dis.*, 2008, **40**, 174.
45. R. Cheng, F. Feng, F. Meng, C. Deng, J. Feijen and Z. Zhong, *J. Controlled Release*, 2011, **152**, 2.
46. J. Li, M. Huo, J. Wang, J. Zhou, J. M. Mohammad, Y. Zhang, Q. Zhu, A. Y. Waddad and Q. Zhang, *Biomaterials*, 2012, **33**, 2310.
47. N. Song, W. Liu, Q. Tu, R. Liu, Y. Zhang and J. Wang, *Colloids Surf., B*, 2011, **87**, 454.
48. Q. Yin, J. Shen, Z. Zhang, H. Yu and Y. Li, *Adv. Drug Delivery Rev.*, 2013, **65**, 1699.
49. L. Brulisauer, M. A. Gauthier and J. C. Leroux, *J. Controlled Release*, 2014, **195**, 147.
50. M. Cai, K. Zhu, Y. Qiu, X. Liu, Y. Chen and X. Luo, *Colloids Surf., B*, 2014, **116**, 424.
51. W. Sun and P. B. Davis, *J. Controlled Release*, 2010, **146**, 118.
52. J. Liu, Y. Pang, W. Huang, Z. Zhu, X. Zhu, Y. Zhou and D. Yan, *Biomacromolecules*, 2011, **12**, 2407.
53. Y. C. Wang, F. Wang, T. M. Sun and J. Wang, *Bioconjugate Chem.*, 2011, **22**, 1939.
54. H. Cho, J. Bae, V. K. Garripelli, J. M. Anderson, H. W. Jun and S. Jo, *Chem. Commun.*, 2012, **48**, 6043.

55. T. Musacchio, O. Vaze, G. D'Souza and V. P. Torchilin, *Bioconjugate Chem.*, 2010, **21**, 1530.
56. G. Salzano, R. Riehle, G. Navarro, F. Perche, G. De Rosa and V. P. Torchilin, *Cancer Lett.*, 2014, **343**, 224.
57. R. Sheng, T. Luo, Y. Zhu, H. Li, J. Sun, S. Chen, W. Sun and A. Cao, *Biomaterials*, 2011, **32**, 3507.
58. M. S. Shim, S. H. Bhang, K. Yoon, K. Choi and Y. Xia, *Angew. Chem., Int. Ed. Engl.*, 2012, **51**, 11899.
59. U. L. Rahbek, A. F. Nielsen, M. Dong, Y. You, A. Chauchereau, D. Oupicky, F. Besenbacher, J. Kjems and K. A. Howard, *J. Drug Targeting*, 2010, **18**, 812.
60. Z. Q. Yu, J. J. Yan, Y. Z. You and Q. H. Zhou, *Int. J. Nanomed.*, 2012, **7**, 5819.
61. W. Xia, P. Wang, C. Lin, Z. Li, X. Gao, G. Wang and X. Zhao, *J. Controlled Release*, 2012, **157**, 427.
62. S. Matsumoto, R. J. Christie, N. Nishiyama, K. Miyata, A. Ishii, M. Oba, H. Koyama, Y. Yamasaki and K. Kataoka, *Biomacromolecules*, 2009, **10**, 119.
63. M. Oishi, T. Hayama, Y. Akiyama, S. Takae, A. Harada, Y. Yamasaki, F. Nagatsugi, S. Sasaki, Y. Nagasaki and K. Kataoka, *Biomacromolecules*, 2005, **6**, 2449.
64. M. Kumagai, S. Shimoda, R. Wakabayashi, Y. Kunisawa, T. Ishii, K. Osada, K. Itaka, N. Nishiyama, K. Kataoka and K. Nakano, *J. Controlled Release*, 2012, **160**, 542.
65. F. Perche, S. Biswas, T. Wang, L. Zhu and V. P. Torchilin, *Angew. Chem., Int. Ed. Engl.*, 2014, **53**, 3362.
66. J. P. Gillet and M. M. Gottesman, *Methods Mol. Biol.*, 2010, **596**, 47.
67. M. Egeblad and Z. Werb, *Nat. Rev. Cancer*, 2002, **2**, 161.
68. A. R. Folgueras, A. M. Pendas, L. M. Sanchez and C. Lopez-Otin, *Int. J. Dev. Biol.*, 2004, **48**, 411.
69. K. Kessenbrock, V. Plaks and Z. Werb, *Cell*, 2010, **141**, 52.
70. C. Gialeli, A. D. Theocharis and N. K. Karamanos, *FEBS J.*, 2011, **278**, 16.
71. R. Yamada, M. B. Kostova, R. K. Anchoori, S. Xu, N. Neamati and S. R. Khan, *Cancer Biol. Ther.*, 2010, **9**, 192.
72. L. Zhu, T. Wang, F. Perche, A. Taigind and V. P. Torchilin, *Proc. Natl. Acad. Sci. U. S. A.*, 2013, **110**, 17047.
73. L. Zhu, P. Kate and V. P. Torchilin, *ACS Nano*, 2012, **6**, 3491.
74. W. H. Chen, X. D. Xu, H. Z. Jia, Q. Lei, G. F. Luo, S. X. Cheng, R. X. Zhuo and X. Z. Zhang, *Biomaterials*, 2013, **34**, 8798.
75. R. Li, W. Wu, Q. Liu, P. Wu, L. Xie, Z. Zhu, M. Yang, X. Qian, Y. Ding, L. Yu, X. Jiang, W. Guan and B. Liu, *PLoS One*, 2013, **8**, e69643.
76. S. Huang, K. Shao, Y. Liu, Y. Kuang, J. Li, S. An, Y. Guo, H. Ma and C. Jiang, *ACS Nano*, 2013, **7**, 2860.
77. C. Zhang, D. Pan, K. Luo, W. She, C. Guo, Y. Yang and Z. Gu, *Adv. Healthcare Mater.*, 2014, **3**, 1299.
78. L. Zhu and V. P. Torchilin, *Integr. Biol.*, 2013, **5**, 96.

79. S. Dutz and R. Hergt, *Int. J. Hyperthermia*, 2013, **29**, 790.
80. D. Ortega and Q. A. Pankhurst, in *Nanoscience*, ed. P. O'Brien, Royal Society of Chemistry, Cambridge, 2012, vol. 1, p. 60.
81. A. Chilkoti, M. R. Dreher, D. E. Meyer and D. Raucher, *Adv. Drug Delivery Rev.*, 2002, **54**, 613.
82. G. R. Anyarambhatla and D. Needham, *J. Liposome Res.*, 1999, **9**, 491.
83. D. Needham, G. Anyarambhatla, G. Kong and M. W. Dewhirst, *Cancer Res.*, 2000, **60**, 1197.
84. D. Needham and M. W. Dewhirst, *Adv. Drug Delivery Rev.*, 2001, **53**, 285.
85. K. Kono, *Adv. Drug Delivery Rev.*, 2001, **53**, 307.
86. W. T. Al-Jamal, Z. S. Al-Ahmady and K. Kostarelos, *Biomaterials*, 2012, **33**, 4608.
87. A. A. Manzoor, L. H. Lindner, C. D. Landon, J. Y. Park, A. J. Simnick, M. R. Dreher, S. Das, G. Hanna, W. Park, A. Chilkoti, G. A. Koning, T. L. ten Hagen, D. Needham and M. W. Dewhirst, *Cancer Res.*, 2012, **72**, 5566.
88. J. Lapointe and S. Martel, *Conf. Proc. IEEE Eng. Med. Biol. Soc.*, 2009, **2009**, 4246.
89. S. R. Abulateefeh, S. G. Spain, J. W. Aylott, W. C. Chan, M. C. Garnett and C. Alexander, *Macromol. Biosci.*, 2011, **11**, 1722.
90. H. G. Schild, *Prog. Polym. Sci.*, 1992, **17**, 163.
91. X. Wang, S. Li, Z. Wan, Z. Quan and Q. Tan, *Int. J. Pharm.*, 2014, **463**, 81.
92. X. L. Yang, Y. L. Luo, F. Xu and Y. S. Chen, *Pharm. Res.*, 2014, **31**, 291.
93. G. Li, Y. Meng, L. Guo, T. Zhang and J. Liu, *J. Biomed. Mater. Res., Part A*, 2014, **102**, 2163.
94. J. R. McDaniel, M. W. Dewhirst and A. Chilkoti, *Int. J. Hyperthermia*, 2013, **29**, 501.
95. J. R. McDaniel, J. Bhattacharyya, K. B. Vargo, W. Hassouneh, D. A. Hammer and A. Chilkoti, *Angew. Chem., Int. Ed. Engl.*, 2013, **52**, 1683.
96. J. R. McDaniel, S. R. Macewan, M. Dewhirst and A. Chilkoti, *J. Controlled Release*, 2012, **159**, 362.
97. V. S. Zaitsev, D. S. Filimonov, I. A. Presnyakov, R. J. Gambino and B. Chu, *J. Colloid Interface Sci.*, 1999, **212**, 49.
98. Y. Zhang, N. Kohler and M. Zhang, *Biomaterials*, 2002, **23**, 1553.
99. R. Jurgons, C. Seliger, A. Hilpert, L. Trahms, S. Odenbach and C. Alexiou, *J. Phys.: Condens. Matter*, 2006, **18**, 2893.
100. A. Nacev, C. Beni, O. Bruno and B. Shapiro, *J. Magn. Magn. Mater.*, 2011, **323**, 651.
101. T. Neuberger, B. Schöpf, H. Hofmann, M. Hofmann and B. von Rechenberg, *J. Magn. Magn. Mater.*, 2005, **293**, 483.
102. H. Y. Acar, R. S. Garaas, F. Syud, P. Bonitatebus and A. M. Kulkarni, *J. Magn. Magn. Mater.*, 2005, **293**, 1.
103. P. B. Santhosh and N. P. Ulrih, *Cancer Lett.*, 2013, **336**, 8.
104. M. K. Yu, Y. Y. Jeong, J. Park, S. Park, J. W. Kim, J. J. Min, K. Kim and S. Jon, *Angew. Chem., Int. Ed. Engl.*, 2008, **47**, 5362.

105. E. K. Lim, Y. M. Huh, J. Yang, K. Lee, J. S. Suh and S. Haam, *Adv. Mater.*, 2011, **23**, 2436.
106. S. Dandamudi, V. Patil, W. Fowle, B. A. Khaw and R. B. Campbell, *Cancer Sci.*, 2009, **100**, 1537.
107. X. Chao, L. Guo, Y. Zhao, K. Hua, M. Peng, C. Chen and Y. Cui, *J. Drug Targeting*, 2011, **19**, 161.
108. C. Wang, C. Ding, M. Kong, A. Dong, J. Qian, D. Jiang and Z. Shen, *Biochem. Biophys. Res. Commun.*, 2011, **410**, 537.
109. Y. Cui, Q. Xu, P. K. Chow, D. Wang and C. H. Wang, *Biomaterials*, 2013, **34**, 8511.
110. E. Gultepe, F. J. Reynoso, A. Jhaveri, P. Kulkarni, D. Nagesha, C. Ferris, M. Harisinghani, R. B. Campbell and S. Sridhar, *Nanomedicine*, 2010, **5**, 1173.
111. S. Laurent, S. Dutz, U. O. Hafeli and M. Mahmoudi, *Adv. Colloid Interface Sci.*, 2011, **166**, 8.
112. T. Neuberger, B. Schöpf, H. Hofmann, M. Hofmann and B. von Rechenberg, *J. Magn. Magn. Mater.*, 2005, **293**, 483.
113. E. Cheraghipour and S. Javadpour, *Int. J. Hyperthermia*, 2013, **29**, 511.
114. D. Komljenovic and T. Bäuerle, in *Cancer Theranostics*, ed. X. Chen and S. Wong, Elsevier, 2014, p. 127.
115. G. A. Husseini, G. D. Myrup, W. G. Pitt, D. A. Christensen and N. Y. Rapoport, *J. Controlled Release*, 2000, **69**, 43.
116. A. Marin, M. Muniruzzaman and N. Rapoport, *J. Controlled Release*, 2001, **71**, 239.
117. N. Rapoport, *Int. J. Hyperthermia*, 2012, **28**, 374.
118. A. Marin, H. Sun, G. A. Husseini, W. G. Pitt, D. A. Christensen and N. Y. Rapoport, *J. Controlled Release*, 2002, **84**, 39.
119. G. A. Husseini, M. A. Diaz de la Rosa, T. Gabuji, Y. Zeng, D. A. Christensen and W. G. Pitt, *J. Nanosci. Nanotechnol.*, 2007, **7**, 1028.
120. M. Ugarenko, C. K. Chan, A. Nudelman, A. Rephaeli, S. M. Cutts and D. R. Phillips, *Oncol. Res.*, 2009, **17**, 283.
121. G. A. Husseini, D. Stevenson-Abouelnasr, W. G. Pitt, K. T. Assaleh, L. O. Farahat and J. Fahadi, *Colloids Surf., A*, 2010, **359**, 18.
122. Z. G. Gao, H. D. Fain and N. Rapoport, *J. Controlled Release*, 2005, **102**, 203.
123. G. A. Husseini, D. Velluto, L. Kherbeck, W. G. Pitt, J. A. Hubbell and D. A. Christensen, *Colloids Surf., B*, 2013, **101**, 153.
124. R. Deckers, A. Paradissis, C. Oerlemans, M. Talelli, G. Storm, W. E. Hennink and J. F. Nijsen, *Langmuir*, 2013, **29**, 9483.
125. Y. S. Li, C. N. Reid and A. P. McHale, *Cancer Lett.*, 2008, **266**, 156.
126. B. Geers, H. Dewitte, S. C. De Smedt and I. Lentacker, *J. Controlled Release*, 2012, **164**, 248.
127. I. Lentacker, B. Geers, J. Demeester, S. C. De Smedt and N. N. Sanders, *Mol. Ther.*, 2010, **18**, 101.
128. N. Nomikou and A. P. McHale, *Cancer Lett.*, 2010, **296**, 133.

129. S. Fokong, B. Theek, Z. Wu, P. Koczera, L. Appold, S. Jorge, U. Resch-Genger, M. van Zandvoort, G. Storm, F. Kiessling and T. Lammers, *J. Controlled Release*, 2012, **163**, 75.
130. N. Rapoport, K. H. Nam, R. Gupta, Z. Gao, P. Mohan, A. Payne, N. Todd, X. Liu, T. Kim, J. Shea, C. Scaife, D. L. Parker, E. K. Jeong and A. M. Kennedy, *J. Controlled Release*, 2011, **153**, 4.
131. Z. Gao, A. M. Kennedy, D. A. Christensen and N. Y. Rapoport, *Ultrasonics*, 2008, **48**, 260.
132. H. Y. Huang, S. H. Hu, S. Y. Hung, C. S. Chiang, H. L. Liu, T. L. Chiu, H. Y. Lai, Y. Y. Chen and S. Y. Chen, *J. Controlled Release*, 2013, **172**, 118.
133. R. H. Perera, L. Solorio, H. Wu, M. Gangolli, E. Silverman, C. Hernandez, P. M. Peiris, A. M. Broome and A. A. Exner, *Pharm. Res.*, 2014, **31**, 1407.
134. Y. Wang, X. Li, Y. Zhou, P. Huang and Y. Xu, *Int. J. Pharm.*, 2010, **384**, 148.
135. R. Suzuki, E. Namai, Y. Oda, N. Nishiie, S. Otake, R. Koshima, K. Hirata, Y. Taira, N. Utoguchi, Y. Negishi, S. Nakagawa and K. Maruyama, *J. Controlled Release*, 2010, **142**, 245.
136. G. Myhr and J. Moan, *Cancer Lett.*, 2006, **232**, 206.
137. Z. Deng, F. Yan, Q. Jin, F. Li, J. Wu, X. Liu and H. Zheng, *J. Controlled Release*, 2014, **174**, 109.
138. T. Yin, P. Wang, J. Li, Y. Wang, B. Zheng, R. Zheng, D. Cheng and X. Shuai, *Biomaterials*, 2014, **35**, 5932.
139. M. Javadi, W. G. Pitt, D. M. Belnap, N. H. Tsosie and J. M. Hartley, *Langmuir*, 2012, **28**, 14720.
140. C. Y. Lin, M. Javadi, D. M. Belnap, J. R. Barrow and W. G. Pitt, *Nanomedicine*, 2014, **10**, 67.
141. P. Shum, J. M. Kim and D. H. Thompson, *Adv. Drug Delivery Rev.*, 2001, **53**, 273.
142. G. Y. Liu, C. H. Chen, D. D. Li, S. S. Wang and J. Ji, *J. Mater. Chem.*, 2012, **22**, 16865.
143. S. Kumar, J. F. Allard, D. Morris, Y. L. Dory, M. Lepage and Y. Zhao, *J. Mater. Chem.*, 2012, **22**, 7252.
144. C. Oerlemans, W. Bult, M. Bos, G. Storm, J. F. Nijsen and W. E. Hennink, *Pharm. Res.*, 2010, **27**, 2569.
145. N. Fomina, J. Sankaranarayanan and A. Almutairi, *Adv. Drug Delivery Rev.*, 2012, **64**, 1005.
146. Q. Yan, D. Han and Y. Zhao, *Polym. Chem.*, 2013, **4**, 5026.
147. A. Higuchi, Q. D. Ling, S. S. Kumar, Y. Chang, T. C. Kao, M. A. Munusamy, A. A. Alarfaj, S. T. Hsu and A. Umezawa, *Prog. Polym. Sci.*, 2014, **39**, 1585.
148. A. Yavlovich, A. Singh, R. Blumenthal and A. Puri, *Biochim. Biophys. Acta*, 2011, **1808**, 117.
149. A. M. Jhaveri and V. P. Torchilin, *Front. Pharmacol.*, 2014, **5**, 77.
150. C. Alvarez-Lorenzo, L. Bromberg and A. Concheiro, *Photochem. Photobiol.*, 2009, **85**, 848.
151. A. Yavlovich, B. Smith, K. Gupta, R. Blumenthal and A. Puri, *Mol. Membr. Biol.*, 2010, **27**, 364.

152. Y. Zhao, *Chem. Rec.*, 2007, **7**, 286.
153. L. Sun, X. Ma, C.-M. Dong, B. Zhu and X. Zhu, *Biomacromolecules*, 2012, **13**, 3581.
154. Y. T. Chang, P. Y. Liao, H. S. Sheu, Y. J. Tseng, F. Y. Cheng and C. S. Yeh, *Adv. Mater.*, 2012, **24**, 3309.
155. K. Niikura, N. Iyo, Y. Matsuo, H. Mitomo and K. Ijiro, *ACS Appl. Mater. Interfaces*, 2013, **5**, 3900.
156. C. Wang, J. Mallela, U. S. Garapati, S. Ravi, V. Chinnasamy, Y. Girard, M. Howell and S. Mohapatra, *Nanomedicine*, 2013, **9**, 903.
157. H. J. Seo and J. C. Kim, *J. Nanosci. Nanotechnol.*, 2013, **13**, 1727.
158. D. Liu, W. Wu, X. Chen, S. Wen, X. Zhang, Q. Ding, G. Teng and N. Gu, *Nanoscale*, 2012, **4**, 2306.
159. J. You, P. Zhang, F. Hu, Y. Du, H. Yuan, J. Zhu, Z. Wang, J. Zhou and C. Li, *Pharm. Res.*, 2014, **31**, 554.
160. S. Mordon, C. Cochrane, J. B. Tylcz, N. Betrouni, L. Mortier and V. Koncar, *Photodiagn. Photodyn. Ther.*, 2015, **12**, 1.
161. N. Barbero, S. Visentin and G. Viscardi, *J. Photochem. Photobiol., A*, 2015, **299**, 38.
162. K. Nawalany, A. Rusin, M. Kepczynski, A. Mikhailov, G. Kramer-Marek, M. Snietura, J. Poltowicz, Z. Krawczyk and M. Nowakowska, *J. Photochem. Photobiol., B*, 2009, **97**, 8.
163. J. P. Longo, S. C. Leal, A. R. Simioni, M. de Fatima Menezes Almeida-Santos, A. C. Tedesco and R. B. Azevedo, *Lasers Med. Sci.*, 2012, **27**, 575.
164. J. Zhang, L. Deng, J. Yao, P. Gu, F. Yang, X. Wang, W. Liu, Y. Zhang, X. Ke, X. Jing and J. Chen, *Bioorg. Med. Chem.*, 2011, **19**, 5520.
165. C. K. Lim, J. Heo, S. Shin, K. Jeong, Y. H. Seo, W. D. Jang, C. R. Park, S. Y. Park, S. Kim and I. C. Kwon, *Cancer Lett.*, 2013, **334**, 176.
166. Q. Yin, J. Shen, L. Chen, Z. Zhang, W. Gu and Y. Li, *Biomaterials*, 2012, **33**, 6495.
167. Z. Zha, S. Zhang, Z. Deng, Y. Li, C. Li and Z. Dai, *Chem. Commun.*, 2013, **49**, 3455.
168. P. J. Pedersen, S. K. Adolph, A. K. Subramanian, A. Arouri, T. L. Andresen, O. G. Mouritsen, R. Madsen, M. W. Madsen, G. H. Peters and M. H. Clausen, *J. Med. Chem.*, 2010, **53**, 3782.
169. M. F. Maitz, U. Freudenberg, M. V. Tsurkan, M. Fischer, T. Beyrich and C. Werner, *Nat. Commun.*, 2013, **4**, 2168.
170. J. Wen, S. M. Anderson, J. Du, M. Yan, J. Wang, M. Shen, Y. Lu and T. Segura, *Adv. Mater.*, 2011, **23**, 4549.
171. N. S. Rejinold, T. Baby, K. P. Chennazhi and R. Jayakumar, *Colloids Surf., B*, 2014, **114**, 209.
172. J. Shi, X. Yu, L. Wang, Y. Liu, J. Gao, J. Zhang, R. Ma, R. Liu and Z. Zhang, *Biomaterials*, 2013, **34**, 9666.
173. S. M. Kim, G. H. Im, D. G. Lee, J. H. Lee, W. J. Lee and I. S. Lee, *Biomaterials*, 2013, **34**, 8941.
174. T. Paik, T. R. Gordon, A. M. Prantner, H. Yun and C. B. Murray, *ACS Nano*, 2013, **7**, 2850.

175. M. B. Martins, M. L. Corvo, P. Marcelino, H. S. Marinho, G. Feio and A. Carvalho, *Nanomedicine*, 2014, **10**, 207.
176. Y. Yang, B. Velmurugan, X. Liu and B. Xing, *Small*, 2013, **9**, 2937.
177. N. Z. Knezevic, B. G. Trewyn and V. S. Lin, *Chemistry*, 2011, **17**, 3338.
178. L. Sun, Y. Yang, C. M. Dong and Y. Wei, *Small*, 2011, **7**, 401.
179. M. J. Shieh, C. L. Peng, W. L. Chiang, C. H. Wang, C. Y. Hsu, S. J. Wang and P. S. Lai, *Mol. Pharm.*, 2010, **7**, 1244.
180. N. Rapoport, Z. Gao and A. Kennedy, *J. Natl. Cancer Inst.*, 2007, **99**, 1095.
181. N. Y. Rapoport, A. M. Kennedy, J. E. Shea, C. L. Scaife and K. H. Nam, *J. Controlled Release*, 2009, **138**, 268.
182. N. Rapoport, A. M. Kennedy, J. E. Shea, C. L. Scaife and K. H. Nam, *Mol. Pharm.*, 2010, **7**, 22.
183. Y. Endo-Takahashi, Y. Negishi, A. Nakamura, D. Suzuki, S. Ukai, K. Sugimoto, F. Moriyasu, N. Takagi, R. Suzuki, K. Maruyama and Y. Aramaki, *Biomaterials*, 2013, **34**, 2807.
184. T. Yin, P. Wang, J. Li, R. Zheng, B. Zheng, D. Cheng, R. Li, J. Lai and X. Shuai, *Biomaterials*, 2013, **34**, 4532.
185. F. Yan, L. Li, Z. Deng, Q. Jin, J. Chen, W. Yang, C. K. Yeh, J. Wu, R. Shandas, X. Liu and H. Zheng, *J. Controlled Release*, 2013, **166**, 246.
186. F. Yan, L. Li, Z. Deng, Q. Jin, J. Chen, W. Yang, C. K. Yeh, J. Wu, R. Shandas, X. Liu and H. Zheng, *J. Controlled Release*, 2013, **166**, 246.
187. P. Mohan and N. Rapoport, *Mol. Pharm.*, 2010, **7**, 1959.
188. B. Al-Lazikani, U. Banerji and P. Workman, *Nat. Biotechnol.*, 2012, **30**, 679.
189. J. P. Gillet and M. M. Gottesman, *Methods Mol. Biol.*, 2010, **596**, 47.
190. C. M. Hu, S. Aryal and L. Zhang, *Ther. Delivery*, 2010, **1**, 323.
191. Y. Zhang, C. Xiao, M. Li, J. Chen, J. Ding, C. He, X. Zhuang and X. Chen, *Macromol. Biosci.*, 2013, **13**, 584.
192. X. Ma, Y. Zhao and K. W. Ng, *Chemistry*, 2013, **19**, 15593.
193. H. Yu, Z. Xu, X. Chen, L. Xu, Q. Yin, Z. Zhang and Y. Li, *Macromol. Biosci.*, 2014, **14**, 100.
194. X. Duan, J. Xiao, Q. Yin, Z. Zhang, H. Yu, S. Mao and Y. Li, *ACS Nano*, 2013, **7**, 5858.
195. Y. Mi, J. Zhao and S. S. Feng, *J. Controlled Release*, 2013, **169**, 185.
196. H. Xiao, W. Li, R. Qi, L. Yan, R. Wang, S. Liu, Y. Zheng, Z. Xie, Y. Huang and X. Jing, *J. Controlled Release*, 2012, **163**, 304.
197. Q. Yin, J. Shen, L. Chen, Z. Zhang, W. Gu and Y. Li, *Biomaterials*, 2012, **33**, 6495.
198. K. Nam, H. Y. Nam, P.-H. Kim and S. W. Kim, *Biomaterials*, 2012, **33**, 8122.
199. X. Ma, K. T. Nguyen, P. Borah, C. Y. Ang and Y. Zhao, *Adv. Healthcare Mater.*, 2012, **1**, 690.
200. F. Zhao, H. Yin and J. Li, *Biomaterials*, 2014, **35**, 1050.
201. S. Y. Qin, J. Feng, L. Rong, H. Z. Jia, S. Chen, X. J. Liu, G. F. Luo, R. X. Zhuo and X. Z. Zhang, *Small*, 2014, **10**, 599.

202. M. Calderon, R. Graeser, F. Kratz and R. Haag, *Bioorg. Med. Chem. Lett.*, 2009, **19**, 3725.
203. L. Zhu, F. Perche, T. Wang and V. P. Torchilin, *Biomaterials*, 2014, **35**, 4213.
204. D.-D. Guo, S.-H. Hong, H.-L. Jiang, J.-H. Kim, A. Minai-Tehrani, J.-E. Kim, J.-Y. Shin, T. Jiang, Y.-K. Kim, Y.-J. Choi, C.-S. Cho and M.-H. Cho, *Biomaterials*, 2012, **33**, 2272.
205. B. Zhang, F. Jia, M. Q. Fleming and S. K. Mallapragada, *Int. J. Pharm.*, 2012, **427**, 88.
206. Q. Liu, J. Zhang, W. Sun, Q. R. Xie, W. Xia and H. Gu, *Int. J. Nanomed.*, 2012, 7, 999.
207. A. Singh, F. Dilnawaz, S. Mewar, U. Sharma, N. R. Jagannathan and S. K. Sahoo, *ACS Appl. Mater. Interfaces*, 2011, **3**, 842.
208. S. H. Hu, B. J. Liao, C. S. Chiang, P. J. Chen, I. W. Chen and S. Y. Chen, *Adv. Mater.*, 2012, **24**, 3627.

CHAPTER 2

Materials and Chemistry of Stimuli-responsive Drug Delivery Systems

MEGHANA RAWAL

Allexcel, Inc., 1 Controls Drive, Shelton, Connecticut, USA
*E-mail: meghana.rawal@allexcel.com

2.1 Introduction

Stimuli-responsive nanomaterials have been used as drug delivery systems (DDS) to target specific sites of a disease and can even be tailored to achieve a desired release rate based on the disease progression. These drug delivery systems allow for great advances in treatment where the drug itself is toxic to healthy tissue or can cause severe side effects. Precise drug release depends on a clear understanding of the required stimuli and release mechanism of the drug delivery system. The field of stimuli responsive materials covers a broad range of stimuli that produce a desired reaction in nanomaterials. The structure property relationship between the stimuli and the response determines the nature of the trigger. A change in pH or the application of specific wavelengths of light can cause a drug carrying nanoparticle to deliver its payload at specific sites or times. The trigger causes a change in the DDS, such as hydrolysis of specific bonds or phase separation through a chemical or physical reaction. Nanomaterial based DDS have a wide range of medical

Biomaterials Science Series No. 1
Stimuli-responsive Drug Delivery Systems
Edited by Amit Singh and Mansoor M. Amiji

Published by the Royal Society of Chemistry, www.rsc.org

applications. Therefore, depending on the application and the mode of the trigger desired, the stimulus may be physical (photo- or thermo-responsive), chemical (pH) or biochemical (enzymes). Consequently, the chemistry of the response is dependent on the type of the stimulus. This chapter will go over a wide range of stimuli-responsive chemistry representing the three types of triggers described above. One must note that as newer materials are developed and new targets are discovered, so does the chemistry and the mechanisms of release expand.

2.2 Physical Stimuli

Physical stimuli include light, temperature, ultrasound, magnetic fields or mechanical stimuli. Physical stimuli typically cause an electronic, structural or a chemical change in the nanomaterial at the molecular level, resulting in significant changes at the macroscopic level.

2.2.1 Thermoresponsive Materials

Thermoresponsive materials are typically grouped into three different categories: liposomes, polymeric micelles or nanoparticles. The trigger for thermoresponsive properties originates from materials exhibiting a lower critical solution temperature (LCST). LCST is the temperature below which the components of a polymer mixture remain miscible. Therefore, at temperatures above the LCST the polymer undergoes phase separation and this property is exploited to allow release of a drug. An example of such a polymer is poly(*N*-isopropylacrylamide) or PANIAm with an LCST of 32 °C. Which means that below 32 °C the polymer remains solubilized in it's aqueous surroundings with the polymer chains in an expanded coil state, and above 32 °C it undergoes coil–globule transition where the chains collapse and aggregate from the solution. Ideally for a DDS, the LCST of the nanosystem should be around 40 to 41 °C and it should be stable at physiological conditions at 37 °C.

An example of a temperature sensitive liposomal systems is Thermodox®, a lyso-lipid temperature sensitive liposome (LTSL) developed by Celsion. Thermodox® consists of three components: dipalmitoylphosphatidylcholine (DPPC) a phospholipid, monosteroylphosphatidylcholine (MSPC), a lyso-lipid and 1,2-distearoyl-*sn*-*glycero*-3-phosphatidylethanolamine-*n*-[methoxy(polyethylene glycol)-2000 (DPSE–PEG_{2000}).[1] The Van der Waals interactions between the lipid molecules determine the phase transition temperature of the liposome, and in turn the Van der Waals forces are determined by the length of the phospholipid and the degree of unsaturation of the phospholipid. At the transition temperature the lipid bilayer changes from an ordered, solid like phase to a fluid, liquid like phase. The lyso-lipid MSPC is similar to a phospholipid, except with one less acyl group. This results in a more porous liposome and allows the drug to be released faster once the desired temperature is reached. Finally the DPSE–PEG_{2000} was used to increase the circulation time of the drug. This thermoresponsive liposomal system has a transition temperature of 43 °C and has been used for delivery of Doxorubicin for the treatment of cancer.

Some modifications of this strategy include the generation of bubbles within the liposome by inclusion of compounds such as ammonium bicarbonate that decompose at higher temperatures and release CO_2.[2]

A triblock copolymer of polylactic-*co*-glycolic acid (PLGA) and polyethylene glycol (PEG) (PLGA–PEG–PLGA) marketed by BTG, known in the UK as ReGel®, has been used as an injectable drug depot for insulin.[3] Upon injection, the polymer undergoes gelation at 37 °C. The gelation is caused by the loss of bound water from PLGA, which is hydrophobic at physiological temperature. The PLGA blocks phase separate by aggregation and cause release of the drug. In addition, as PLGA has an ester backbone, it hydrolyzes to it's constituent acids (lactic acid and glycolic acid) in the presence of water resulting in accelerated release of the payload.

Li and coworkers[4] have reported a thermoresponsive polymer based on the upper critical solution temperature (UCST) for use as a temperature-responsive DDS. The use of such a polymer *in vivo* would require hyperthermia conditions at the point of treatment, such as the use of microwave therapy. They prepared a copolymer of acrylamide and acrylonitrile, poly(acrylamide-*co*-acrylonitrile) followed by a PEG graft on the free amides. This poly(AAm-*co*-AN)-*g*-PEG formed micelles in aqueous solutions. However, as the temperature increased, the micelle size began to shrink until at temperatures above 43 °C the micellar form of the polymer was completely disassembled. This is a polymer that transitioned from being hydrophobic to hydrophilic, caused by an increase in temperature.

2.2.2 Photoresponsive Materials

Photoresponsive materials react to specific wavelengths of light causing either a chemical or conformational modification of the photoresponsive element in the DDS. This mechanism is typically used as a "gate" valve or a switch. The trigger may be a one-time release or reversible depending on the chemistry of the reaction. There is a vast repertoire of molecules that are photoresponsive. This section contains a representation of a few typical reactions employed in this field.

2.2.2.1 Photoisomerizable Groups

An example of a reversible system is the use of the dimerization thymine reported by He and coworkers.[5] In this case mesoporous silica nanoparticles were functionalized with thymine and then loaded with a drug. Irradiation of the nanomaterial at 365 nm caused the thymine molecules to dimerize, thus trapping the payload while irradiation at 240 nm reversed the dimerization, opening the gate and releasing the drug (Figure 2.1).

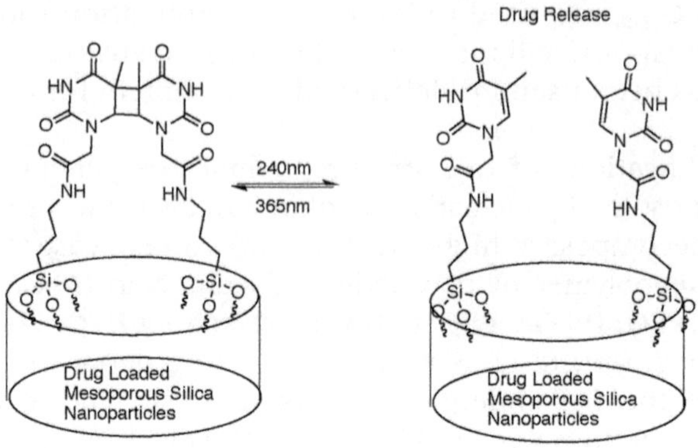

Figure 2.1 Reversible photo-dimerization of thymine.

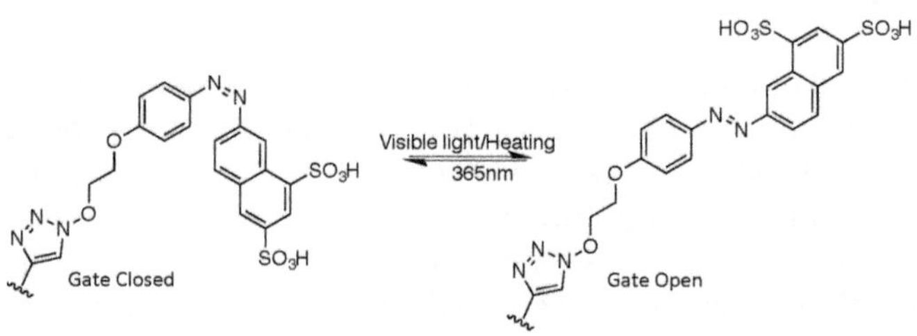

Figure 2.2 *Cis*–trans photo-isomerization of azobenzene systems.

Cis–trans photoisomerization of azobenzene ring derivatives occurs between 300–380 nm and is reversible. A [2]rotaxane system consisting of α-cyclodextrin and an azobenzene derivative was used by Yan and coworkers[6] to create reversible on–off gate functionalized mesoporous silica nanoparticles (Figure 2.2).

Photo-responsive catanionic vesicles were prepared by Liu *et al.*[7] by mixing azobenzene based cationic species and anionic surfactants, either sodium dodecylbenzene sulfonate (SDBS) or sodium dodecyl sulfonate (SDS) in a 7 : 3 ratio. These vesicles were developed for the photo-triggered release of DNA molecules after cellular uptake by endocytosis. Upon exposure to UV light of wavelength 350 nm, the azobenzene species that make up the vesicle isomerize from the *trans* to *cis* conformation resulting in release of the DNA. The disruption of the vesicle is a reversible system with the *cis* to *trans* isomerization occurring *via* visible light or thermal relaxation.

Spiropyrans are reversible photoswitching type molecules that undergo ring opening and isomerize to their merocyanine form upon UV exposure. The ring-closed spiropyran form is uncharged and hydrophobic, whereas the ring opened merocyanine form is zwitterionic and hydrophilic. This change in hydrophilic/hydrophobic character was exploited by Tong and coworkers[8] to develop a nanoparticle DDS that changed size upon being triggered by light. The nanoparticles consisted of a hybrid spyropyran–lipidic PEG system. The PEG was introduced to improve the circulation time and increase drug-loading capacity of the DDS, as well as to affect the microenvironment of the photoswitching molecule. The spiropyran form, being hydrophobic, moves into the core of the nanoparticle, disrupting the lipids and expanding the whole system. Application of UV light results in a change from the hydrophobic spiropyran to the hydrophilic merocyanine form, moving towards the outside of the nanosystem and allowing the lipidic core to collapse, shrinking the nanoparticle size. The process is reversible by leaving the system in the dark, as the less stable merocyanine form naturally reverts back to the more stable spiropyran form (Figure 2.3).

2.2.2.2 Photo-cleavable Groups

An example of a one-time release strategy is the use of a photocleavable linkage to release a drug. Agasti and coworkers[9] prepared gold nanoparticles conjugated to the anticancer drug 5-fluorouracil through a photoresponsive *o*-dinitrobenzyl linker. Irradiation with 365 nm light results in the photolytic cleavage of 5-fluorouracil from through a radical mechanism.

Polymerization of hydroxyethylacrylate (HEA) and a crosslinker containing a nitrobenzene group prepared by Azagarsamy *et al.*[10] produced a photodegradable nanogel. A protein was encapsulated within the nanogel during the polymerization process. Protein release was achieved by application of UV light (365 nm) causing degradation of the nanogel, resulting from the cleavage of the polymer at the nitrobenzene group (Figure 2.4).

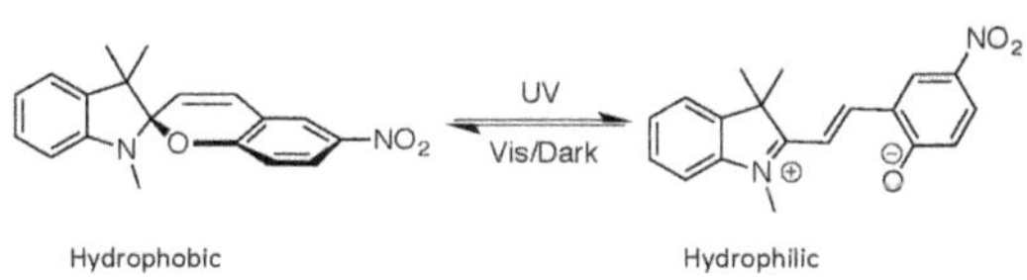

Figure 2.3 Light mediated reversible ring opening of spyropyrans.

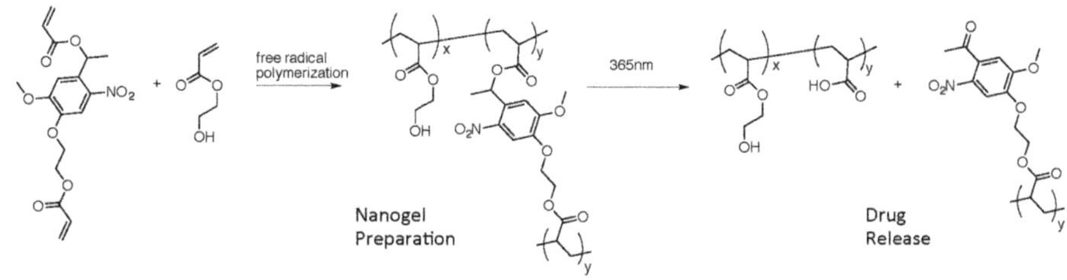

Figure 2.4 Synthesis and photo-cleavage of nanogel containing photo-labile nitrobenzene group.

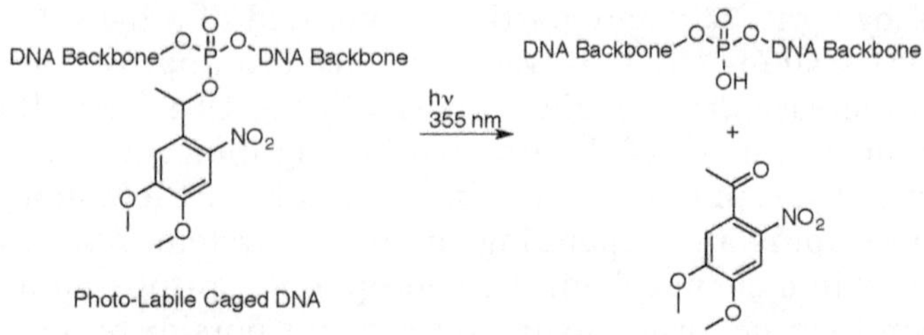

Figure 2.5 Structure of caged DNA and its light mediated release by cleavage of the dimethoxy nitrophenyl group.

Schroeder and coworkers[11] used (4,5-dimethoxy-2-nitrophenyl)diazoethane (DMNPE) based photo-labile caging and silencing of plasmid DNA previously reported by Monroe *et al.*[12] to develop a phospholipid vesicle capable of protein synthesis when triggered. The dimethoxy nitrophenyl group was attached to the phosphate on the DNA backbone. This bond is cleaved when irradiated with 300–400 nm light releasing active DNA, resulting in targeted protein expression. The Friedman group[13] has published a detailed protocol for the DMNPE functionalization of interference RNA (RNAi) followed by light triggered release using a UV lamp (Figure 2.5).

The photo-labile *p*-hydroxyphenacyl (*p*HP) group and peroxide sensitive boronate group were used by the Singh group[14] to develop a DDS for anticancer drugs. This nanoprodrug is a two-step detection and release system. The presence of high levels of hydrogen peroxide in cancerous environments triggers the release of a benzothiazole-*p*HP-drug conjugate resulting in green fluorescence. The application of visible light ≥410 nm releases the drug itself from the *p*HP portion of the molecule and also causes the residual fluorescent moiety to change to blue fluorescence, thus indicating the release of the drug (Figure 2.6).

This group has also shown the use of coumarin as a photo-cleavable group for the release of chlorambucil towards the treatment of prostate cancer.[15] The molecule also contained di-(2-picolyl)amine as a zinc sensor to distinguish healthy prostrate cells, which are zinc rich, from cancerous cells. The efficiency of the photo-triggered drug release was higher in the absence of Zn^{2+} than when the di-(2-picolyl) amine moiety formed a complex with Zn^{2+}. A 90% drug release was observed within 50 to 60 minutes in the former case, compared to 5 hours in the latter case. In addition, the photo release could be triggered either by one photon absorption (1PA) at 365 nm or by a two photon absorption (2PA) at 675 nm (Figure 2.7). The advantage of 2PA is that the use of near infrared (NIR) wavelengths have deeper tissue penetration and are therefore more realistic for use in phototherapy.

Photolabile groups may also include protecting groups that are photosensitive. While not targeted at drug delivery, the reader is directed to an extensive review by Klan and coworkers[16] on photocleavable protecting groups, that contains in depth information on the chemistry of these moieties.

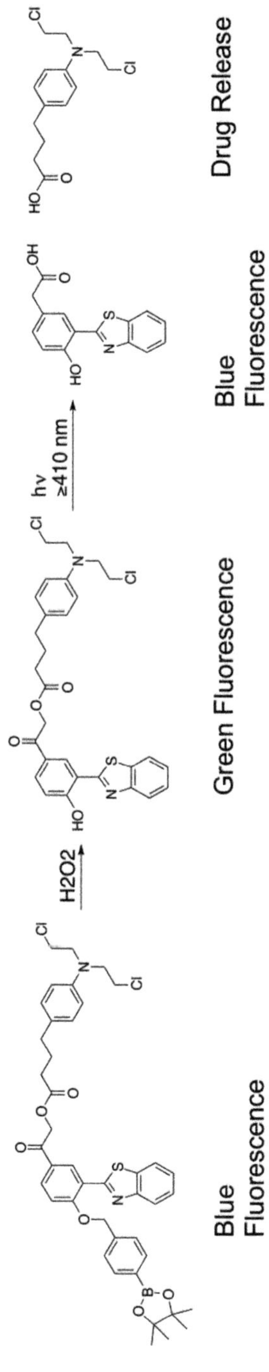

Figure 2.6 Two step peroxide mediated cancer detection and light mediated drug release.

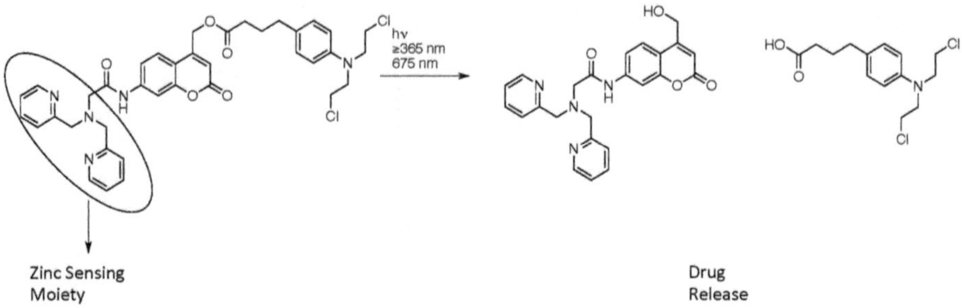

Figure 2.7 Coumarin based photo-cleavable drug release with di-(2-picolyl)amine as a zinc sensor.

2.2.3 Magnetically Responsive Materials

Magnetically responsive materials are usually made of core–shell systems in which the core is magnetic, while the shell may be a polymer or ligand/polymer decorated silica or liposomes. While the magnetic properties of these nanomaterials are usually used for guiding the DDS towards the target, the use of an alternating magnetic field can cause an increase in temperature resulting in the release of the drug once the target is approached.

In a static magnetic field approach, Qin *et al.*[17] described the use of a mixture of oleic acid coated superparamagnetic iron oxide nanoparticles (SPIONs) and a drug loaded Pluronic F127 (PF127) copolymer for use as a magnetically triggered DDS. Here, the application of a magnetic field causes the SPIONs to align with the magnetic field. It also causes SPION–SPION alignment and attraction. This movement of the ferrogel results in squeezing of the drug loaded PF127 micelles.

In a similar approach, a polyethylenimine (PEI) coated SPION was used for the transfection of malaria DNA vaccine using an external magnetic field.[18] In this case the drug was coated on top of the nanoparticles and was attached to the PEI electrostatically. The trigger mechanism for magnetofection in this case was a combination of the magnetic field, resulting in SPION–SPION attraction and squeezing of the plasmid DNA off the nanoparticles. The SPION also immobilized near the cells, allowing close proximity of the DNA to the cells and speeding up the transfection.

PEG functionalized magnetite nanoparticle clusters (PMNC) were reported by the Park group,[19] whereby magnetite nanoparticles were functionalized with a 3,4-dihydroxy-L-phenylalanine (DOPA)–polyethylenimine (PEI) conjugate (PEI-DOPA). Anionic siRNA was loaded into the nanoparticle system through electrostatic interaction with the cationic PEI. These clusters were targeted towards transfection of cancer cells and triggered using a magnetic field. This system of nanoclusters exhibited very little cellular uptake after incubation with cells for 120 min. However, upon exposure to a magnetic field (70 to 250 mT) for 15 min the cells showed cellular uptake of the PMNC's at 6.6 times that of the unexposed cells. The PMNC exhibit superparamagnetic properties without hysterysis. Therefore, exposure to a gradient magnetic field caused the magnetite cores to cluster quickly within the vicinity of the cells. Therefore, increased transfection was simply a result of increased interaction time between the cells and the siRNA loaded PMNC.

2.3 Chemical Stimuli

2.3.1 pH Responsive Materials

Mucoadhesive polymers made of lightly crosslinked high molecular weight polyacrylc acid or polymethacrylic acid have been incorporated in formulations for drug delivery to mucosal surfaces such as the eyes, nose, stomach and vagina. Mucin polymers present on these surfaces are high in COOH and OH groups and thus easily hydrogen bond to the polymers. Mahalingam *et al.*[20] demonstrated the use of pH triggered reversible crosslinking in a copolymer mixture. This was used as a microbicide to prevent the transmission of HIV through semen. Individual copolymers were prepared by reacting *N*-[3-(2-Methyl-acryloylamino)-propyl]-4-amidophenylboronic acid (APMAmPBA) and 4-[(2-Methyl-acryloylamino)-methyl]-salicylhydroxamic acid (MAAmSHA) with 2-hydroxypropyl methacrylamide (HPMAm). The two copolymers (HPMAm-*co*-APMAmPBA and HPMAm-*co*-MAAmSHA) when mixed together form a flowable lightly crosslinked gel. Upon coming into contact with cervical mucus at the vaginal pH of 4–5.5 the boronic acid weakly binds to the mucins. Upon coming into contact with semen, the pH is closer to neutral and this results in dense crosslinking between the boronic acid and the salicylhydroxamic acid preventing HIV transmission (Figure 2.8).

The guest host complex created by the cation–dipole interaction of cucurbit[7]uril (CB[7]) with ethylene diamine endcapped poly(glycidyl methacrylate) (PGMA) can act as a pH responsive gate when functionalized onto nanoparticles.[21,22] At physiological pH one CB[7] molecule is bound to two amines. Reducing the pH to below 5 causes competitive

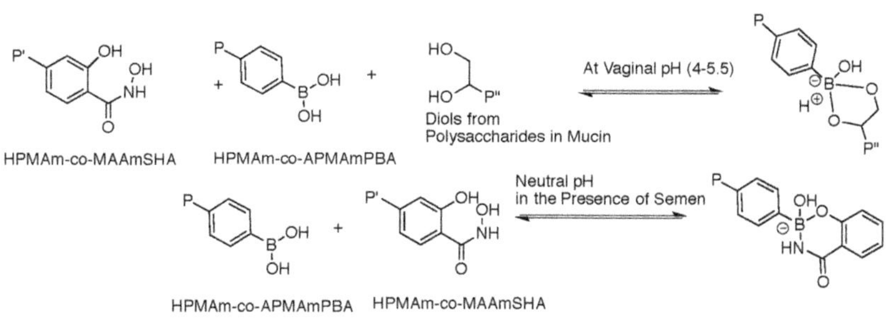

Figure 2.8 pH sensitive thickening of boronic acid and salicylhydroxamic acid based polymer for protection against HIV.

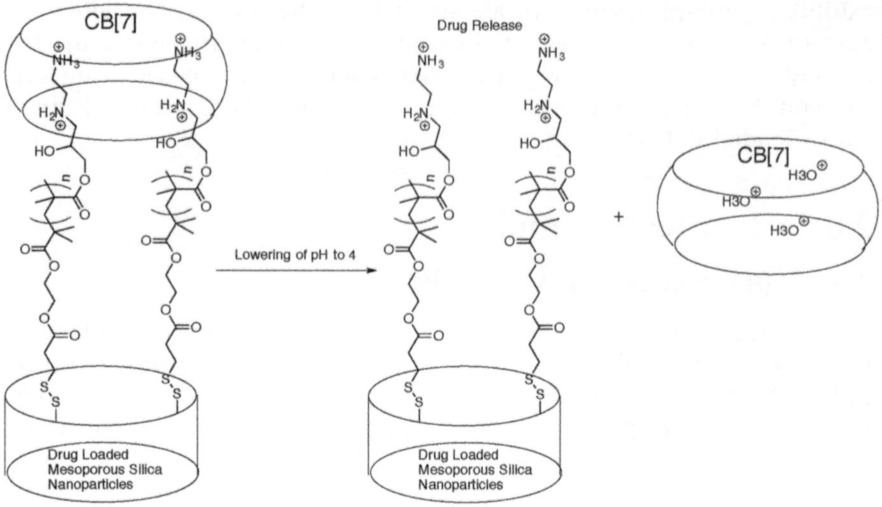

Figure 2.9 pH mediated drug release observed in a guest host complex of cucurbit[7]uril (CB[7]) with ethylene diamine endcapped poly(glycidyl methacrylate) (PGMA).

complexation of CB[7] with hydronium ions instead of the amines. As the CB[7] dissociates from the polymer bound amine, the drug payload is released (Figure 2.9).

Wang *et al.* describe a complex "plug" gate based strategy involving a conformational change caused by a change in pH.[23] In this system, a mono-benzimidazole functionalized β-cyclodextrin based pseudo[1]rotaxane (BZI-β-CD) was covalently bonded to mesoporous silica nanoparticles containing *p*-coumaric acid embedded in it's pores. At neutral pH the benzimidazole sits in the β-CD cavity blocking the release of *p*-coumaric acid. However, at acidic pH benzimidazole becomes protonated and unplugs the β-CD cavity releasing the *p*-coumaric acid (Figure 2.10).[24]

In a multimodal strategy the same group created a DDS that was triggered by two stimuli, pH and voltage, that allowed them to incorporate multiple drugs into one DDS.[25] Once again, mesoporous silica nanoparticles were used as the vehicle to carry the anti-cancer drugs, doxorubicin and gemcitabine. The silica was functionalized with an alkyne end-group, prop-2-yn-1-yl (3-(triethoxysilyl) propyl) carbamate (PTPC). An acid cleavable linker, 2-(2-azidoethoxy)-2-(2-bromoethoxy) propane (ABP) was reacted with the alkyne using click chemistry. The acid cleaving group is a ketal functional group within the ABP and it also contains a bromo end group to allow for attachment of an amino functionalized mono-ferrocenyl β-cyclodextrin. This mono-ferrocenyl β-cyclodextrin acted as a reversible gate that was closed at −1.5 V with the ferrocenyl group sitting inside the β-CD cavity and opened at +1.5V to release the gemcitabine. Upon reducing the pH of the DDS environment, the ketal group becomes hydrolyzed to release the ferrocenyl

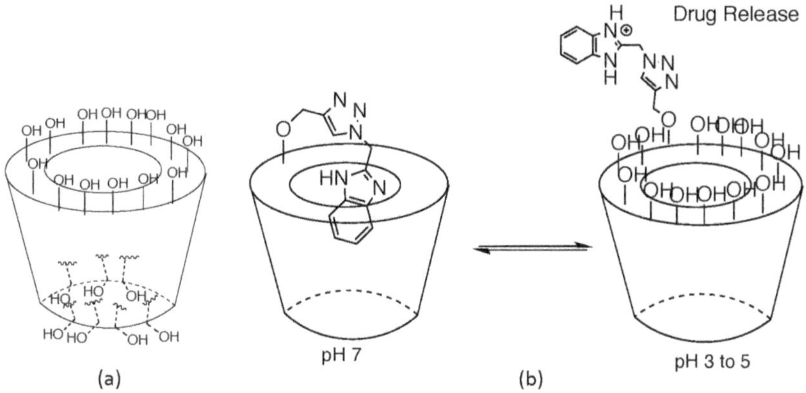

Figure 2.10 (a) Representative structure of β-cyclodextrin (β-CD) showing the hydroxyl groups and the cavity formed. (b) Reversible pH sensitive gate opening/closing using a benzimidazole molecule and β-CD.

β-cyclodextrin and detaching it from the silica surface resulting in the release of doxorubicin. Thus, two drugs could be released sequentially or concurrently for use in combination chemotherapy against cancer (Figure 2.11).

Mesoporous silica nanoparticle (MSN) supported lipid bilayers (protocells) have been used to carry and deliver siRNAs targeted against cancer.[26–28] The supported lipid bilayer was decorated with a peptide that would recognize hepatocellular carcinomas and a second endosomolytic peptide that would allow for the release of the protocell once inside the endosome. These protocells are stable at physiological pH but are triggered to release therapeutic agents under acidic pH. At acidic pH, there is a decrease in the Van der Waals forces between the payload (siRNA in this case), the MSN and the lipid bilayer, resulting in the disruption of the protocell membrane and release of the cargo.

Porous hollow iron oxide (Fe_2O_3) nanoparticles (2–4 nm pore size) derivatized with Herceptin to target breast cancer and loaded with cisplatin were reported by Cheng and coworkers. These nanoparticles were stable at a pH above 7 but at acidic pH (<6) the nanoparticle pores were etched and resulted in release of cisplatin.[29]

2.3.2 Thiol-responsive Materials

2-(propyldisulfanyl)ethylamine functionalized mesoporous silica nanospheres were prepared by Lai *et al.*[30] These were loaded with drugs such as vancomycin or adenosine triphosphate (ATP) and then capped with mercaptoacetic acid-derivatized cadmium sulfide (CdS) nanocrystals using an amidation procedure. The CdS nanoparticles consequently become attached to the silica nanospheres through disulfide links and are therefore susceptible to thiol-reducing agents such as dithiothreitol or mercaptoethanol. The DDS is leak proof, until it comes into contact with a glutathione (GSH) rich

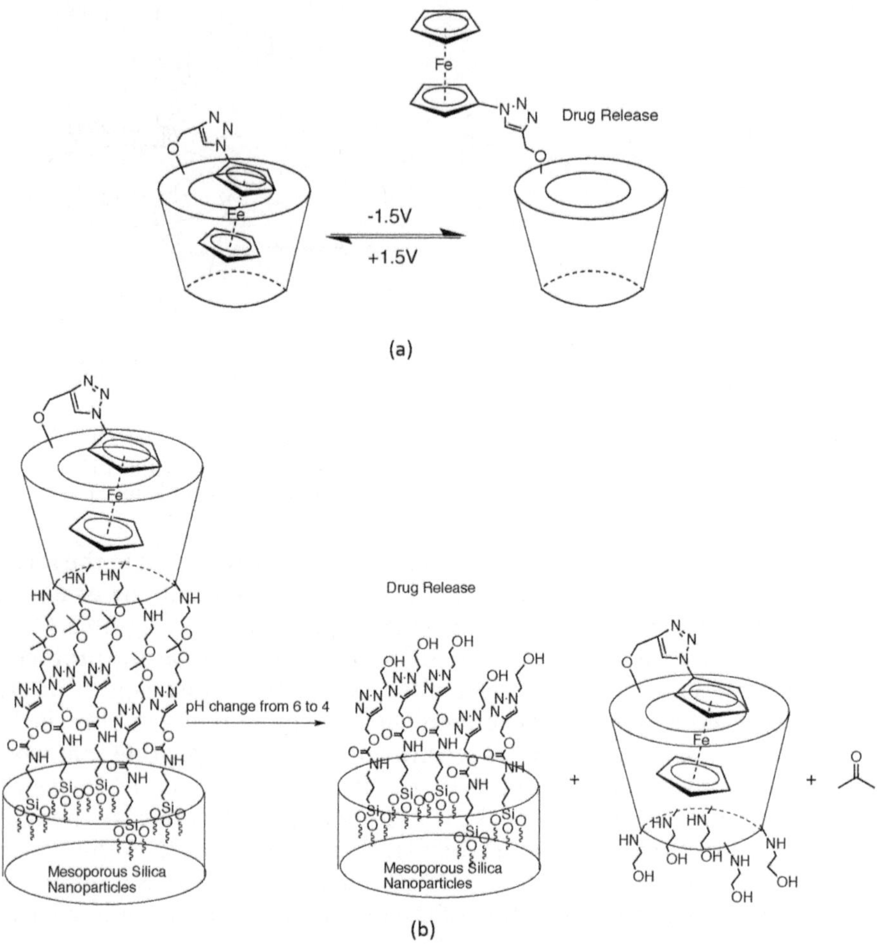

Figure 2.11 (a) Voltage triggered reversible gate opening and closing of β-CD using a ferrocenyl group. (b) A two drug system including the voltage triggered β-CD system and the pH triggered drug release by hydrolysis of ketal groups on mesoporous silica.

environment of a tumor where the disulfide bonds break releasing the CdS caps and discharging the drug from the silica pores (Figure 2.12). The use of disulfide linkers has shown great versatility in creating "plug" type gates and has been used to derivatize silica nanoparticles with various materials such as iron oxide,[31] ZnO quantum dots,[32] second generation polyamidoamines (G2-PAMAM)[33] and gold nanoparticles.[34]

2.4 Biological/Biochemical Stimuli

The use of biologically triggered nanomaterials exploits disease specific microenvironments to enable drug release. We have already seen some examples of bioresponsive DDS in the case of pH responsive materials and

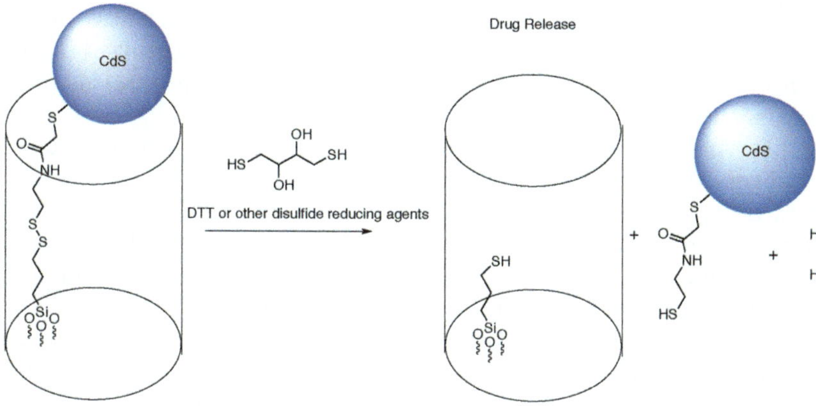

Figure 2.12 Drug release from mesoporous silica by reduction of disulfide linker in thiol rich environments.

thiol-responsive materials. In addition, nanomaterials can be created that are responsive to other biological molecules such as enzymes. Since enzymes work specifically by cleaving specific bonds in molecules, nanomaterials can be made enzyme specific by using targeting-ligands and the incorporation of specific enzyme cleavable groups.

Law and coworkers[35] designed a gel matrix sensitive to urokinase-type plasminogen activator (uPA), a serine protease by using self-assembling peptide units containing two β-sheets and a uPA sensitive sequence. They introduced a therapeutic peptide into the gel and upon incubation with uPA they were able to achieve uPA mediated release of the drug from the matrix.

Kang *et al.*[36] developed a peptide they named Alphatomega (HFKKQGS-FAKKK-NH2)[37] which was selective to a cancer specific signal transduction protein (STP) known as protein kinase Cα (PKCα). The drug delivery system consisted of a copolymer of methacryloyl peptide and acrylamide. Plasmid DNA (pDNA) was loaded on to the DDS *via* polyplex formation of the pDNA and the copolymer. When injected into B16 melanoma cells, the PKCα, which is over expressed in the cells, caused an enzymatic degradation of the polymer containing the peptide and released the plasmid DNA into the cells.

An activatable low molecular weight protamine (E10- PLGLAG-VSRRRRR-RGGRRRR) (ALMWP) was developed by Gu *et al.*,[38] It was conjugated to a matrix of poly(ethylene glycol)-*block*-poly(ε-caprolactone) (PEG–PCL)to construct a matrix metalloprotease(MMP)-responsive DDS. The design of the ALMWP consists of three sections: a polycationic cell penetrating peptide (CPP) sequence on one end, a polyanionic sequence E10 that masks the CPP on the other end and a peptide sequence PLGLAG, that is cleaved by matrix metalloproteases MMP-2 and MMP-9 in the middle. Studies have shown that two matrix metalloproteases MMP-2 and MMP-9 are over-expressed in glioblastomas. The authors prepared a DDS by conjugating the PEG–PCL with the ALMWP, loaded with paclitaxel for treating glioblastomas through targeted MMP-triggered drug therapy. Many other MMP sensitive peptide sequences have been identified in the literature that may be utilized in a similar manner.[39–41] Vartak and Gemeinhart describe the use of MMP as targets for various pathological conditions.[42]

Phospholipases are another group of enzymes that may be targeted for drug delivery. For example, a liposomal DDS stabilized by gold–chitosan complex designed to be hydrolyzed in the presence of phospholipase A2 (PLA2) was used by Thamphiwatana and coworkers to deliver doxycycline to H. pylori infected sites.[43] The rate of release was directly proportional to the extent of infection, as more bacteria secreted more PLA2.

Harnoy *et al.*[44] synthesized an enzyme responsive micellar system consisting of hydrophilic PEG and hydrophobic dendrons with phenyl acetamide end groups. The phenyl acetamide groups were hydrolyzable by an enzyme called penicillin G amidase. The micelle self-assembled, forming a hydrophilic shell made of PEG chains and a hydrophobic core made up of the dendrons incorporating a hydrophobic payload. Upon treatment with the enzyme the micelles disassembled due to breakdown of the core and released the payload.

2.5 Summary

A vast array of stimuli-responsive materials have been developed for the delivery of drugs for various diseases such as cancer or diabetes. It is pertinent to find specific substrates in order for the drug delivery to be a truly targeted system. While certain chemo- and bioresponsive materials are targeted towards specific diseases through changes in pH or over-expression of enzymes, physically triggered materials such as thermo-, photo- or magneto-responsive systems can be physically activated at the point of interest. Thus, it becomes imperative to understand how the structure of the DDS material affects the chemistry that results in the release of the payload. Our understanding of their chemistry allows for the creation of systems that are highly tunable and will go far in helping develop successful therapies for many difficult to treat diseases.

References

1. D. Needham, J.-Y. Park, A. M. Wright and J. Tong, Materials characterization of the low temperature sensitive liposome (LTSL): effects of the lipid composition (lysolipid and DSPE–PEG_{2000}) on the thermal transition and release of doxorubicin, *Faraday Discuss.*, 2013, **161**, 515–534.
2. K.-J. Chen, H.-F. Liang, H.-L. Chen, Y. Wang, P.-Y. Cheng and H.-L. Liu, *et al.*, A thermoresponsive bubble-generating liposomal system for triggering localized extracellular drug delivery, *ACS Nano*, 2013, 7(1), 438–446.

3. J. Zhang, S. S. Desale and T. K. Bronich, Polymer-based vehicles for therapeutic peptide delivery, *Ther. Delivery*, 2015, **6**(11), 1279–1296.
4. W. Li, L. Huang, X. Ying, Y. Jian, Y. Hong and F. Hu, *et al.*, Antitumor drug delivery modulated by a polymeric micelle with an upper critical solution temperature, *Angew. Chem., Int. Ed.*, 2015, **54**(10), 3126–3131.
5. D. He, X. He, K. Wang, J. Cao and Y. Zhao, A light-responsive reversible molecule-gated system using thymine-modified mesoporous silica nanoparticles, *Langmuir*, 2012, **28**(8), 4003–4008.
6. H. Yan, C. Teh, S. Sreejith, L. Zhu, A. Kwok and W. Fang, *et al.*, Functional mesoporous silica nanoparticles for photothermal-controlled drug delivery *in vivo*, *Angew. Chem., Int. Ed.*, 2012, **51**(33), 8373–8377.
7. Y.-C. Liu, A.-L. M. Le Ny, J. Schmidt, Y. Talmon, B. F. Chmelka and C. T. Lee, Photo-assisted gene delivery using light-responsive catanionic vesicles, *Langmuir*, 2009, **25**(10), 5713–5724.
8. R. Tong, H. D. Hemmati, R. Langer and D. S. Kohane, Photoswitchable nanoparticles for triggered tissue penetration and drug delivery, *J. Am. Chem. Soc.*, 2012, **134**(21), 8848–8855.
9. S. S. Agasti, A. Chompoosor, C.-C. You, P. Ghosh, C. K. Kim and V. M. Rotello, Photoregulated release of caged anticancer drugs from gold nanoparticles, *J. Am. Chem. Soc.*, 2009, **131**(16), 5728–5729.
10. M. A. Azagarsamy, D. L. Alge, S. J. Radhakrishnan, M. W. Tibbitt and K. S. Anseth, Photocontrolled nanoparticles for on-demand release of proteins, *Biomacromolecules*, 2012, **13**(8), 2219–2224.
11. A. Schroeder, M. S. Goldberg, C. Kastrup, Y. Wang, S. Jiang and B. J. Joseph, *et al.*, Remotely activated protein-producing nanoparticles, *Nano Lett.*, 2012, **12**(6), 2685–2689.
12. W. T. Monroe, M. M. McQuain, M. S. Chang, J. S. Alexander and F. R. Haselton, Targeting expression with light using caged DNA, *J. Biol. Chem.*, 1999, **274**(30), 20895–20900.
13. A. Kala, P. K. Jain, D. Karunakaran, S. Shah and S. H. Friedman, The synthesis of tetra-modified RNA for the multidimensional control of gene expression *via* light-activated RNA interference, *Nat. Protoc.*, 2014, **9**(1), 11–20.
14. S. Biswas, J. Das, S. Barman, B. Rao Pinninti, K. T. Maiti and N. D. P. Singh, Environment activatable nanoprodrug: two-step surveillance in the anticancer drug release, *ACS Appl. Mater. Interfaces*, 2017, **9**(34), 28180–28184.
15. S. Biswas, J. Das, S. Barman, S. S. Shah, M. Gangopadhyay and T. K. Maiti, *et al.*, Single component image guided "On-demand" drug delivery system for early stage prostate cancer, *Sens. Actuators, B*, 2017, **244**, 327–333.
16. P. Klán, T. Šolomek, C. G. Bochet, A. Blanc, R. Givens and M. Rubina, *et al.*, Photoremovable protecting groups in chemistry and biology: reaction mechanisms and efficacy, *Chem. Rev.*, 2013, **113**(1), 119–191.

17. J. Qin, I. Asempah, S. Laurent, A. Fornara, R. N. Muller and M. Muhammed, Injectable superparamagnetic ferrogels for controlled release of hydrophobic drugs, *Adv. Mater.*, 2009, **21**(13), 1354–1357.
18. F. N. Al-Deen, J. Ho, C. Selomulya, C. Ma and R. Coppel, Superparamagnetic nanoparticles for effective delivery of malaria DNA vaccine, *Langmuir*, 2011, **27**(7), 3703–3712.
19. J. W. Park, K. H. Bae, C. Kim and T. G. Park, Clustered magnetite nanocrystals cross-linked with PEI for efficient siRNA delivery, *Biomacromolecules*, 2011, **12**(2), 457–465.
20. A. Mahalingam, J. I. Jay, K. Langheinrich, S. Shukair, M. D. McRaven and L. C. Rohan, *et al.*, Inhibition of the transport of HIV *in vitro* using a pH-responsive synthetic mucin-like polymer system, *Biomaterials*, 2011, **32**(33), 8343–8355.
21. Q.-L. Li, Y. Sun, Y.-L. Sun, J. Wen, Y. Zhou and Q.-M. Bing, *et al.*, Mesoporous silica nanoparticles coated by layer-by-layer self-assembly using cucurbit[7]uril for *in vitro* and *in vivo* anticancer drug release, *Chem. Mater.*, 2014, **26**(22), 6418–6431.
22. Q.-L. Li, S.-H. Xu, H. Zhou, X. Wang, B. Dong and H. Gao, *et al.*, pH and glutathione dual-responsive dynamic cross-linked supramolecular network on mesoporous silica nanoparticles for controlled anticancer drug release, *ACS Appl. Mater. Interfaces*, 2015, **7**(51), 28656–28664.
23. T. Wang, M. Wang, C. Ding and J. Fu, Mono-benzimidazole functionalized [small beta]-cyclodextrins as supramolecular nanovalves for pH-triggered release of *p*-coumaric acid, *Chem. Commun.*, 2014, **50**(83), 12469–12472.
24. P. Díez, A. Sánchez, M. Gamella, P. Martínez-Ruíz, E. Aznar and C. de la Torre, *et al.*, Toward the design of smart delivery systems controlled by integrated enzyme-based biocomputing ensembles, *J. Am. Chem. Soc.*, 2014, **136**(25), 9116–9123.
25. T. Wang, G. Sun, M. Wang, B. Zhou and J. Fu, Voltage/pH-driven mechanized silica nanoparticles for the multimodal controlled release of drugs, *ACS Appl. Mater. Interfaces*, 2015, **7**(38), 21295–21304.
26. P. N. Durfee, Y.-S. Lin, D. R. Dunphy, A. J. Muñiz, K. S. Butler and K. R. Humphrey, *et al.*, Mesoporous silica nanoparticle-supported lipid bilayers (protocells) for active targeting and delivery to individual leukemia cells, *ACS Nano*, 2016, **10**(9), 8325–8345.
27. C. E. Ashley, E. C. Carnes, K. E. Epler, D. P. Padilla, G. K. Phillips and R. E. Castillo, *et al.*, Delivery of small interfering RNA by peptide-targeted mesoporous silica nanoparticle-supported lipid bilayers, *ACS Nano*, 2012, **6**(3), 2174–2188.
28. C. E. Ashley, E. C. Carnes, G. K. Phillips, D. Padilla, P. N. Durfee and P. A. Brown, *et al.*, The targeted delivery of multicomponent cargos to cancer cells by nanoporous particle-supported lipid bilayers, *Nat. Mater.*, 2011, **10**(5), 389–397.
29. K. Cheng, S. Peng, C. Xu and S. Sun, Porous hollow Fe_3O_4 nanoparticles for targeted delivery and controlled release of cisplatin, *J. Am. Chem. Soc.*, 2009, **131**(30), 10637–10644.

30. C.-Y. Lai, B. G. Trewyn, D. M. Jeftinija, K. Jeftinija, S. Xu and S. Jeftinija, *et al.*, A mesoporous silica nanosphere-based carrier system with chemically removable CdS nanoparticle caps for stimuli-responsive controlled release of neurotransmitters and drug molecules, *J. Am. Chem. Soc.*, 2003, **125**(15), 4451–4459.
31. S. Giri, B. G. Trewyn, M. P. Stellmaker and V. S.-Y. Lin, Stimuli-responsive controlled-release delivery system based on mesoporous silica nanorods capped with magnetic nanoparticles, *Angew. Chem., Int. Ed.*, 2005, **44**(32), 5038–5044.
32. S. Wu, X. Huang and X. Du, pH- and redox-triggered synergistic controlled release of a ZnO-gated hollow mesoporous silica drug delivery system, *J. Mater. Chem. B*, 2015, **3**(7), 1426–1432.
33. D. R. Radu, C.-Y. Lai, K. Jeftinija, E. W. Rowe, S. Jeftinija and V. S.-Y. Lin, A polyamidoamine dendrimer-capped mesoporous silica nanosphere-based gene transfection reagent, *J. Am. Chem. Soc.*, 2004, **126**(41), 13216–13217.
34. X. Sun, Y. Zhao, V. S.-Y. Lin, I. I. Slowing and B. G. Trewyn, Luciferase and luciferin co-immobilized mesoporous silica nanoparticle materials for intracellular biocatalysis, *J. Am. Chem. Soc.*, 2011, **133**(46), 18554–18557.
35. B. Law, R. Weissleder and C.-H. Tung, Peptide-based biomaterials for protease-enhanced drug delivery, *Biomacromolecules*, 2006, **7**(4), 1261–1265.
36. J.-H. Kang, D. Asai, J.-H. Kim, T. Mori, R. Toita and T. Tomiyama, *et al.*, Design of polymeric carriers for cancer-specific gene targeting: utilization of abnormal protein kinase cα activation in cancer cells, *J. Am. Chem. Soc.*, 2008, **130**(45), 14906–14907.
37. J.-H. Kang, D. Asai, S. Yamada, R. Toita, J. Oishi and T. Mori, *et al.*, A short peptide is a protein kinase C (PKC) α-specific substrate, *Proteomics*, 2008, **8**(10), 2006–2011.
38. G. Gu, H. Xia, Q. Hu, Z. Liu, M. Jiang and T. Kang, *et al.*, PEG-*co*-PCL nanoparticles modified with MMP-2/9 activatable low molecular weight protamine for enhanced targeted glioblastoma therapy, *Biomaterials*, 2013, **34**(1), 196–208.
39. H. Nagase and G. B. Fields, Human matrix metalloproteinase specificity studies using collagen sequence-based synthetic peptides, *Pept. Sci.*, 1996, **40**(4), 399–416.
40. S.-J. Deng, D. M. Bickett, J. L. Mitchell, M. H. Lambert, R. K. Blackburn and H. L. Carter, *et al.*, Substrate specificity of human collagenase 3 assessed using a phage-displayed peptide library, *J. Biol. Chem.*, 2000, **275**(40), 31422–31427.
41. S. Netzel-Arnett, G. B. Fields, H. Birkedal-Hansen, H. E. Van Wart and G. Fields, Sequence specificities of human fibroblast and neutrophil collagenases, *J. Biol. Chem.*, 1991, **266**(11), 6747–6755.
42. D. G. Vartak and R. A. Gemeinhart, Matrix metalloproteases: underutilized targets for drug delivery, *J. Drug Targeting*, 2007, **15**(1), 1–20.

43. S. Thamphiwatana, W. Gao, D. Pornpattananangkul, Q. Zhang, V. Fu and J. Li, *et al.*, Phospholipase A2-responsive antibiotic delivery *via* nanoparticle-stabilized liposomes for the treatment of bacterial infection, *J. Mater. Chem. B*, 2014, **2**(46), 8201–8207.
44. A. J. Harnoy, I. Rosenbaum, E. Tirosh, Y. Ebenstein, R. Shaharabani and R. Beck, *et al.*, Enzyme-responsive amphiphilic PEG-dendron hybrids and their assembly into smart micellar nanocarriers, *J. Am. Chem. Soc.*, 2014, **136**(21), 7531–7534.

CHAPTER 3

pH-responsive Drug Delivery Systems

ZHAOHUI WANG, XINYI ZHANG, GANG HUANG AND JINMING GAO*

Department of Pharmacology, Harold C. Simmons Comprehensive Cancer Center, UT Southwestern Medical Center at Dallas, 5323 Harry Hines Blvd, Dallas, TX 75390, USA
*E-mail: jinming.gao@utsouthwestern.edu

3.1 Introduction

Rapid progress in materials science and pharmaceuticals have advanced a variety of drug delivery systems (DDS) to overcome challenges in low drug solubility, poor pharmacokinetics and toxicity in a wide range of pharmaceutical indications.[1–3] In order to maximize the therapeutic efficacy and minimize the side effects of loaded drugs, researchers have explored the use of stimuli-responsive nanocarriers for site-specific drug targeting and release. In these systems, a variety of "triggers" including external physical and endogenous biological stimuli have been employed.[4–6] Among them, pH is a ubiquitous biological stimulus that has been widely used to design novel, responsive drug delivery systems.[7,8]

Biomaterials Science Series No. 1
Stimuli-responsive Drug Delivery Systems
Edited by Amit Singh and Mansoor M. Amiji

Published by the Royal Society of Chemistry, www.rsc.org

3.1.1 pH Environment in Gastrointestinal Organs

pH varies throughout the gastrointestinal (GI) tract, with the lowest pH value of about 1.0–3.0 being in stomach, compared to small intestine pH at 5.5–6.8 and colon at pH 6.4–7.0.[9] Extensive efforts have been devoted in using pH-sensitive polymers as enteric coating materials that protect acid-labile drugs from degradation in the stomach to increase intestinal absorption.[10] Many enteric coated tablets and capsules have been approved for clinical use by the Food and Drug Administration.

3.1.2 Acidic pH Environment in Pathological Tissues

Tumor extracellular environments, infection and inflammatory tissues are acidic, with reported pH values ranging from 6.3 to 7.0, compared to 7.4 in blood and normal tissue interstitium.[11] Tissue acidosis is a hallmark of both the ischemia and inflammation processes. Bacterial infections are generally characterized by low pH values because of anaerobic fermentation and subsequent inflammation. During this process, there are increased levels of acidic products, through massive infiltration of neutrophils and macrophages to the site of infection, leading to a pH to as low as 5.5–7.0.[5,12,13]

Dysregulated pH with acidic extracellular pH (pH_e) is a hallmark of solid tumors regardless of cancer genotypes or phenotypes. Compared with normal tissue, cancer cells have increased uptake of glucose with elevated aerobic glycolysis to produce lactic acids.[14] This phenomenon was first observed by Otto Warburg in the 1930s, also known as the Warburg effect.[15,16] Besides lactic acid, generation of CO_2 by tumor cells *via* oxidative metabolism is another major source of extracellular acidity.[17] In addition, other pH regulatory proteins (*e.g.*, V-ATPase, Na^+/bicarbonate transporter, and anion exchangers) also contribute to acidic pH_e.[11,18] As a result of the increased proton production by the cancer cells and poor proton clearance from solid tumors (*e.g.*, due to impaired lymphatics), pH_e is a ubiquitous signature with a low value ranging from 6–7.[19,20]

3.1.3 Acidic Subcellular Organelles

Apart from mild acidic extracellular pH in diseased tissues, an even greater pH gradient can be found in multiple subcellular compartments. For example, the pH of Golgi apparatus is ~6.4, whereas cytoplasmic pH is 7.4. The most widely known acidic organelles are lysosomes (4–5) and endosomes (5–6.5), which play an important role in cell signaling, endocytic maturation, and protein/lipid catabolism.[21–23] Vacuolar H^+-ATPase plays a major part in the acidification of endocytic organelles.[24]

pH gradients within the organs, tissues, and subcellular organelles can be targeted for the design of drug carriers to achieve site-specific therapeutic delivery. In this chapter, we will review different strategies and material systems for pH-responsive drug delivery (Figure 3.1). At the organ level,

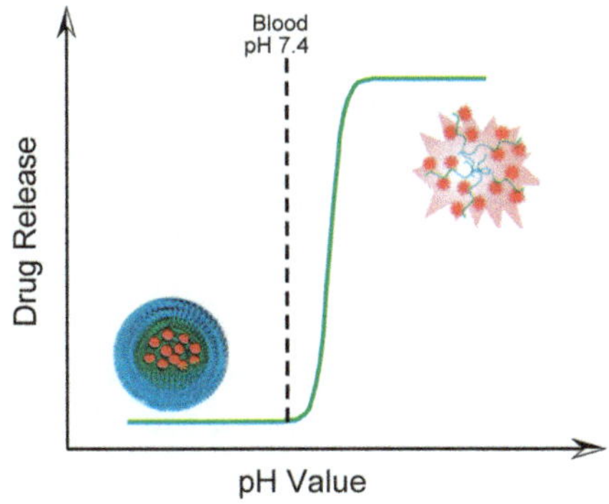

Figure 3.1 Scheme of pH-responsive drug delivery system.

oral delivery systems have been established to differentiate drug release and uptake along the GI tract with multiple clinical products. At the tissue level, various drug carriers have been designed to exploit the pH difference between tumor or infection site and normal tissue to achieve high drug accumulation.[25–27] Finally, at the intracellular level, pH-responsive nanoparticles have been extensively studied to escape acidic endolysosomal compartments for cytosolic drug delivery.[7,28]

3.2 Classification of pH-responsive Materials

We divide pH-responsive materials into three main categories. The first category consists of non-covalent ionizable polymeric systems with variable degrees of ionization in response to pH changes. The second category consists of polymers with acid-labile linkages, where breakage of covalent bonds at acidic pH is employed for the construction of pH-responsive nanocarriers.[29] Lastly, pH-sensitive inorganic nanomaterials, such as silica, calcium phosphate (CaP), and zinc oxide (ZnO), which are relatively insoluble at physiological pH, but can be dissolved in acidic microenvironments for payload release.[30,31] In the following section, we review each category in more detail.

3.2.1 Polymers with Ionizable Functional Groups

Non-covalent pH sensitive polymers can be further divided into two subcategories, namely acid-based polymers (–COOH) and polymers with organic bases (–NR_2). These acidic/basic pendant groups accept protons at low pH and release them at high pH depending on pK_a. In several cases, the pH-induced hydrophobic/hydrophilic transition can be used to control the assembly/disassembly of nanoparticles and swelling/deswelling of hydrogels in the aqueous environment, respectively. These properties can be used to load drugs in the carriers and release them when the environmental pH changes.

3.2.1.1 pH-responsive Acidic Polymers

Carboxylic Acid Groups. The carboxylic acid groups undergo deprotonation and become carboxylate anions at basic pH values (Figure 3.2). Poly(acrylic acid) (PAA), poly(methacrylic acid) (PMAA), poly(ethyl acrylic acid) (PEAA), poly(propyl acrylic acid) (PPAA), and poly(butyl acrylic acid) (PBAA) have been reported as pH-sensitive polyacids. For the copolymers composed of methyl methacrylate (MMA) and MAA, MMA is the hydrophobic segment and MAA is the ionizable segment. MAA becomes hydrophilic at high pH due to deprotonation of the carboxylic acid groups; when the carboxylic acid group is protonated at low pH, it becomes hydrophobic. In several cases, phase transition occurs across the pK_a value of the carboxylic acid groups, which is in the pH range of 4.5–5.5. The hemolytic activity and pH sensitivity of these compounds can be controlled by adjusting the hydrophobicity of alkyl chains. Polycarboxylates have been used

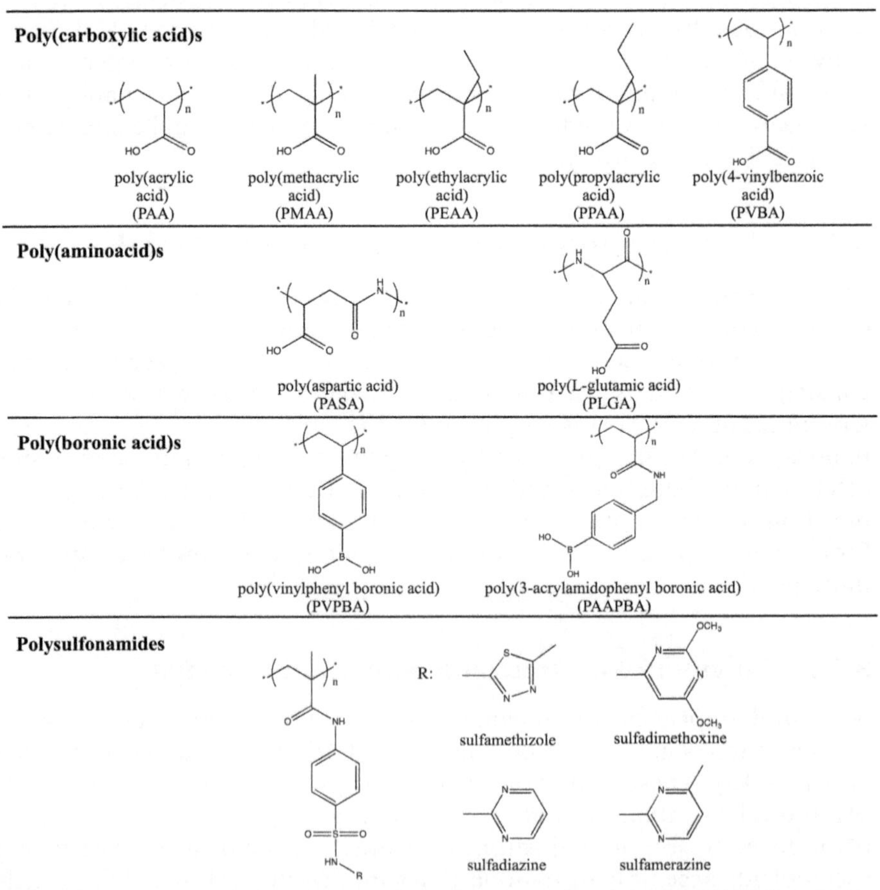

Figure 3.2 Structure of pH-responsive acidic polymers.

to construct pH-responsive vesicles, polymeric micelles, dendrimer and nanospheres.[32–34]

Boronic Acids. The pK_a of a boronic acid is typically 9, but it can form tetrahedral boronate complexes with $pK_a \sim 7$. Most of the polymeric boronic acids reported in the literature are phenylboronic acid moieties (–Ph–B(OH)$_2$).[35] Introduction of substituents on the phenyl ring allows the pK_a to be finely tuned in the physiologically relevant pH range. For example, the pK_a of aryl boronic acid can be modulated by the introduction of electron withdrawing groups (*e.g.*, nitro and fluorine groups) at the *para-* or *ortho-*positions of the boronic acid group, resulting in a lowered pK_a of phenylboronic acid from 7.8 to 7.2 (3-fluoro-4-carboxyphenylboronic acid and 3-nitrophenylboronic acid).[36,37] Polymers containing boronic acid groups are strongly hydrophobic when uncharged but it becomes hydrophilic when it is negatively charged at pH values above its pK_a, offering a pH responsive mechanism for drug delivery.[38]

Alternatively, boronic acids are able to bind diols reversibly, forming boronate esters that exhibit a fast dual response to external pH values and competing diols.[39] Li *et al.* reported a dual-responsive boronate cross-linked micelle for drug delivery based on the self-assembly and *in situ* complexation of boronic acid containing polymers and catechol containing polymers. The micelles retained the encapsulated drug under physiological conditions, while releasing the payload quickly when triggered by the lower pH value of the tumor environment, or when exposed to exogenous competing diols. These polymeric drug carriers have a chemically defined mechanism for drug loading and release through the pH-sensitive catechol-boronic acid interaction.[39,40]

Sulfonamide. Another type of pH-sensitive polymers contain arylsulfonamide groups.[41] The acidic protons of sulfonamide groups are readily dissociated as the pH increases, and the behavior of these polymers can be controlled over a narrower pH range close to physiological conditions, rendering them more pH sensitive than conventional carboxylic acid polymers.[42] Bae *et al.* investigated the weakly acidic sulfonamide as a trigger for extracellular delivery of the anticancer drug doxorubicin (DOX). Sulfonamide-carrying nanoparticles (NP) were stable in neutral solution, while the sulfonamide was deionized in acidic conditions and the NP collapsed, leading to the release of the encapsulated DOX. Since then, various sulfonamide-based pH-sensitive polymeric systems, including micelles, hydrogels, nanoparticles and dendrimers have been developed as drug carriers to enhance therapeutic effect at targeted sites.[43]

3.2.1.2 *pH-responsive Basic Polymers*

Polybases, which undergo protonation transitions from pH ~5–11, are employed as pH-responsive basic polymers. The amine groups accept protons at low pH values by forming cationic ammonium groups and subsequently releasing them under basic conditions.[44] The (meth)acrylate, (meth)acrylamide, and vinylic polymers containing tertiary amine, morpholino, pyrrolidine, imidazole, piperazine, and pyridine groups are shown in Figure 3.3.

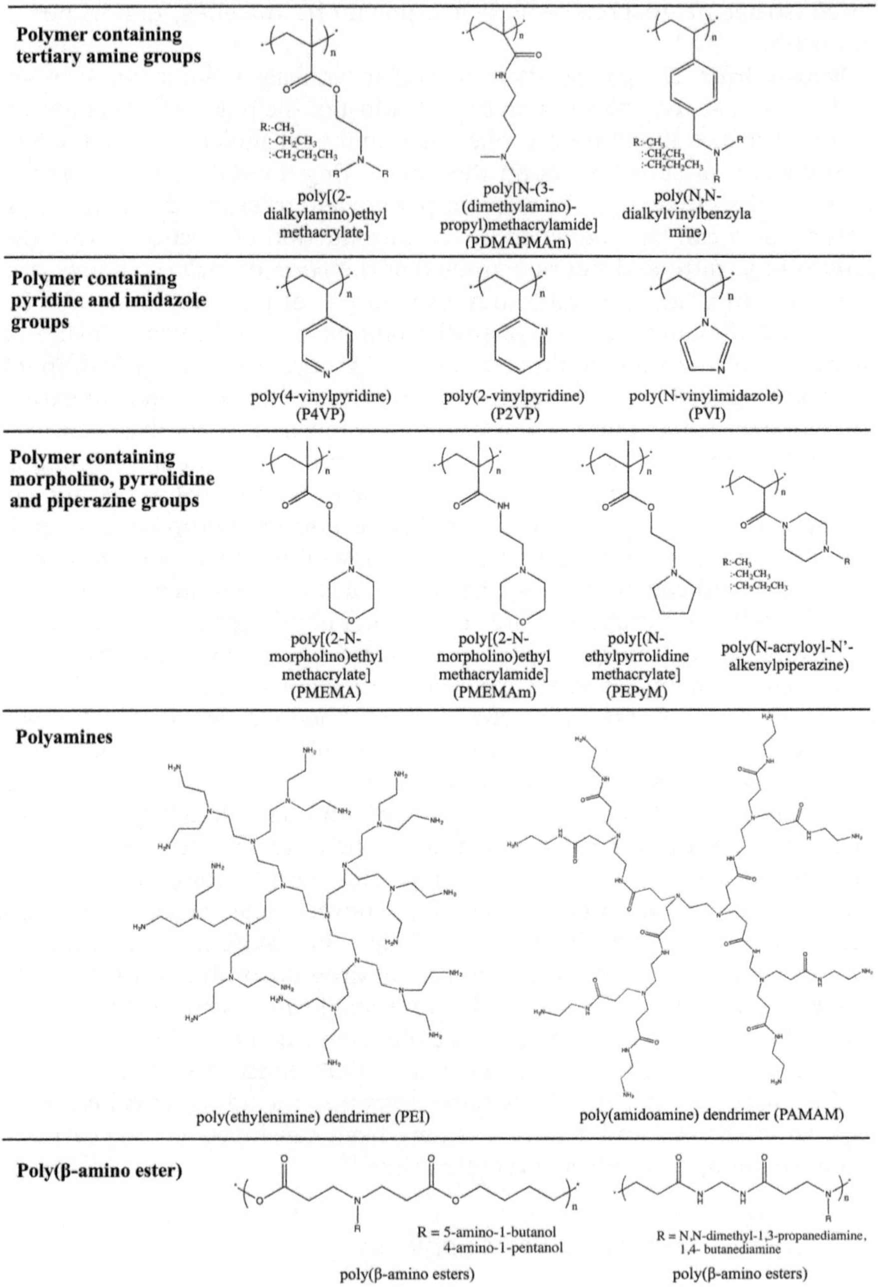

Figure 3.3 Structure of pH-responsive basic polymers.

Tertiary amine-containing polymethacrylate based polymers such as poly[(2-diethylamino)ethyl methacrylate] (PDEA), and poly[(2-diisopropylamino)ethyl methacrylate] (PDPA) are a representative class of basic polymers.[45] PDPA has a pK_a of 6.2. At pH 7.4, PDPA is hydrophobic due to deionization, however, at acidic tumor pH$_e$, it could be protonated and becomes hydrophilic. Poly(vinyl pyridine) based polymers are also widely used polymers, including poly(4-vinylpyridine) and poly(2-vinylpyridine). These polymers undergo a phase transition above pH 5 owing to deprotonation of the pyridine groups.[46,47] Poly (L-histidine) with imidazole side groups has a pK_a of 5.5–6.5 exhibiting potential for pH-activated phase transition.[48] Poly(ethylene imine) (PEI) and its derivatives are commercially available and widely used to deliver macromolecules into the cell cytosol. Importantly, dendrimers such as branched PEI, poly(propylene imine) (PPI), and poly(amidoamine) (PAMAM) have also been developed as pH-responsive polymers and used as DDS.

3.2.1.3 pH-responsive Natural Polymers

In recent years, the use of natural biodegradable polymers has also gained great attention owing to their good biocompatibility and easy modification by simple chemistry. Several natural polymers are also used in the pH-responsive self-healing gel system.[49] Dextran, hyaluronic acid (HA), heparin, alginic acid, cellulose, chitosan, and gelatine are among the most widely used natural polymers (Figure 3.4). By proper chemical modification, these polymers can exhibit pH-sensitivity for DDS.[50,51] For example, chitosan is a polycationic biopolymer (pK_a ~ 6.3) which is soluble in acidic solution and widely used for drug and gene delivery. The gelation mechanism of chitosan occurs through electrostatic attraction between the ammonium group of the chitosan and a counter ion, hydrogen bonding between the chitosan chains

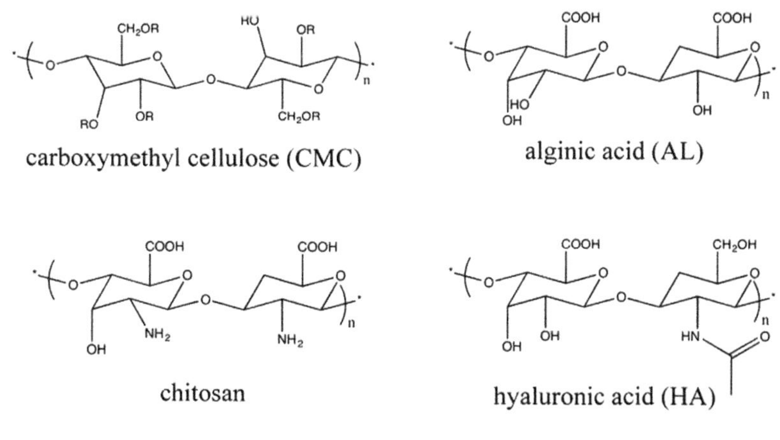

Figure 3.4 Structure of pH-responsive natural polymers.

and hydrophobic interactions.[52] Swelling of chitosan leads to the release of encapsulated tumor necrosis factor-α in the local acidic environment of tumor tissues.[53]

HA is also commonly used in drug delivery formulations. The presence of one carboxylic acid group per repeat unit imparts a pH-responsiveness which is enhanced in a crosslinked hydrogel network. Pitarresi *et al.* demonstrated the pH-responsive behavior of photo-crosslinked HA hydrogels for the release of thrombin.[54] Sharpe *et al.* introduced more carboxylic groups to optimize the pH-responsive of HA for drug delivery to the colon.[55]

3.2.2 Acid-labile pH-sensitive Nanocarriers

Acyl hydrazone, acetal/ketal, *cis*-acotinyl, imine, substituted trityl, orthoester linkers are pH-labile bonds that have been used in the syntheses of pH sensitive polymer structures.[26] These acid-labile chemical bonds are stable at neutral pH but are hydrolyzed in acidic media. Their chemical structures and degradation products are illustrated in Table 3.1. Ketal-containing poly(β-amino esters) (PBAE) are acid-degradable polymers that possess ketal linkages, as well as ester bonds, along with secondary amines in the backbone. Hydrolysis of the ketal linkage occurs approximately 1000 times faster at the phagosomal pH 4.5 *versus* the pH 7.4 environment of the blood.[56]

The acyl hydrazone group also received much attention in the design of pH-sensitive DDS. For example, Yu *et al.* developed a novel drug carrier *via* (PEG)-hyperbranched polyacylhydrazone for controlled release of paclitaxel (PTX) in a pH-dependent manner.[57] Alternatively, the conjugation of DOX to a polymer *via* the hydrazine is able to achieve rapid release at low pH,

Table 3.1 Examples of acid-labile chemical bonds and their degradation products.

Name	Acid-labile bonds	Degradation product
Orthoester	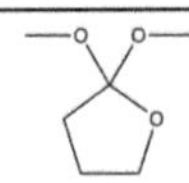	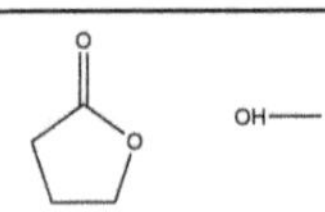
Hydrazone	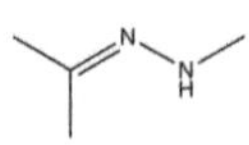	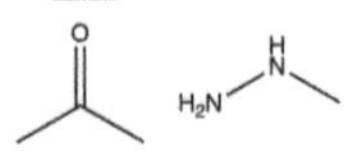
Ester	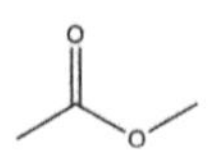	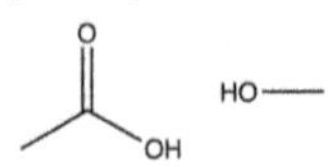
Acetal/ketal		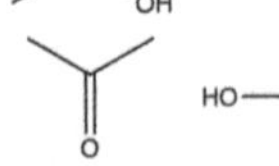
Imine	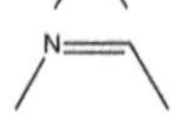	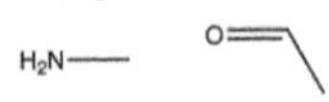
Cis-aconityl	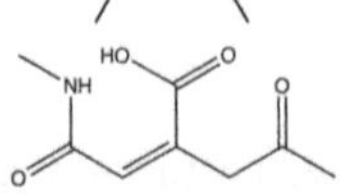	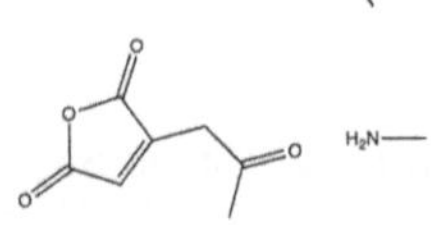

while being stable at neutral pH.[58] Kratz and coworkers developed a novel albumin-binding prodrug of DOX with acid-sensitive properties (INNO-206). INNO-206 can selectively release DOX in the acidic tumor tissue, which demonstrated superior antitumor efficacy in murine tumor models and a favorable toxicity profile in mice, rats and dogs, including significantly reduced cardiotoxicity. INNO-206 is under phase II clinical evaluation in advanced soft tissue sarcoma.[59]

3.2.3 pH-responsive Inorganic Nanoscale Materials

There have been rapid advances in the construction of pH-sensitive inorganic nanosystems for biomedical applications.[60] The most reported inorganic materials include Au nanostructures, carbon-based nanostructures, mesoporous silica, and calcium phosphate-based nanostructured materials. For example, CaP, regardless of Ca : P ratio, crystallinity, or phase, are relatively insoluble at physiological pH (pH 7.4) and become increasingly soluble below pH 6.5, yielding nontoxic calcium and phosphate ions in acidic microenvironments.[61] This unique property provides a multifunctional platform with high biocompatibility and pH sensitivity for the delivery of drugs, fluorescent dyes, or other cargos into cells. Kester *et al.* encapsulated both fluorophores and hydrophobic chemotherapeutics into CaP nanoparticles. The pH changes during endocytosis leads to dissolution of nanoparticles and subsequent cytosolic or perinuclear release of encapsulated agents.[61]

Besides CaP nanoparticles, ZnO quantum dots (QDs) also decompose in acidic aqueous solution. As a known biocompatible material with mesoporous structures, ZnO QDs are employed as platforms for targeted and pH-responsive intracellular delivery of drugs. Muhammad *et al.* demonstrated that ZnO QDs dissolve rapidly in the acidic environment of intracellular compartments in cancer cells, releasing DOX into the cytosol and killing the cells.[62]

3.3 Release Mechanism of pH-responsive DDS

pH-sensitive DDS is designed to be stable at neutral pH during circulation, but rapidly release drugs under mildly acidic/basic conditions. For example, a pH-sensitive liposomal system constructed by incorporation of pH-sensitive lipid palmitoyl homocysteine demonstrated pH-dependent drug release.[63] Generally, pH-responsive drug release can be exploited by several strategies. First, pH-sensitive DDS have been designed by using molecular moieties with specific pK_a values. In this case, pH change causes the protonation of multiple functional groups, disrupting the hydrophilic–hydrophobic balance and triggering a dramatic structural or conformational transformation, leading to rapid drug release. The second strategy is based on pH-labile linkages that are stable at neutral pH, but are cleaved under acidic conditions.[64] Other

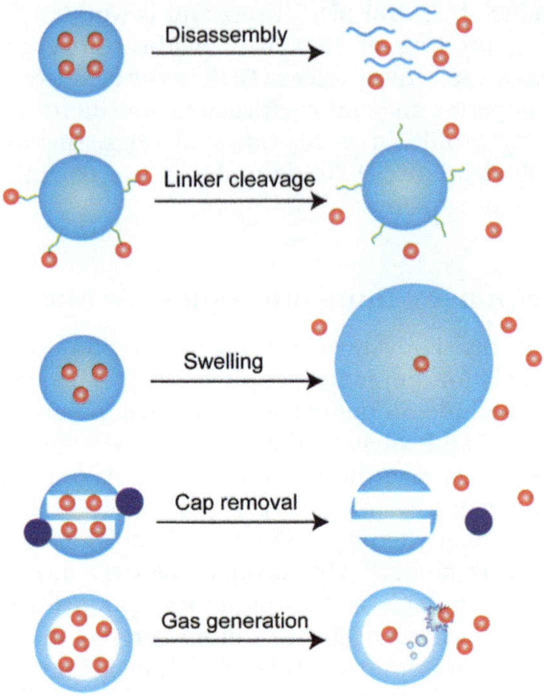

Figure 3.5 Schematic illustration of pH-activated drug release by various strategies.

strategies to induce pH-activated drug release include cap/coating removal from the nanoparticle, swelling/deswelling of the polymer and disintegration of nanoparticles by gas-generation (Figure 3.5).

3.3.1 Hydrophobic–Hydrophilic Transition Induced Disassembly

The types of polymers inducing drug carrier destabilization are materials that could be ionized upon a change in pH value. For example, anionic polymers containing carboxylate groups, such as poly(acrylic acid) and poly(methacrylic acid) have been utilized to construct pH-responsive DDS. At physiological pH, they are negatively charged and hydrophilic. At low pH, they are protonated to a neutral acid form to induce hydrophobic characteristics, which result in phase transition and the destabilization of vehicles. In contrast, poly(4-vinylpyridine), PDEA, poly(amino esters) and poly(L-histidine), which contain tertiary amine or pyridine groups, could be protonated as the pH decreases and they undergo hydrophobic to hydrophilic phase transition that induces the drug release.

Poly(L-histidine) (pHis) has a pK_a value near pH 6.5–7.0 and shows reversible hydrophilic to hydrophobic transitions in accordance with its protonated and deprotonated states, respectively.[65,66] Lee *et al.* developed mixed micelles by blending pHis-*b*-PEG diblock copolymer with poly(L-lactic acid) (PLLA)-*b*-PEG. The mixed micelles are stable at a pH above 7.4, but become gradually destabilized below pH 7.0 due to the protonation of the pHis block in the micelle core. Increased electrostatic repulsions which arise from the progressive protonation of imidazole rings overcome the hydrophobic interactions among the uncharged neutral blocks, leading to destabilization and swelling of the micellar structure at pH 6.8. When the pH further decreases to 6.0, dissociation of p(His)/PEG from the micelle core occurs. The mixed micelles can encapsulate the anticancer drug DOX through hydrophobic interactions between DOX and deprotonated pHis segments. At acidic tumor sites, the mixed micelles dissociate and selectively release the drug, leading to an improvement in antitumor efficacy as compared with their pH-insensitive counterparts. A relatively higher mixing ratio of PLLA/PEG can demonstrate a higher colloidal stability and a lower triggering pH of the micelles against pH drops and *vice versa*. The mechanism of pH-induced micelle destabilization may provide higher local concentrations of drugs after the micelles were accumulated at tumor sites.[67]

Cationic polymers containing tertiary amine groups are also developed as pH sensitive drug delivery systems. Peng *et al.* synthesized PEG methacrylate-*co*-DPA (PEGMA-*co*-DPA) and prepared NPs encapsulating hydrophobic photosensitizer, *meso*-tetra(hydroxyphenyl) chlorin (*m*-THPC). *In vitro* release of *m*-THPC at pH 6.0 was faster than the neutral or basic environment at pH 7.0 or 8.0.[68] Gao and coworkers reported ultra-pH sensitive (UPS) nanoparticles using a block copolymer design (PEO-*b*-PR, where PEO is poly(ethylene oxide) and PR is an ionizable tertiary amine block) (Figure 3.6). At low pH, the nanoparticles dissociate into free unimers with protonated ammonium groups, where conjugated fluorophores emit strong fluorescence. When the pH increases, the neutralized PR segments become hydrophobic and self-assemble into the micelles, which results in fluorescence quenching. They discovered that hydrophobic micellization is responsible for the cooperativity in pH transitions (on/off states is <0.25 pH unit, compared to 2 pH units for small molecular pH sensors as dictated by the Henderson-Hasselbalch equation).[69,70] This UPS and nonlinear fluorescence response to pH opens up many exciting opportunities for basic biological research (*e.g.*, endosome/lysosome biology) as well as for translational studies in tumor imaging, image-guided surgery and drug delivery.[71,72]

3.3.2 Cleavage of Acid-labile Linkers for Drug Release

pH-labile linkers represent another strategy for the development of pH-responsive drug delivery systems.[73] Kim *et al.* developed biodegradable polyacetal-based block copolymers that underwent pH-dependent degradation *via* an acid-catalyzed hydrolysis for controlled drug delivery. The polymeric micelles released the loaded PTX in a pH-dependent manner.[74]

Polymer-drug conjugates using a pH-sensitive spacer/linker are also broadly reported to achieve pH-responsive drug release. Acyl hydrazone

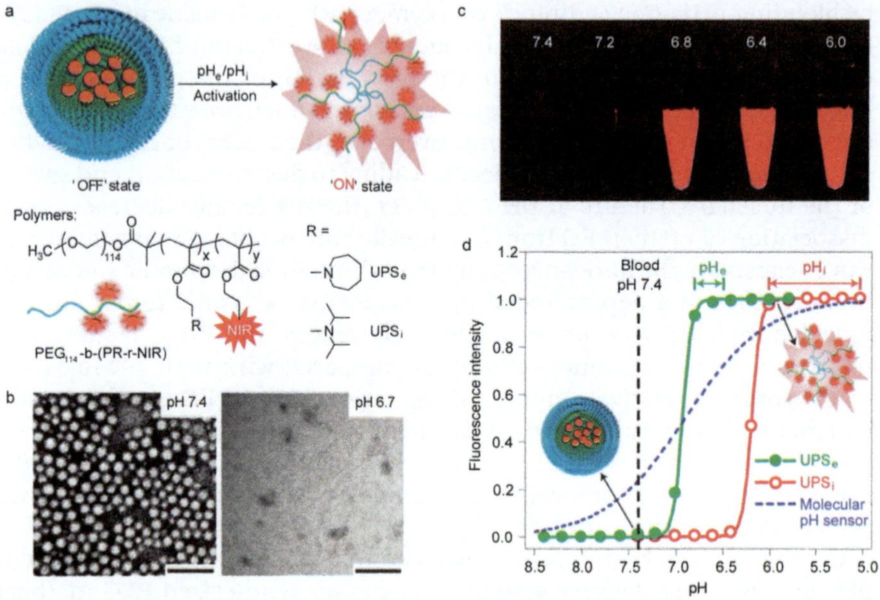

Figure 3.6 (a) Structures of two types of nanoprobe, UPS_e and UPS_i, with pH transitions at 6.9 and 6.2, respectively. (b) Transmission electron micrographs of UPS_e nanoprobes at pH 7.4 and 6.7 (scale bars, 100 nm). (c) Fluorescent images of UPS_e-Cy5.5 nanoprobe in different pH buffers ($\lambda_{ex}/\lambda_{em}$ = 675/710 nm). (d) Normalized fluorescence intensity as a function of pH for UPS_e and UPS_i nanoprobes. For UPS, the pH response ($\Delta pH_{10-90\%}$) is sharp (<0.25 pH unit between ON/OFF states) with >100-fold signal amplification. In contrast, small molecular pH sensors require 3 pH units for a comparable signal change. Reproduced with permission from ref. 71, Copyright 2013 The Authors, rights managed by Nature Publishing Group.

is one of the commonly used covalent linkages that are selectively hydrolyzed in an acidic environment. Patil *et al.* conjugated DOX *via* a pH-sensitive hydrazone linkage to a biodegradable poly(β-L-malic acid) polymer. The DOX-nanoconjugates were found stable under physiological conditions and shown to successfully inhibit *in vitro* cell growth of several invasive breast carcinoma cell lines.[75] An acyl hydrazone bond was also studied in hydroxy propyl methyl cellulose-DOX, PAMAM dendrimer-DOX, HA-DOX and poly(aspartate hydrazide)-PTX conjugates. In addition, Das *et al.* developed an ultra-small super-paramagnetic iron oxide nanocore conjugating the anticancer drug methotrexate *via* ether linkage. The drug release rate was much higher at a lower pH, whereas negligible drug release was observed at pH 7.4.[76]

3.3.3 Gel Swelling

pH-dependent swelling/deswelling of the gel matrix is frequently used to achieve pH-controlled drug release.[77,78] For example, when using acrylic acid-based polymers (*e.g.* PMAA), NPs retain a hydrophobic, collapsed state in the stomach because of protonation of carboxylates. After passing though the gastric lumen, increasing pH results in the NP swelling due to the formation of carboxylate groups and hydrogen bond breakage, leading to enhanced drug release.[79] PMAA-PEG diblock copolymers were able to achieve swelling ratios as high as (mass of swollen polymer/mass of dry polymer) of 40–90 fold. About 90% of the insulin was released at pH 7.4 within two hours, whereas only 10% was released at pH 1.2, as in the collapsed state.[80]

3.3.4 Cap/Coating Removal

Hollow mesoporous channels are "capped" by pH-sensitive linkers, acid-decomposable quantum dots, and polyelectrolytes to render pH responsive drug release.[81,82] For example, Rafi *et al.* developed a core–shell system composed of mesoporous silica nanoparticles and a pH-sensitive polymer (Poly4-vinylpyridine) as the "gatekeeper" to cover pores for controlled release of anti-cancer drugs. At higher pH, the gatekeepers are in their closed state, but they switch to the open state as a result of repulsive forces between positively charged polymer chains in acidic media, causing the release of the capped drugs. The release profiles show that the release rates are pH-dependent and it proceeds with a decrease in pH. This study demonstrated that drug release can be controlled by intracellular pH due to the degradation of the gatekeeper.[83]

The use of pH-sensitive polymers to coat the surface of nanoparticles is another widely used strategy to control the drug release. Eudragit S100 is an anionic copolymer based on methacrylic acid and methyl methacrylate. Subudhi *et al.* developed a colon specific delivery system by the coating of citrus pectin nanoparticles with Eudragit S100. The nanoparticles are able to protect the drug leakage in the upper GI tract because of the deterrent nature of Eudragit S100 to the strongly acidic milieu of the upper GI tract and to deliver fluorouracil upon reaching the colon due to the amiable pH of the colonic fluid. The design showed the promise of nanoparticulate DDS for site-specific delivery of drugs for the effective treatment of colorectal cancer.[84]

3.3.5 Gas Generation

Recently, a type of pH-sensitive DDS that can generate CO_2 gas in the acidic environment has been developed. It is well known that HCO_3^- reacts with acids to produce carbonic acid, which can be converted to CO_2 gas and H_2O. Nanoparticles containing HCO_3^- have the potential to trigger disintegration of the carrier and selective drug release in an acidic environment. Common CO_2-generating agents include sodium bicarbonate ($NaHCO_3$) and ammonium bicarbonate (NH_4HCO_3). Sung and co-workers reported a DDS based on hollow microspheres that could deliver anticancer drugs into tumor cells and quickly release the drug in acidic organelles, such as endosomes and lysosomes. In an acidic environment, $NaHCO_3$ reacts with hydrogen ions to generate CO_2 bubbles, which causes the microsphere wall to burst and thereby release the anticancer drug.[85] In a different approach, Liu *et al.* prepared a pH-sensitive liposome encapsulating DOX by a NH_4HCO_3 gradient method. The liposomes were able to release the drug at pH 5.0 by the production of CO_2 bubbles. The cumulative amount of DOX released at pH 7.4 in 48 h was in the range of 25%, while at pH 5.0, about 60% of the drug was released in the first 2 h. More importantly, the drug-loaded liposome effectively inhibited the breast cancer cells resistant to DOX.[86]

3.4 pH-responsive DDS for Targeted Delivery

Taking advantage of the unique stimuli of pH, pH responsive DDS can be implemented for targeted therapy, as compared to conventional DDS.[87] These DDS can remain stable during circulation and respond to the pH stimulation upon reaching the disease sites and change the physicochemical properties to enhance targeting efficiency, eventually releasing the encapsulated drugs.[88,89] Overall, several responsive strategies are reported including PEG shell detachment, targeting ligand re-exposure, surface charge reversal, particle-size shrinkage and membrane fusion (Figure 3.7).[64,90] Smart DDS with pH-responsive drug release have been gaining more attention in recent years. In this section, different responsive modes are discussed.

3.4.1 PEG Shedding

The surface coating of nanoparticles with PEG has become the most preferable strategy to ensure prolonged blood circulation by preventing protein adsorption and nanoparticle clearance.[91] On the other hand, for nanoparticle–cell interactions, the PEG coating hinders subsequent cellular uptake in tumor cells, compromising the efficiency of drug delivery into targeted cells and tissues and, even after endocytosis, interfering with the interaction of nanoparticles with the endosomal membrane for endosomal escape. This paradox is called the 'PEG dilemma'.[92,93] PEG-sheddable DDSs were established as a strategy to overcome the negative effects of PEG after achieving prolonged circulation time.[94] The PEG shedding process is usually achieved *via* two mechanisms, covalent bond breakage of pH-labile linkers and dissociation of ionic interactions.

Typical pH-labile degradable linkages, such as acyl hydrazone, acetal, imine and succinate bonds, have been investigated as a PEG shedding system. Fella *et al.* constructed an acid-labile pyridylhydrazone PEG to modify PEI NPs. When NPs accumulate in the acidic tumor environment, the hydrazine linker can be hydrolyzed, resulting in the removal of the PEG shield and re-exposure of the PEI NPs. It was demonstrated that this dynamic shielding strategy can improve the biophysical and biological characteristics of non-viral PEI polyplexes.[95] Sun *et al.* reported a tumor-pH-responsive

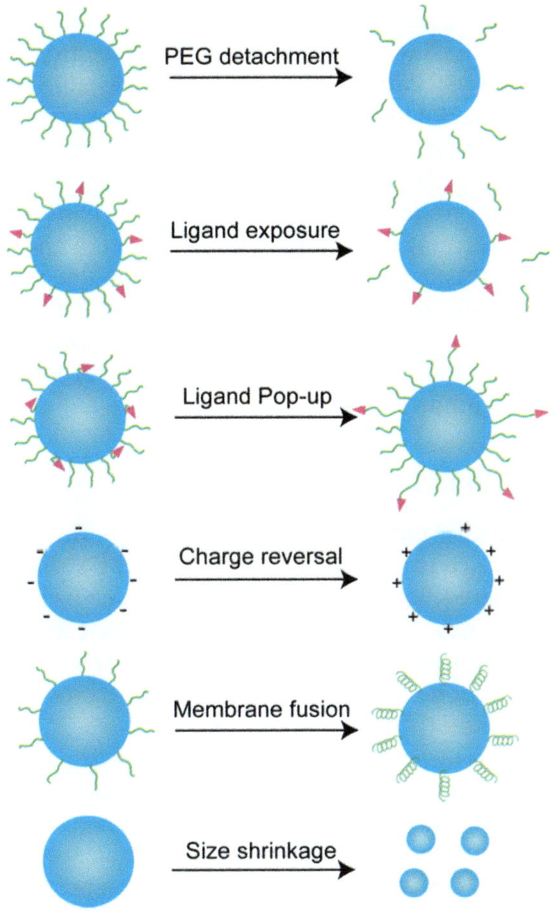

Figure 3.7 Schematic illustration of pH-responsive DDS to enhance drug delivery efficiency.

polymeric nanoparticle based on a bridged PEG and poly(D,L-lactide) (PEG-Dlinkm-PDLLA) block copolymer for improved cancer therapy. The acid-labile copolymer, denoted as PEG-Dlinkm-PDLLA, was covalently connecting the PEG and PDLLA segments with an acid-degradable amide bond. This amphiphilic copolymer assembles into PEGylated nanoparticles, which would accumulate preferentially at the tumor site. Upon arriving at the tumor site, the PEG-Dlinkm-PDLLA nanoparticles will lose the PEG layer and increase zeta potential by responding to tumor acidity, both of which significantly enhances cellular uptake and improves the tumor growth inhibition rate to 78.1% in comparison to 47.8% of the non-responsive control.[96]

In the second strategy, Bae *et al.* employed pH-sensitive polymers to develop a PEG sheddable nanoparticle system for gene delivery. Plasma DNA was complexed with PEI and further modified with a pH-sensitive diblock copolymer, poly(methacryloyl sulfadimethoxine)-*b*-PEG (PSD-*b*-PEG). The nanoparticles showed low cytotoxicity and transfection at pH 7.4 due to shielding of PEI by PSD-*b*-PEG. The PSD-*b*-PEG was able to differentiate the small difference in pH between normal and tumor tissues and be detached from the complexes when exposed to acidic environment (pH 6.6) and permit PEI to interact with cells, leading to high cytotoxicity and transfection. The PSD-*b*-PEG can show a sharp pH-sensitive phase transition, from being in the solution phase to forming aggregates, between pH 6.9 and 6.8. While typical pH-sensitive polymers based on carboxylic acid moieties show a broad transition (over 1 pH unit) in a pH range too low to be exploited for tumor pH targeting, the sulfonamide-based polymers can demonstrate very sharp pH transitions between the physiological and tumor pH range, making themselves suitable for tumor targeting.[43]

3.4.2 Ligand Shielding/Deshielding

For a long time, active targeting based on the specific interaction between receptors on tumor cells and related ligands on the surface of nanocarriers have been considered as a major strategy to selectively deliver therapeutics into tumor tissues.[97–100] However, concerns were raised about the role of active targeting for tumor accumulation because of non-specific uptake of nanocarriers by non-cancerous cells, tumor heterogeneity and variable expression levels of receptors in solid tumors.[101]

To achieve tumor-targeting delivery efficiency with reduced nonspecific uptake, Han *et al.* designed DGL-PEG-T7-hydrazonediethylenetriamine pentaacetic acid (DTPA)/pDNA NPs by a shielding/deshielding mechanism. DTPA, a hydrophilic and negatively charged molecule, which could prevent the contact of NPs with normal tissue cell membrane in neutral conditions, was chosen as the shielding molecule. Tumor-targeting peptide T7 was masked by DTPA *via* a pH-sensitive hydrazone linkage. In neutral conditions, targeting ligand T7 is shielded to avoid nonspecific binding, while in the acidic tumor environment, the hydrazone bond is hydrolyzed to expose ligands for specific tumor targeting.[102] Quan *et al.* developed a peptide containing Arg-Gly-Asp sequence as a target ligand to improve the cell uptake efficacy, while PEGylated technology was employed *via* benzoic-imine bonds to protect the ligands in normal tissues and body fluid. It was found that the targeting property of nanoparticles was switched off before reaching the tumor sites and switched on after removing the PEG segment in the tumor sites.[103]

3.4.3 Ligand Pop-up Targeting

Ligand exposure by pop-up mechanism is a common strategy in designing a pH-sensitive DDS. Aiming to reduce nonspecific uptake and retain the tumor-targeting delivery efficiency, Bae and coworkers developed

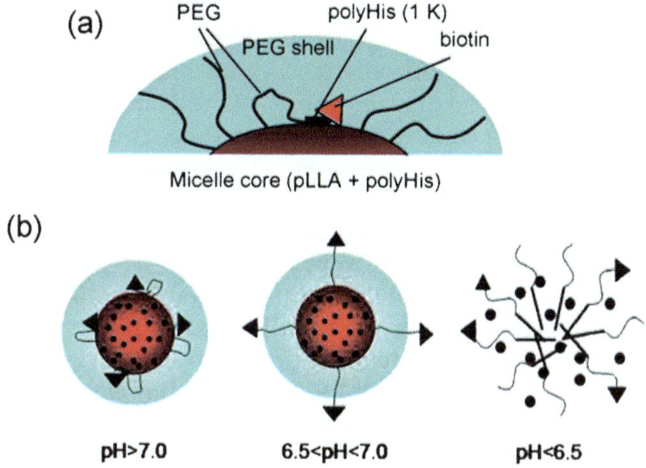

Figure 3.8 (a) Schematic diagram depicting the concept of pH-induced vitamin repositioning on the micelle surface. (b) Above pH 7.0, biotin is shielded by the PEG shell of the micelle; when 6.5 < pH < 7.0, biotin is exposed on the micelle surface and can interact with cells, which facilitates biotin receptor-mediated endocytosis. At pH < 6.5, the micelle dissociates, resulting in enhanced drug release. Reproduced with permission from ref. 104 E. S. Lee, K. Na and Y. H. Bae, *Nano Lett*, 2005, **5**, 325. Copyright 2005 American Chemical Society.

novel pH-sensitive pop-up polymeric micelle systems that present cell-interacting moieties (*e.g.*, targeting ligand (Biotin) and cell penetrating peptide (TAT)) on their surface in response to a small pH change (Figure 3.8).[104,105] These micelles were constructed by mixing of pHis(5 kDa)-*b*-PEG(3.4 kDa) with poly(L-lactic acid) (3 kDa)-*b*-PEG(2 kDa)-*b*-pHis(2 kDa)-TAT or PLLA-*b*-PEG-pHis (2 kDa)-TAT (PHSM^{pop-up}). During blood circulation, the cell-interacting moieties are anchored on the inside of the polymeric micelles *via* a short pHis chain as a pH-sensitive actuator, thus being shielded by the PEG shell of the micelle. When the micelles accumulate in acidic tumor sites, pHis becomes charged and exposes TAT or Biotin to the outside of the PEG shell, facilitating specific cell penetration in tumor cells.

In addition to the pop-up mechanism, the micelles could be designed to become physically destabilized with a further decrease in pH (pH < 6.5), which results in enhanced drug release, disruption of the cell's endosomal membrane and thus increased intracellular drug concentration. At pH 7.4, PHSM^{pop-up} TAT micelle uptake by human breast tumor MCF-7 cells was minimal. At pH 6.8, cellular uptake increased 70-fold compared with that at pH 7.4. These pH-sensitive multifunctional micelle systems are universal for various solid tumors by effectively repositioning and activating cell targeting moieties in a pH-dependent manner.[106]

3.4.4 Charge Reversal

The surface charge of nanoparticles plays an important role in stability, cellular uptake and *in vivo* distribution.[107] On the one hand, a negatively or neutrally charged surface is necessary for nanoparticles to remain stable and circulate in the bloodstream after intravenous administration, while cationic particles have a strong non-specific cellular uptake, and can be rapidly cleared from the plasma compartment.[108] On the other hand, positively charged nanoparticles are superior due to the electrostatic interactions with tumor cell membranes that possess an overall negative charge, leading to significantly improved cellular uptake in tumor tissues.[109,110] Therefore, surface charge-switchable drug delivery design, which have a negative or neutral charge in the bloodstream and can convert to a positive surface once inside the acidic microenvironment, are effective for enhanced targeting efficiency.[111] For example, Zhou *et al.* utilized the acid labile amides with β-carboxylic acid groups to reversibly modify the amine groups and developed charge-reversal nanoparticles for nuclear drug delivery. Owing to the cationic poly(L-lysine)'s primary amine groups were amidized by β-carboxylic acid (PLL/amide), the resulting nanocarrier was negatively charged at neutral pH, which inhibited the PLL's non-specific cellular uptake and clearance from the blood compartment. Once in cancer cells' lysosomes, the amides quickly hydrolyzed to regenerate the positively charged PLL, which then escaped from the lysosomes and traversed to the nucleus. Camptothecin (CPT) conjugated to the PLL/amide shows much enhanced cytotoxicity over free CPT.[112]

3.4.5 Size Change

Particle size plays an important role in controlling the penetration of nanoparticles into tumor tissue.[113] Larger nanoparticles (~100 nm) are usually suitable for exploiting the enhanced permeability and retention effects and hold high propensity of extravasation across tumor vasculature, but have poor penetration due to the high interstitial fluid pressure and dense extracellular matrix.[114] As a result, limited diffusion into the tumor mass is a common shortcoming of nanotherapeutics, which achieve only modest clinical efficacy. Smaller nanoparticles typically show greater tumor penetration because of their reduced diffusional hindrance, but generally suffer from a short half-life and insufficient tumor accumulation because of their rapid clearance.[115] The enhanced tumor accumulation and poor tumor penetration are two conflicting attributes of nanoparticles for cancer treatment.[116,117] This dilemma can be solved by a size-changeable delivery system that could maintain larger initial size for prolonged blood circulation and selective extravasation, while transforming to smaller nanoparticles at tumor sites for increased penetration and tumor distribution.[118]

Li *et al.* reported a pH-activatable clustered nanoparticle (iCluster/Pt) by responding to the acidic pH of the tumor microenvironment (Figure 3.9). The nanoparticles were constructed through molecular assembly of a platinum prodrug-conjugated poly(amidoamine)-*graft*-polycaprolactone (PCL-CDM-PAMAM/Pt) with a PCL homopolymer and poly(ethylene glycol)-*b*-poly(ε-caprolactone) (PEG-*b*-PCL) copolymer. The CDM anhydride residue is acid labile and will be cleaved at low pH. At physiological pH, the clustered nanoparticles hold the size around 100 nm and have a high propensity for long blood circulation and enhanced tumor accumulation. Once accumulating at tumor sites, the tumor extracellular acidity can trigger

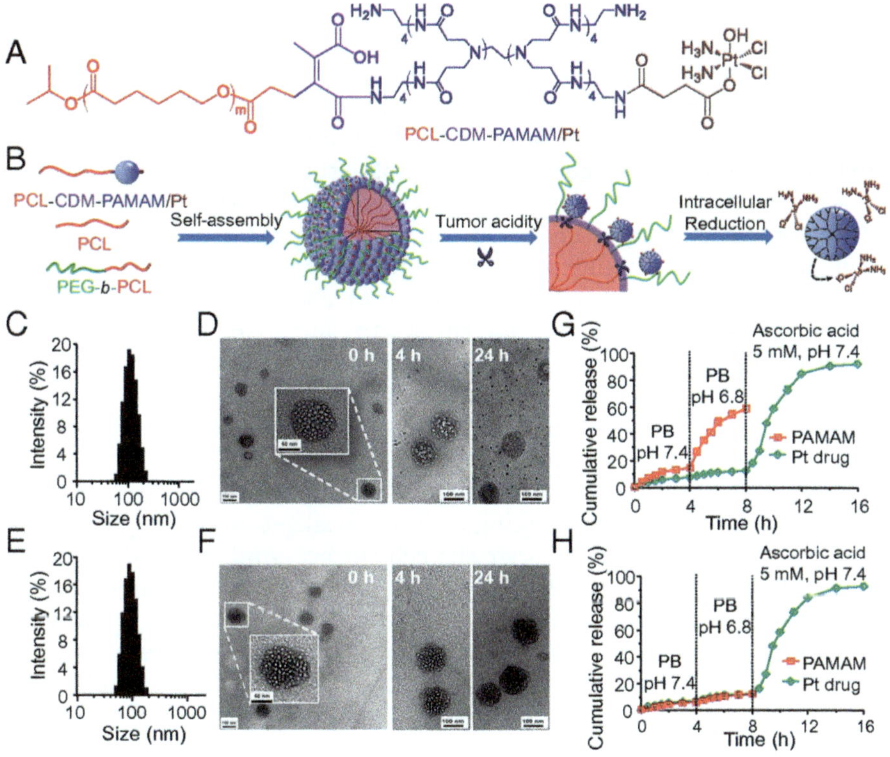

Figure 3.9 Preparation and physicochemical properties of the clustered nanoparticles. (A) Chemical structure of PCL-CDM-PAMAM/Pt. (B) Self-assembly and structural change of iCluster/Pt in response to tumor acidity and intracellular reductive environment. (C and E) Size of iCluster/Pt (C) and Cluster/Pt (E). (D and F) TEM images of iCluster/Pt (D) and Cluster/Pt (F) treated in PB at pH 6.8 for 0, 4, and 24 h, respectively. (Scale bar, 100 nm and for the Inset images, 50 nm.) (G and H) PAMAM and platinum drug release from iCluster (G) and Cluster (H) under three different conditions. Reproduced with permission from ref. 119 H. J. Li, J. Z. Du, X. J. Du, C. F. Xu, C. Y. Sun, H. X. Wang, Z. T. Cao, X. Z. Yang, Y. H. Zhu, S. M. Nie and J. Wang, *Proc Natl Acad Sci U S A*, 2016, **113**, 4164.

the discharge of the platinum prodrug-conjugated poly(amidoamine) dendrimers (diameter ~5 nm). Such a structural alteration greatly facilitates tumor penetration and cell internalization of the therapeutics. The internalized dendrimer prodrugs are further reduced intracellularly to release cisplatin to kill cancer cells. The superior antitumor activities of the iCluster are validated in varying intractable tumor models, including poorly permeable pancreatic cancer, drug-resistant cancer, and metastatic cancer, demonstrating its versatility and broad applicability.[119]

3.4.6 Membrane Fusion by pH-sensitive Peptides

3.4.6.1 pH-(Low) Insertion Peptide (pHLIP)

In the 1990s, Engelman and colleagues discovered a novel class of pH-responsive transmembrane peptides, pHLIPs. pHLIP was found to insert into cell membranes to form a stable transmembrane-helix, which is driven by a pH drop at the cell surface.[120] At acidic pH_e, the protonatable residues and negatively charged C-terminal carboxyl group of the pHLIP become neutral. The protonation leads to an increase in the hydrophobicity of the pHLIP peptide, increasing its affinity to the hydrophobic membrane and triggering the pHLIP to spontaneously fold into a helix and insert across the membrane.[121] When the C-terminal residue and carboxyl group are exposed to the normal intracellular pH of the cell, they become negatively charged, and anchor the pHLIP in the membrane. The insertion is predominantly unidirectional as the C-terminus usually propagates across the bilayer into the cytosol, while the N-terminus remains in the extracellular space.[122,123]

pHLIP insertion occurs with a favorable change in Gibbs free energy. Cargoes associated with a N-terminus will be localized to the targeted tissues and upon insertion will decorate the cell surface. Andreev *et al.* conjugated near-infrared fluorescent dyes to pHLIP and intravenously injected them into a breast adenocarcinoma mouse model. The peptide-dye constructs were found to light up tumors of various sizes. The fluorescence signals achieved from tagged tumor tissues were approximately five-fold higher than in healthy tissues and remained stable for more than 4 days.[124] Clinical imaging modalities such as positron emission tomography and single-photon emission computed tomography imaging agent have also been reported by using pHLIPs.[125,126]

The insertion of the C-terminus facilitates the delivery of drugs across cell membranes. Cytosolic delivery has been achieved for a number of agents, including peptides, chemotherapeutic agents, phalloidin toxin, and even peptide-nucleic acids and microRNAs.[127–129] As the first example of *in vivo* efficacy, Cheng *et al.* attached peptide nucleic acid antimiRs to pHLIP that could target the tumor microenvironment, transport antimiRs across plasma membranes under acidic conditions, and effectively inhibit the miR-155 oncomiR in a mouse model of lymphoma.[129] In addition, pHLIP peptide technology can also assist in the tumor targeted delivery of gold nanoparticles, liposomes and mesoporous silica nanoparticles.[130–132]

A potential challenge in using the insertion peptides to target the tumor microenvironment is that the transition pH of the wild-type peptide is around 6.1, while most solid tumors have a pH_e of 6.5–7.0. To close this gap, Onyango *et al.* carried out structure-activity relationship studies for a series of peptide variants with improved pH response. They found that replacing Asp25 with α-aminoadipic acid (Aad) increases the pH_{50} to 6.74, matching average tumor acidity. Replacing Asp14 with γ-carboxyglutamic acid (Gla) increases the sharpness of the pH response (transition over 0.5 instead of 1 pH unit). These effects are additive: the Asp14Gla/Asp25Aad double variant shows a pH_{50} of 6.79, with sharper transition than Asp25Aad. Furthermore, the advantage of the double variant over WT pHLIP in terms of cargo delivery was demonstrated in fluorescence assays and anti-proliferation studies (using PTX as drug) in A549 lung cancer cells. This progress further increased the usefulness of insertion peptides, with the potential to deliver more drugs into the cytoplasm of cancer cells *via* plasma membrane insertion.[133]

3.4.6.2 pH-activatable Cell-penetrating Peptides CPPs

The positive charges of CPP non-specifically allow transmembrane delivery of therapeutic payloads to a variety of cells.[134] To improve tumor specificity, Jin *et al.* demonstrated a molecular modification approach for inhibiting nonspecific interactions of CPPs in the bloodstream while reactivating their functions in the targeted tissues or cells (Figure 3.10). In their work, the amines of TAT lysines were amidized to succinyl amides (aTAT), preventing nonspecific interactions in the blood compartment; once in the acidic tumor interstitium or internalized into cell endo/lysosomes, the succinyl amides of aTAT were quickly hydrolyzed, fully restoring the functions of TAT for fast cellular uptake or fast endo/lysosomal escape and nuclear targeting. The aTAT-functionalized PEG-*b*-PCL micelles achieved long circulation in the blood compartment and efficiently accumulated and delivered DOX to tumor tissues, giving rise to high antitumor activity and low cardiotoxicity.[135]

Zhang *et al.* developed a new type of CPP by replacing all the lysines with histidine residues. The resulting acid-activatable CPP cannot bind to the cell membrane at physiological pH because of the neutral states of histidines. In the tumor microenvironment, the histidines are protonated and become cationic, which allowed more cellular uptake at pH 6.5 than that at pH 7.4. After attachment of CPT, this conjugate exhibited remarkable cytotoxicity to cancer cells in a pH-dependent manner compared with free CPT.[136]

3.5 pH-responsive DDS for Intracellular Delivery

Macromolecular therapeutics such as peptides, proteins, antisense oligonucleotides, and nucleic acids have received considerable attention.[137,138] Biomacromolecules are typically taken up by endocytosis or phagocytosis, but these processes confine the internalized compounds to endosomes or

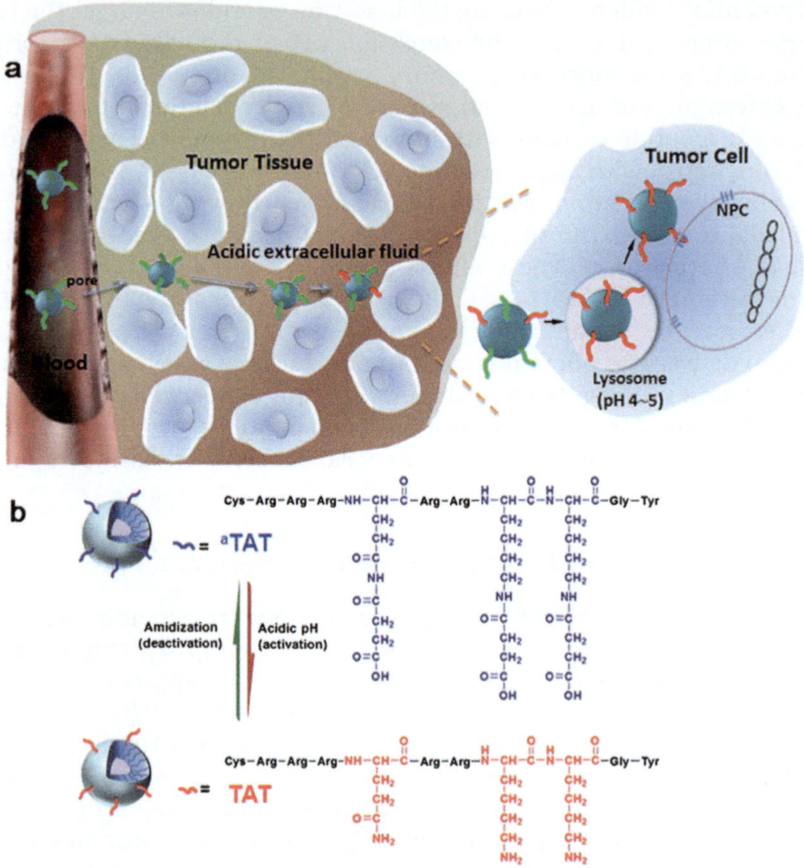

Figure 3.10 (a) Illustration of the concept of CPP deactivation in the blood compartment and its activation in the tumor interstitium or cells for tumor-targeted drug delivery. (b) Amidization of TAT's primary amines to succinyl amides and their pH-triggered hydrolysis to produce intact TAT for cell-targeted delivery. Reproduced with permission from ref. 135 E. L. Jin, B. Zhang, X. R. Sun, Z. X. Zhou, X. P. Ma, Q. H. Sun, J. B. Tang, Y. Q. Shen, E. Van Kirk, W. J. Murdoch and M. Radosz, *J. Am. Chem. Soc.*, 2013, **135**, 933. Copyright 2013 American Chemical Society.

phagosomes. The fusion of these vesicles with lysosomes often leads to rapid destruction of therapeutic molecules with little or no release into the cytosol.[139,140] Thus, a limiting step in achieving an effective biologics-based therapy is to facilitate the endosomal escape and ensure cytosolic delivery of the therapeutics.[28,141–143]

To address the above challenges, a variety of systems have been developed to efficiently overcome the endosomal barriers. Among these, pH-sensitive delivery systems can take advantage of the "proton sponge effect", which is mediated by agents with a high buffering capacity and the ability to swell when protonated.[144] PEI and poly (amidoamine) have been shown to accumulate in the endosomes and increase the osmotic pressure that results in disruption of the membrane.[145] Despite these polymers exhibiting high efficiency in gene transfection *in vitro*, the *in vivo* therapeutic translation is hampered by the relatively high cytotoxicity. The development of polymers which can demonstrate endosomal pH dependent lytic activity could offer a more favorable approach for reducing toxic side-effects by confining the function of the membrane disruption only within the endosomes.

Recently, Ma *et al.* synthesized a series of diblock copolymers with varying molar fractions of the two tertiary amine containing methacrylate monomers with different hydrophobicity.[146] The resulting copolymers displayed tunable pH transitions. A library of pH-sensitive polymers was generated with a 0.3 pH increment of pK_a that span the entire physiologic range of pH (4.0–7.4). The copolymers show a strong pH-specific buffering effect at their pK_a values.[147] Wang *et al.* demonstrated that polymers with a higher pK_a induce stronger membrane disruption to allow efficient protein delivery into cell cytosol, indicating buffering of the organelle pH at an early stage of endocytic maturation is more effective for cytosolic delivery.[148] Gao *et al.* adopted the library to discover a minimalist nanovaccine, comprising a simple physical mixture of an antigen and a synthetic polymeric nanoparticle, PC7A NP, which generated robust T cell responses *in vivo* over another 20 polymer counterparts and conventional innate stimulants (*e.g.* Alum, lipopolysaccharides, CpG, Poly(I:C)) (Figure 3.11). PC7A NP achieved efficient cytosolic delivery of tumor antigens to antigen presenting cells in draining lymph nodes leading to increased surface presentation while simultaneously activating type I interferon-stimulated genes. The nanovaccine produced potent tumor growth inhibition in melanoma, colon cancer, and human papilloma virus-E6/E7 tumor models.[149]

Biodegradable PBAE polymers have been used for pH buffering and gene delivery applications.[150–152] Anderson *et al.* synthesized polymers with 70 different primary structures using monomers previously identified for effective gene delivery. In this library, polymers with 20 of the 70 primary structures possess transfection efficiencies as good as, or better than, Lipofectamine 2000. In general, the most effective polymers condense DNA into sub-150 nm complexes with a positive surface charge. Among this group, the 2 most effective polymers condensed DNA to the smallest particle sizes (71 and 79 nm). The top 9 polymers were formed from amino alcohols, and the structure of the 3 top performing polymers differs by only one carbon. This convergence in structure suggests a common mode of action and provides a framework with which future polymers can be designed.[153]

Alternatively, lipid nanoparticles (LNPs) from ionizable amino lipids are the leading delivery technology for siRNA, with several products in clinical trials.[154] They are composed of an ionizable amino lipid (*e.g.*, heptatriaconta-6,9,28,31-tetraen-19-yl 4-(dimethylamino)butanoate, DLin-MC3-DMA), a phosphatidylcholine (1,2-distearoyl-*sn*-glycero-3-phosphocholine, DSPC), cholesterol and a coat lipid (polyethylene glycol-dimyristolglycerol,

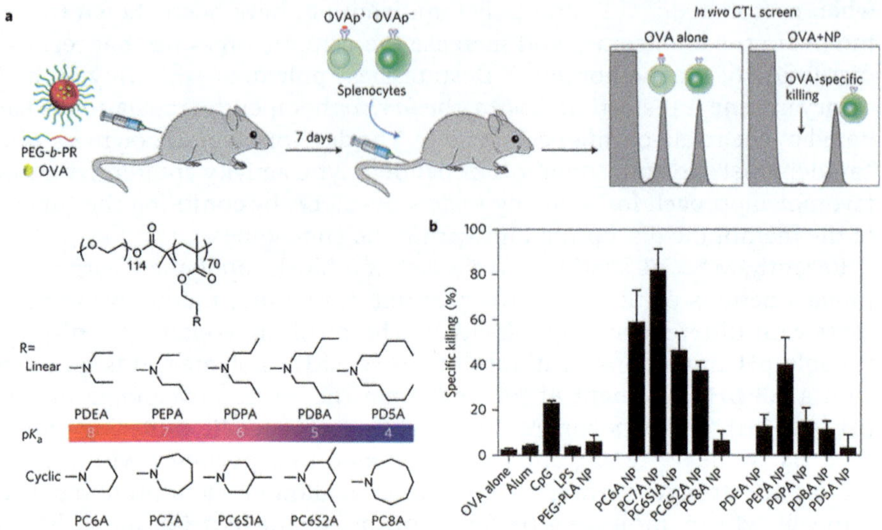

Figure 3.11 (a) Schematic of the carboxy fluorescein succinimidyl ester method to screen for polymer structures that generate a strong antigen-specific CTL response. Ovalbumin (OVA) was used as a model antigen (10 μg) and loaded into different polymer NPs (30 μg). (b) Quantitative comparison of OVA-specific CTL responses in different NP groups (n = 3 for each group), identifying the PC7A NP as the best candidate. Reproduced with permission from ref. 149, rights managed by Nature Publishing Group.

PEG-DMG) (Figure 3.12). LNP siRNA systems have an interior lipid core containing siRNA complexed with ionizable cationic lipid, as shown by the absence of a ^{31}P NMR signal from the free phosphorothioate in the siRNA and the complete protection of siRNA degradation by external RNases. In physiological environments where the pH is above the pK_a of the ionizable lipid (*e.g.*, pH 7.4), the surface of the LNP has an almost neutral charge, which improves circulation and reduces toxicity. Subsequently, in the acidic environment of endosomes, the amino group of the ionizable lipid is positively charged and associates with the anionic endosomal lipids. This interaction enables the destabilization of the endosomal membranes and promotes the release of siRNA into the cytosol.[155,156]

Besides siRNA, LNP was also used to deliver other biomacromolecules (DNA, mRNA, protein, *etc.*) for gene therapy and genome editing. To date, several mRNA LNPs have been developed that can induce functional protein production in specific tissues (*e.g.*, liver, spleen) following systemic administration in mice.[157,158] Fenton *et al.* developed degradable LNP formulations capable of delivering mRNA to the spleen, transfecting B lymphocytes, and inducing more than 60 pg of protein expression per million B cells. Importantly, this LNP induces more than 85% of total protein production in the spleen, despite LNPs being observed transiently in the liver and other organs.

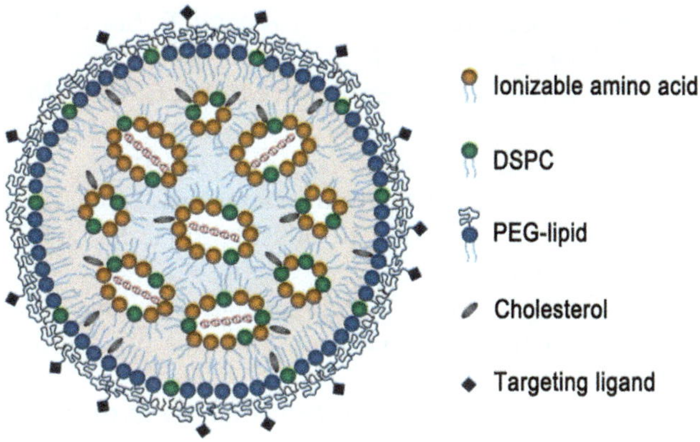

Figure 3.12 A schematic of LNP delivery of siRNA showing a nanostructured core. Reproduced from ref. 154, http://dx.doi.org/10.3390/pharmaceutics5030498. Copyright 2013 The Authors. Published under the terms of the CC BY 3.0 licence, https://creativecommons.org/licenses/by/3.0/.

These results demonstrate that LNP composition can be used to modulate the site of protein induction *in vivo*, highlighting the importance of designing and synthesizing new nanomaterials for nucleic acid delivery.[159]

3.6 Conclusions

pH differences among different organs, tissues, and cellular compartments can serve as a biological stimulus for controlled drug release and delivery. These pH-responsive DDS have the potential to improve tissue targeting and cellular internalization, resulting in higher therapeutic efficacy with decreased toxicity to normal tissues. In this chapter, we summarized the recent advances in the pH-responsive DDS, including the progress in the synthesis of pH-sensitive materials, pH-induced drug release and targeting, as well as related biomedical applications. Mechanistic understanding of these processes in the context of pathophysiology is important to adopting the pH-sensitive technology to tackle different disease indications. Strategies to improve the robustness in function and reduce the barriers in manufacturing and quality control will be important to propel the clinical evaluation and translation of these novel DDS.

References

1. L. Zhang, F. X. Gu, J. M. Chan, A. Z. Wang, R. S. Langer and O. C. Farokhzad, *Clin. Pharmacol. Ther.*, 2008, **83**, 761.
2. Z. Wang, Y. Yu, W. Dai, J. Lu, J. Cui, H. Wu, L. Yuan, H. Zhang, X. Wang, J. Wang, X. Zhang and Q. Zhang, *Biomaterials*, 2012, **33**, 8451.

3. D. Peer, J. M. Karp, S. Hong, O. C. Farokhzad, R. Margalit and R. Langer, *Nat. Nanotechnol.*, 2007, **2**, 751.
4. M. A. Stuart, W. T. Huck, J. Genzer, M. Muller, C. Ober, M. Stamm, G. B. Sukhorukov, I. Szleifer, V. V. Tsukruk, M. Urban, F. Winnik, S. Zauscher, I. Luzinov and S. Minko, *Nat. Mater.*, 2010, **9**, 101.
5. S. Mura, J. Nicolas and P. Couvreur, *Nat. Mater.*, 2013, **12**, 991.
6. J. R. Lopes, G. Santos, P. Barata, R. Oliveira and C. M. Lopes, *Curr. Pharm. Des.*, 2013, **19**, 7169.
7. W. Gao, J. M. Chan and O. C. Farokhzad, *Mol. Pharm.*, 2010, **7**, 1913.
8. S. Ganta, H. Devalapally, A. Shahiwala and M. Amiji, *J. Controlled Release*, 2008, **126**, 187.
9. N. W. Read and K. Sugden, *Crit. Rev. Ther. Drug Carrier Syst.*, 1988, **4**, 221.
10. L. Liu, W. Yao, Y. Rao, X. Lu and J. Gao, *Drug Delivery*, 2017, **24**, 569.
11. Y. Kato, S. Ozawa, C. Miyamoto, Y. Maehata, A. Suzuki, T. Maeda and Y. Baba, *Cancer Cell Int.*, 2013, **13**, 89.
12. W. H. Evans and M. L. Karnovsky, *Biochemistry*, 1962, **1**, 159.
13. A. F. Radovic-Moreno, T. K. Lu, V. A. Puscasu, C. J. Yoon, R. Langer and O. C. Farokhzad, *ACS Nano*, 2012, **6**, 4279.
14. R. A. Gatenby and R. J. Gillies, *Nat. Rev. Cancer*, 2004, **4**, 891.
15. J. W. Kim and C. V. Dang, *Cancer Res.*, 2006, **66**, 8927.
16. M. G. Vander Heiden, L. C. Cantley and C. B. Thompson, *Science*, 2009, **324**, 1029.
17. D. Neri and C. T. Supuran, *Nat. Rev. Drug Discovery*, 2011, **10**, 767.
18. A. Gorbatenko, C. W. Olesen, E. Boedtkjer and S. F. Pedersen, *Front. Physiol.*, 2014, **5**, 130.
19. T. Volk, E. Jahde, H. P. Fortmeyer, K. H. Glusenkamp and M. F. Rajewsky, *Br. J. Cancer*, 1993, **68**, 492.
20. B. A. Webb, M. Chimenti, M. P. Jacobson and D. L. Barber, *Nat. Rev. Cancer*, 2011, **11**, 671.
21. Y. B. Hu, E. B. Dammer, R. J. Ren and G. Wang, *Transl. Neurodegener.*, 2015, **4**, 18.
22. F. R. Maxfield and D. J. Yamashiro, *Adv. Exp. Med. Biol.*, 1987, **225**, 189.
23. R. F. Murphy, S. Powers and C. R. Cantor, *J. Cell Biol.*, 1984, **98**, 1757.
24. I. Mellman, R. Fuchs and A. Helenius, *Annu. Rev. Biochem.*, 1986, **55**, 663.
25. D. Chen and H. Wang, *J. Nanosci. Nanotechnol.*, 2014, **14**, 983.
26. X. He, J. Li, S. An and C. Jiang, *Ther. Delivery*, 2013, **4**, 1499.
27. J. Liu, Y. Huang, A. Kumar, A. Tan, S. Jin, A. Mozhi and X. J. Liang, *Biotechnol. Adv.*, 2014, **32**, 693.
28. K. A. Whitehead, R. Langer and D. G. Anderson, *Nat. Rev. Drug Discovery*, 2009, **8**, 129.
29. B. Mathew, C. A. D'Angelis, S. Lakshminrusimha, P. A. Nickerson, J. J. Sokolowski, V. H. S. Kumar, H. Wang, K. A. Wynn, B. A. Holm and R. M. Ryan, *Pediatr. Res.*, 2017, **82**, 349.

30. M. Liu, H. Du, W. Zhang and G. Zhai, *Mater. Sci. Eng., C*, 2017, **71**, 1267.
31. K. N. Yang, C. Q. Zhang, W. Wang, P. C. Wang, J. P. Zhou and X. J. Liang, *Cancer Biol. Med.*, 2014, **11**, 34.
32. W. Dong, Y. Zhou, D. Yan, H. Li and Y. Liu, *Phys. Chem. Chem. Phys.*, 2007, **9**, 1255.
33. A. E. Felber, M. H. Dufresne and J. C. Leroux, *Adv. Drug Delivery Rev.*, 2012, **64**, 979.
34. D. Avichezer, B. Schechter and R. Arnon, *React. Funct. Polym.*, 1998, **36**, 59.
35. Y. Guan and Y. Zhang, *Chem. Soc. Rev.*, 2013, **42**, 8106.
36. A. Matsumoto, T. Ishii, J. Nishida, H. Matsumoto, K. Kataoka and Y. Miyahara, *Angew. Chem., Int. Ed.*, 2012, **51**, 2124.
37. V. Yesilyurt, M. J. Webber, E. A. Appel, C. Godwin, R. Langer and D. G. Anderson, *Adv. Mater.*, 2016, **28**, 86.
38. W. L. Brooks and B. S. Sumerlin, *Chem. Rev.*, 2016, **116**, 1375.
39. Y. J. Tseng, S. W. Chou, J. J. Shyue, S. Y. Lin, J. K. Hsiao and P. T. Chou, *ACS Nano*, 2016, **10**, 5809.
40. Y. Li, W. Xiao, K. Xiao, L. Berti, J. Luo, H. P. Tseng, G. Fung and K. S. Lam, *Angew. Chem., Int. Ed.*, 2012, **51**, 2864.
41. K. Na, K. H. Lee and Y. H. Bae, *J. Controlled Release*, 2004, **97**, 513.
42. S. I. Kang and Y. H. Bae, *J. Controlled Release*, 2003, **86**, 115.
43. V. A. Sethuraman, K. Na and Y. H. Bae, *Biomacromolecules*, 2006, **7**, 64.
44. G. Kocak, C. Tuncer and V. Butun, *Polym. Chem.*, 2017, **8**, 144.
45. H. Lomas, J. Du, I. Canton, J. Madsen, N. Warren, S. P. Armes, A. L. Lewis and G. Battaglia, *Macromol. Biosci.*, 2010, **10**, 513.
46. P. Midoux, C. Pichon, J. J. Yaouanc and P. A. Jaffres, *Br. J. Pharmacol.*, 2009, **157**, 166.
47. H. L. Tang and N. V. Tsarevsky, *Polym. Chem.*, 2016, **54**, 3683.
48. A. L. Hansen and L. E. Kay, *Proc. Natl. Acad. Sci. U. S. A.*, 2014, **111**, E1705.
49. A. S. Hoffman, *Adv. Drug Delivery Rev.*, 2002, **54**, 3.
50. M. V. Risbud, A. A. Hardikar, S. V. Bhat and R. R. Bhonde, *J. Controlled Release*, 2000, **68**, 23.
51. Z. Amoozgar, J. Park, Q. Lin and Y. Yeo, *Mol. Pharm.*, 2012, **9**, 1262.
52. H. Priya James, R. John, A. Alex and K. R. Anoop, *Acta Pharm. Sin. B*, 2014, **4**, 120.
53. Z. Deng, Z. Zhen, X. Hu, S. Wu, Z. Xu and P. K. Chu, *Biomaterials*, 2011, **32**, 4976.
54. G. Pitarresi, P. Pierro, G. Giammona, F. Iemma, R. Muzzalupo and N. Picci, *Biomaterials*, 2004, **25**, 4333.
55. L. A. Sharpe, A. M. Daily, S. D. Horava and N. A. Peppas, *Expert Opin. Drug Delivery*, 2014, **11**, 901.
56. S. Lee, S. C. Yang, M. J. Heffernan, W. R. Taylor and N. Murthy, *Bioconjugate Chem.*, 2007, **18**, 4.
57. J. Yu, H. Deng, F. Xie, W. Chen, B. Zhu and Q. Xu, *Biomaterials*, 2014, **35**, 3132.

58. X. Guo, C. Shi, J. Wang, S. Di and S. Zhou, *Biomaterials*, 2013, **34**, 4544.
59. B. Elsadek, R. Graeser, A. Warnecke, C. Unger, T. Saleem, N. El-Melegy, H. Madkor and F. Kratz, *ACS Med. Chem. Lett.*, 2010, **1**, 234.
60. J. Panyam and V. Labhasetwar, *Adv. Drug Delivery Rev.*, 2003, **55**, 329.
61. M. Kester, Y. Heakal, T. Fox, A. Sharma, G. P. Robertson, T. T. Morgan, E. I. Altinoglu, A. Tabakovic, M. R. Parette, S. M. Rouse, V. Ruiz-Velasco and J. H. Adair, *Nano Lett.*, 2008, **8**, 4116.
62. F. Muhammad, M. Y. Guo, Y. J. Guo, W. X. Qi, F. Y. Qu, F. X. Sun, H. J. Zhao and G. S. Zhu, *J. Mater. Chem.*, 2011, **21**, 13406.
63. M. B. Yatvin, W. Kreutz, B. A. Horwitz and M. Shinitzky, *Science*, 1980, **210**, 1253.
64. B. L. Chen, W. B. Dai, B. He, H. Zhang, X. Q. Wang, Y. G. Wang and Q. Zhang, *Theranostics*, 2017, **7**, 538.
65. E. S. Lee, H. J. Shin, K. Na and Y. H. Bae, *J. Controlled Release*, 2003, **90**, 363.
66. H. Yin, E. S. Lee, D. Kim, K. H. Lee, K. T. Oh and Y. H. Bae, *J. Controlled Release*, 2008, **126**, 130.
67. E. S. Lee, K. T. Oh, D. Kim, Y. S. Youn and Y. H. Bae, *J. Controlled Release*, 2007, **123**, 19.
68. C. L. Peng, L. Y. Yang, T. Y. Luo, P. S. Lai, S. J. Yang, W. J. Lin and M. J. Shieh, *Nanotechnology*, 2010, **21**, 155103.
69. K. Zhou, Y. Wang, X. Huang, K. Luby-Phelps, B. D. Sumer and J. Gao, *Angew. Chem., Int. Ed.*, 2011, **50**, 6109.
70. X. Ma, Y. Wang, T. Zhao, Y. Li, L. C. Su, Z. Wang, G. Huang, B. D. Sumer and J. Gao, *J. Am. Chem. Soc.*, 2014, **136**, 11085.
71. Y. G. Wang, K. J. Zhou, G. Huang, C. Hensley, X. N. Huang, X. P. Ma, T. Zhao, B. D. Sumer, R. J. DeBerardinis and J. M. Gao, *Nat. Mater.*, 2014, **13**, 204.
72. Y. G. Wang, C. S. Wang, Y. Li, G. Huang, T. Zhao, X. P. Ma, Z. H. Wang, B. D. Sumer, M. A. White and J. M. Gao, *Adv. Mater.*, 2017, **29**, 1603794.
73. E. R. Gillies, A. P. Goodwin and J. M. J. Frechet, *Bioconjugate Chem.*, 2004, **15**, 1254.
74. J. K. Kim, V. K. Garripelli, U. H. Jeong, J. S. Park, M. A. Repka and S. Jo, *Int. J. Pharm.*, 2010, **401**, 79.
75. R. Patil, J. Portilla-Arias, H. Ding, B. Konda, A. Rekechenetskiy, S. Inoue, K. L. Black, E. Holler and J. Y. Ljubimova, *Int. J. Mol. Sci.*, 2012, **13**, 11681.
76. M. Das, D. Mishra, P. Dhak, S. Gupta, T. K. Maiti, A. Basak and P. Pramanik, *Small*, 2009, **5**, 2883.
77. A. P. Griset, J. Walpole, R. Liu, A. Gaffey, Y. L. Colson and M. W. Grinstaff, *J. Am. Chem. Soc.*, 2009, **131**, 2469.
78. B. S. Kim and Y. C. Shin, *J. Appl. Polym. Sci.*, 2007, **105**, 3656.
79. P. Colombo, F. Sonvico, G. Colombo and R. Bettini, *Pharm. Res.*, 2009, **26**, 601.
80. N. A. Peppas, *Int. J. Pharm.*, 2004, **277**, 11.
81. H. Zhou, X. Wang, J. Tang and Y. W. Yang, *Polymers*, 2016, **8**, 277.
82. X. Ma, Y. Zhao, K. W. Ng and Y. Zhao, *Chemistry*, 2013, **19**, 15593.

83. A. A. Rafi, M. Mahkam, S. Davaran and H. Hamishehkar, *Eur. J. Pharm. Sci.*, 2016, **93**, 64.
84. M. B. Subudhi, A. Jain, A. Jain, P. Hurkat, S. Shilpi, A. Gulbake and S. K. Jain, *Materials*, 2015, **8**, 832.
85. C. J. Ke, T. Y. Su, H. L. Chen, H. L. Liu, W. L. Chiang, P. C. Chu, Y. Xia and H. W. Sung, *Angew. Chem., Int. Ed.*, 2011, **50**, 8086.
86. J. Liu, H. L. Ma, T. Wei and X. J. Liang, *Chem. Commun.*, 2012, **48**, 4869.
87. C. Alvarez-Lorenzo and A. Concheiro, *Chem. Commun.*, 2014, **50**, 7743.
88. R. M. Sawant, J. P. Hurley, S. Salmaso, A. Kale, E. Tolcheva, T. S. Levchenko and V. P. Torchilin, *Bioconjugate Chem.*, 2006, **17**, 943.
89. K. Jiang, T. Chi, T. Li, G. R. Zheng, L. L. Fan, Y. J. Liu, X. F. Chen, S. J. Chen, L. Jia and J. W. Shao, *Nanoscale*, 2017, **9**, 9428.
90. J. Z. Du, L. A. Lane and S. M. Nie, *J. Controlled Release*, 2015, **219**, 205.
91. Y. Matsumura and H. Maeda, *Cancer Res.*, 1986, **46**, 6387.
92. S. Mishra, P. Webster and M. E. Davis, *Eur. J. Cell Biol.*, 2004, **83**, 97.
93. H. Hatakeyama, H. Akita and H. Harashima, *Adv. Drug Delivery Rev.*, 2011, **63**, 152.
94. A. M. Aldayel, Y. W. Naguib, H. L. O'Mary, X. Li, M. M. Niu, T. B. Ruwona and Z. R. Cui, *Mol. Ther.–Nucleic Acids*, 2016, **5**, e340.
95. C. Fella, G. F. Walker, M. Ogris and E. Wagner, *Eur. J. Pharm. Sci.*, 2008, **34**, 309.
96. C. Y. Sun, Y. Liu, J. Z. Du, Z. T. Cao, C. F. Xu and J. Wang, *Angew. Chem., Int. Ed.*, 2016, **55**, 1010.
97. Z. H. Wang, Y. Yu, J. Ma, H. R. Zhang, H. Zhang, X. Q. Wang, J. C. Wang, X. Zhang and Q. Zhang, *Mol. Pharm.*, 2012, **9**, 2646.
98. M. M. Galagudza, D. V. Korolev, D. L. Sonin, I. V. Alexandrov, S. M. Minasian, V. N. Postnov, E. B. Kirpicheva, G. V. Papayan and I. S. Uskov, *Bull. Exp. Biol. Med.*, 2011, **152**, 105.
99. F. Danhier, O. Feron and V. Preat, *J. Controlled Release*, 2010, **148**, 135.
100. J. D. Byrne, T. Betancourt and L. Brannon-Peppas, *Adv. Drug Delivery Rev.*, 2008, **60**, 1615.
101. Y. H. Bae and K. Park, *J. Controlled Release*, 2011, **153**, 198.
102. L. Han, Y. B. Guo, H. J. Ma, X. He, Y. Y. Kuang, N. Zhang, E. Lim, W. J. Zhou and C. Jiang, *Small*, 2013, **9**, 3647.
103. C. Y. Quan, J. X. Chen, H. Y. Wang, C. Li, C. Chang, X. Z. Zhang and R. X. Zhuo, *ACS Nano*, 2010, **4**, 4211.
104. E. S. Lee, K. Na and Y. H. Bae, *Nano Lett.*, 2005, **5**, 325.
105. E. S. Lee, Z. G. Gao, D. Kim, K. Park, I. C. Kwon and Y. H. Bae, *J. Controlled Release*, 2008, **129**, 228.
106. K. M. Huh, H. C. Kang, Y. J. Lee and Y. H. Bae, *Macromol. Res.*, 2012, **20**, 224.
107. C. B. He, Y. P. Hu, L. C. Yin, C. Tang and C. H. Yin, *Biomaterials*, 2010, **31**, 3657.
108. M. Longmire, P. L. Choyke and H. Kobayashi, *Nanomedicine*, 2008, **3**, 703.
109. S. F. Ma, M. Nishikawa, H. Katsumi, F. Yamashita and M. Hashida, *J. Controlled Release*, 2005, **102**, 583.

110. D. Fischer, Y. X. Li, B. Ahlemeyer, J. Krieglstein and T. Kissel, *Biomaterials*, 2003, **24**, 1121.
111. Y. Huang, Z. H. Tang, X. F. Zhang, H. Y. Yu, H. Sun, X. Pang and X. S. Chen, *Biomacromolecules*, 2013, **14**, 2023.
112. Z. X. Zhou, Y. Q. Shen, J. B. Tang, M. H. Fan, E. A. Van Kirk, W. J. Murdoch and M. Radosz, *Adv. Funct. Mater.*, 2009, **19**, 3580.
113. M. R. Junttila and F. J. de Sauvage, *Nature*, 2013, **501**, 346.
114. C. H. Heldin, K. Rubin, K. Pietras and A. Ostman, *Nat. Rev. Cancer*, 2004, **4**, 806.
115. E. A. Sykes, Q. Dai, C. D. Sarsons, J. Chen, J. V. Rocheleau, D. M. Hwang, G. Zheng, D. T. Cramb, K. D. Rinker and W. C. W. Chan, *Proc. Natl. Acad. Sci. U. S. A.*, 2016, **113**, E1142.
116. H. Cabral, Y. Matsumoto, K. Mizuno, Q. Chen, M. Murakami, M. Kimura, Y. Terada, M. R. Kano, K. Miyazono, M. Uesaka, N. Nishiyama and K. Kataoka, *Nat. Nanotechnol.*, 2011, **6**, 815.
117. J. Q. Wang, W. W. Mao, L. L. Lock, J. B. Tang, M. H. Sui, W. L. Sun, H. G. Cui, D. Xu and Y. Q. Shen, *ACS Nano*, 2015, **9**, 7195.
118. C. Wong, T. Stylianopoulos, J. A. Cui, J. Martin, V. P. Chauhan, W. Jiang, Z. Popovic, R. K. Jain, M. G. Bawendi and D. Fukumura, *Proc. Natl. Acad. Sci. U. S. A.*, 2011, **108**, 2426.
119. H. J. Li, J. Z. Du, X. J. Du, C. F. Xu, C. Y. Sun, H. X. Wang, Z. T. Cao, X. Z. Yang, Y. H. Zhu, S. M. Nie and J. Wang, *Proc. Natl. Acad. Sci. U. S. A.*, 2016, **113**, 4164.
120. J. F. Hunt, P. Rath, K. J. Rothschild and D. M. Engelman, *Biochemistry*, 1997, **36**, 15177.
121. F. N. Barrera, D. Weerakkody, M. Anderson, O. A. Andreev, Y. K. Reshetnyak and D. M. Engelman, *J. Mol. Biol.*, 2011, **413**, 359.
122. Y. K. Reshetnyak, O. A. Andreev, M. Segala, V. S. Markin and D. M. Engelman, *Proc. Natl. Acad. Sci. U. S. A.*, 2008, **105**, 15340.
123. D. Weerakkody, A. Moshnikova, M. S. Thakur, V. Moshnikova, J. Daniels, D. M. Engelman, O. A. Andreev and Y. K. Reshetnyak, *Proc. Natl. Acad. Sci. U. S. A.*, 2013, **110**, 5834.
124. O. A. Andreev, A. D. Dupuy, M. Segala, S. Sandugu, D. A. Serra, C. O. Chichester, D. M. Engelman and Y. K. Reshetnyak, *Proc. Natl. Acad. Sci. U. S. A.*, 2007, **104**, 7893.
125. S. Macholl, M. S. Morrison, P. Iveson, B. E. Arbo, O. A. Andreev, Y. K. Reshetnyak, D. M. Engelman and E. Johannesen, *Mol. Imaging Biol.*, 2012, **14**, 725.
126. A. L. Vavere, G. B. Biddlecombe, W. M. Spees, J. R. Garbow, D. Wijesinghe, O. A. Andreev, D. M. Engelman, Y. K. Reshetnyak and J. S. Lewis, *Cancer Res.*, 2009, **69**, 4510.
127. P. Daumar, C. A. Wanger-Baumann, N. Pillarsetty, L. Fabrizio, S. D. Carlin, O. A. Andreev, Y. K. Reshetnyak and J. S. Lewis, *Bioconjugate Chem.*, 2012, **23**, 1557.
128. M. An, D. Wijesinghe, O. A. Andreev, Y. K. Reshetnyak and D. M. Engelman, *Proc. Natl. Acad. Sci. U. S. A.*, 2010, **107**, 20246.

129. C. J. Cheng, R. Bahal, I. A. Babar, Z. Pincus, F. Barrera, C. Liu, A. Svoronos, D. T. Braddock, P. M. Glazer, D. M. Engelman, W. M. Saltzman and F. J. Slack, *Nature*, 2015, **518**, 107.
130. L. Yao, J. Danniels, A. Moshnikova, S. Kuznetsov, A. Ahmed, D. M. Engelman, Y. K. Reshetnyak and O. A. Andreev, *Proc. Natl. Acad. Sci. U. S. A.*, 2013, **110**, 465.
131. Z. Zhao, H. Meng, N. Wang, M. J. Donovan, T. Fu, M. You, Z. Chen, X. Zhang and W. Tan, *Angew. Chem., Int. Ed.*, 2013, **52**, 7487.
132. L. Yao, J. Daniels, D. Wijesinghe, O. A. Andreev and Y. K. Reshetnyak, *J. Controlled Release*, 2013, **167**, 228.
133. J. O. Onyango, M. S. Chung, C. H. Eng, L. M. Klees, R. Langenbacher, L. Yao and M. An, *Angew. Chem., Int. Ed.*, 2015, **54**, 3658.
134. D. Sarko, B. Beijer, R. G. Boy, E. M. Nothelfer, K. Leotta, M. Eisenhut, A. Altmann, U. Haberkorn and W. Mier, *Mol. Pharm.*, 2010, **7**, 2224.
135. E. L. Jin, B. Zhang, X. R. Sun, Z. X. Zhou, X. P. Ma, Q. H. Sun, J. B. Tang, Y. Q. Shen, E. Van Kirk, W. J. Murdoch and M. Radosz, *J. Am. Chem. Soc.*, 2013, **135**, 933.
136. W. Zhang, J. J. Song, B. Z. Zhang, L. W. Liu, K. R. Wang and R. Wang, *Bioconjugate Chem.*, 2011, **22**, 1410.
137. A. P. McCaffrey, L. Meuse, T. T. T. Pham, D. S. Conklin, G. J. Hannon and M. A. Kay, *Nature*, 2002, **418**, 38.
138. H. J. Kong and D. J. Mooney, *Nat. Rev. Drug Discovery*, 2007, **6**, 455.
139. U. Lachelt and E. Wagner, *Chem. Rev.*, 2015, **115**, 11043.
140. L. M. Bareford and P. W. Swaan, *Adv. Drug Delivery Rev.*, 2007, **59**, 748.
141. L. M. Kranz, M. Diken, H. Haas, S. Kreiter, C. Loquai, K. C. Reuter, M. Meng, D. Fritz, F. Vascotto, H. Hefesha, C. Grunwitz, M. Vormehr, Y. Husemann, A. Selmi, A. N. Kuhn, J. Buck, E. Derhovanessian, R. Rae, S. Attig, J. Diekmann, R. A. Jabulowsky, S. Heesch, J. Hassel, P. Langguth, S. Grabbe, C. Huber, O. Tureci and U. Sahin, *Nature*, 2016, **534**, 396.
142. H. Yin, R. L. Kanasty, A. A. Eltoukhy, A. J. Vegas, J. R. Dorkin and D. G. Anderson, *Nat. Rev. Genet.*, 2014, **15**, 541.
143. H. Yin, K. J. Kauffman and D. G. Anderson, *Nat. Rev. Drug Discovery*, 2017, **16**, 387.
144. A. E. Nel, L. Madler, D. Velegol, T. Xia, E. M. V. Hoek, P. Somasundaran, F. Klaessig, V. Castranova and M. Thompson, *Nat. Mater.*, 2009, **8**, 543.
145. P. Resnier, T. Montier, V. Mathieu, J. P. Benoit and C. Passirani, *Biomaterials*, 2013, **34**, 6429.
146. X. P. Ma, Y. G. Wang, T. Zhao, Y. Li, L. C. Su, Z. H. Wang, G. Huang, B. D. Sumer and J. M. Gao, *J. Am. Chem. Soc.*, 2014, **136**, 11085.
147. C. Wang, Y. Wang, Y. Li, B. Bodemann, T. Zhao, X. Ma, G. Huang, Z. Hu, R. J. DeBerardinis, M. A. White and J. Gao, *Nat. Commun.*, 2015, **6**, 8524.
148. Z. Wang, M. Luo, C. Mao, Q. Wei, T. Zhao, Y. Li, G. Huang and J. Gao, *Angew. Chem., Int. Ed.*, 2017, **56**, 1319.
149. M. Luo, H. Wang, Z. H. Wang, H. C. Cai, Z. G. Lu, Y. Li, M. J. Du, G. Huang, C. S. Wang, X. Chen, M. R. Porembka, J. Lea, A. E. Frankel, Y. X. Fu, Z. J. J. Chen and J. M. Gao, *Nat. Nanotechnol.*, 2017, **12**, 648.

150. H. Devalapally, D. Shenoy, S. Little, R. Langer and M. Amiji, *Cancer Chemother. Pharmacol.*, 2007, **59**, 477.
151. D. Shenoy, S. Little, R. Langer and M. Amiji, *Pharm. Res.*, 2005, **22**, 2107.
152. D. Shenoy, S. Little, R. Langer and M. Amiji, *Mol. Pharm.*, 2005, **2**, 357.
153. D. G. Anderson, A. Akinc, N. Hossain and R. Langer, *Mol. Ther.*, 2005, **11**, 426.
154. Y. Y. Tam, S. Chen and P. R. Cullis, *Pharmaceutics*, 2013, **5**, 498.
155. I. M. Hafez, N. Maurer and P. R. Cullis, *Gene Ther.*, 2001, **8**, 1188.
156. Y. H. Xu and F. C. Szoka, *Biochemistry*, 1996, **35**, 5616.
157. K. J. Kauffman, J. R. Dorkin, J. H. Yang, M. W. Heartlein, F. DeRosa, F. F. Mir, O. S. Fenton and D. G. Anderson, *Nano Lett.*, 2015, **15**, 7300.
158. O. S. Fenton, K. J. Kauffman, R. L. McClellan, E. A. Appel, J. R. Dorkin, M. W. Tibbitt, M. W. Heartlein, F. DeRosa, R. Langer and D. G. Anderson, *Adv. Mater.*, 2016, **28**, 2939.
159. O. S. Fenton, K. J. Kauffman, J. C. Kaczmarek, R. L. McClellan, S. Jhunjhunwala, M. W. Tibbitt, M. D. Zeng, E. A. Appel, J. R. Dorkin, F. F. Mir, J. H. Yang, M. A. Oberli, M. W. Heartlein, F. DeRosa, R. Langer and D. G. Anderson, *Adv. Mater.*, 2017, **29**, 1606944.

CHAPTER 4

Thermo-responsive Nanomedicines for Drug Delivery in the Gastrointestinal Tract

ANDREIA ALMEIDA[a,b,c], HELENA MACEDO[a,b,c], JOSÉ DAS NEVES[a,b] AND BRUNO SARMENTO*[a,b,d]

[a]Instituto de Investigação e Inovação em Saúde (i3S), Universidade do Porto, Rua Alfredo Allen, 208, 4200-393 Porto, Portugal; [b]Instituto de Engenharia Biomédica (INEB), Universidade do Porto, Rua Alfredo Allen, 208, 4200-393 Porto, Portugal; [c]Instituto Ciências Biomédicas Abel Salazar (ICBAS), Universidade do Porto, Rua de Jorge Viterbo Ferreira 228, 4050-313 Porto, Portugal; [d]CESPU, Instituto de Investigação e Formação Avançada em Ciências e Tecnologias da Saúde, Rua Central de Gandra 1317, 4585-116 Gandra, Portugal
*E-mail: bruno.sarmento@ineb.up.pt

4.1 Introduction

Cancer comprises of a multitude of heterogeneous pathological conditions and has been one of the most studied fields in biomedical sciences. Despite all efforts being made over the last few decades, many aspects of this group of diseases are complex and remain misunderstood. Cancer is generally characterized by the deregulation of the normal cell cycle leading to abnormal cell

Biomaterials Science Series No. 1
Stimuli-responsive Drug Delivery Systems
Edited by Amit Singh and Mansoor M. Amiji

Published by the Royal Society of Chemistry, www.rsc.org

growth and spreading beyond its natural limits.[1] When auto-regulation of cellular division fails, the resulting cells can invade other parts of the body and spread to other tissues and organs through a mechanism called metastasis, a major cause of death by cancer.[2,3]

The number of new cancer cases has been increasing over the years. In 2015, cancer was the second group of diseases with the highest mortality rate worldwide and the numbers are expect to increase by about 70% over the next decades.[3] The probability of developing cancer increases with aging in association with other external factors such as tobacco use, infectious diseases, unhealthy diet or exposure to ultraviolet and ionizing radiation, or by internal factors such as genetic mutations, hormones or immune conditions.[3,4] In addition, it has been established that most cancers are the result of chronic inflammatory diseases related to infection. For example, gastric infection from *Helicobacter pylori* can increase the risk of gastric cancer development by 75%.[4] Also, it has been widely described that inflammation is the primary cause of cancer development. In particular, inflammatory bowel disease (IBD), which is a condition characterized by intestinal inflammation, can increase the risk of developing colorectal cancer (CRC) by twice.[5] IBD includes two main intestinal inflammatory diseases: ulcerative colitis (UC) and Crohn's disease (CD). UC is a mucosal inflammation of the large intestine that always affects the rectum and typically reaches the colon. On the other hand, CD can affect all the layers of the gastrointestinal tract (GIT). Management of IBD and cancer face significant challenges such as (i) non-specific systemic distribution of drugs, in which active molecules reach both normal and abnormal cells/tissues; (ii) unsuitable drug concentrations at affected sites; (iii) and, adverse drug effects. In the specific case of cancer, development of drug resistance may also occur.[6]

One alternative that has been studied for decades to overcome the previous difficulties and improve the effectiveness of treatment is the use of nanomedicines. Delivery systems such as nanoparticles,[7–12] micelles[13–15] or liposomes[16–20] are able to incorporate and protect hydrophobic and/or hydrophilic drugs, preventing premature leakage to sites other than those intended. Also, nanomedicines can promote drug concentration at tumors by taking advantage of the well-described enhanced permeability and retention (EPR) effect.[21,22] Leaky vasculature at cancer sites allows facilitates access of nanomedicines that can further accumulate due to hindrance in their tissue transport. Once at the desired site, drug delivery systems can disassemble and release the associated drug.

Stimuli-responsive materials present variable behavior depending on physical, chemical or biological changes, including those observed in inflammatory and/or tumor environments. The use of stimuli-responsive materials in the design of delivery systems allows the incorporation and protection of the drug of interest in the absence of a specific stimulus. However, when intended environmental changes occur, the drug can be rapidly released. Overall, such systems may be able to improve efficacy by providing targeted drug delivery, while minimizing off-target effects.[16,23] Stimuli triggering drug release can be classified as external or internal. External stimuli are induced by a device located outside of the body, such as light, ultrasound or microwaves. Internal stimuli are inherent to physiopathological conditions, being mostly related with changes in pH, temperature, redox environment, enzymes or proteins.[24] In particular, slight but significant changes in temperature associated with inflammation or cancer have been receiving special attention in the development of stimuli-responsive drug delivery nanosystems.

Thermo-responsive polymers are characterized by their lower critical solution temperature (LCST), for example, the value at which the polymers exhibit coil-to-globule transition in aqueous solution.[24] This shift is characterized by changes in molecular conformation, solubility or hydrophobicity/hydrophilicity.[25,26] One of the most recognizable thermo-responsive polymers used in drug delivery systems is poly(*N*-isopropylacrylamide) (PNIPAM), with the LCST at 32 °C in aqueous solution.[27] The LCST of PNIPAM can be adjusted to a desired temperature for specific drug delivery purposes, for example, above the physiologic temperature, namely by conjugation with other materials, such as chitosan.[28,29]

This chapter provides a review on nanotechnology-based strategies for the management of gastrointestinal inflammatory/tumor diseases, particularly the use of thermo-responsive nanocarriers that are triggered by pathological changes in temperature occurring at inflammatory/tumor tissues or rely on an external heat source to directly release the drug at the site of interest.

4.2 Gastrointestinal Tract

4.2.1 Constitution of the Gastrointestinal Tract

The main function of the GIT is the digestion of food and absorption of nutrients and vitamins. The GIT is designed to facilitate absorption of relatively low molecular weight compounds but also to prevent the entry of macromolecules, toxins or antigens.[30] The GIT begins in the oral cavity and connects the esophagus to the stomach, small intestine (duodenum, jejunum and ileum) and large intestine (colon), ending up at the rectum. Each part of the GIT presents distinct anatomical, biochemical and physiological properties, which determine digestion and absorption. The intestine wall is composed of several layers that are not fully separated from each other, but are joined together by connective tissue and vascular and neuronal elements.[31] There are four main layers: mucosa, submucosa, muscularis propria and serosa. The mucosa contacts directly with the intestinal lumen, being structurally and functionally more complex. Absorption phenomena occur mainly at the mucosa, which can be further subdivided into the epithelium, lamina propria and muscularis mucosae. The intestinal epithelium comprises of epithelial cells that are directly in contact with nutrients, microbiota and mucus, being connected to the basal membrane covering the second layer, namely the lamina propria. This last consists of subepithelial connective tissue featuring blood capillaries and lymph nodes. The next underlying layer is called muscularis mucosae and is constituted of a thin layer of smooth muscle.[31,32] The submucosa is a thin layer of connective tissue that supports the mucosa and joins it to the muscularis propria, this last being constituted of longitudinal and circular muscle cells contributing to peristalsis.[33] Finally, the innermost layer, the serosa, is formed essentially by fat and, also by a layer of epithelial cells called mesothelium.[34]

Two main barriers can be implied in drug absorption and prevention of the entry of undesirable materials. First, the physical barrier consisting of a mucus layer and epithelial cells and, second the biochemical barrier composed of catabolic enzymes and efflux systems, such as P-glycoprotein (P-gp) and multidrug resistance-associated protein 2 (MRP2), which act against drug absorption.[35] However, it has been reported that the mucus can also influence nanocarriers adhesion, thus favoring drug absorption.[36]

The epithelium of the small intestine consists of different cell types that are located in the villi (enterocytes, goblet and enteroendocrine cells) and differentiated cells residing at the bottom of the crypts, such as Paneth cells and microfold or membranous cells (M-cells).[31,37] Crypts are tubular invaginations of the intestinal epithelium and are characterized by presenting basal Paneth cells and stem cells, which continuously divide and are the source of all intestinal epithelial cells.[37] Villi are tiny finger-like projections, which increase the surface area by approximately 30-fold.[38]

The large intestine or colon begins in the right iliac region of the small intestine and is very similar in terms of cellular organization and morphology to the small intestine. Unlike the small intestine, the mucosa of the large intestine is not covered with villous projections but contains deep tubular pits, which increase in depth as they approach the rectum, as well as the muscularis propria that further increase the surface area for digestion and absorption by approximately 600-fold.[31,38]

4.2.2 Differences Between Normal and Diseased Tissues

It is well-known that anticancer or anti-inflammatory drugs have to overcome diverse biological barriers before dispersing in the bloodstream in therapeutically relevant concentrations. Ideally, drugs should preferentially reach damaged cells and increase the intracellular concentrations, leaving normal cells unaffected, thus reducing toxicity. There are several differences between normal and inflammatory and/or tumor tissues that can be useful when considering differential drug delivery using nanomedicines. The main differences between normal and pathological tissues are listed in Table 4.1.

In tumor tissues, the formation of new blood vessels occurs through a process called angiogenesis. Nevertheless, these new vessels display several abnormalities such as vessel tortuosity, deficient pericytes and large gaps between adjacent endothelial cells.[39] These gaps can reach 100–800 nm, whereas in healthy vasculature the endothelial cells are tightly bound with fenestrations around 1–10 nm.[40,41] In this sense, nanomedicines can preferentially bypass tumor vasculature allowing the extravasation of drugs into

Table 4.1 Main biological differences between normal and inflamed and tumor tissues.

Biological features	Normal tissues	Inflammatory and tumor tissues	References
Blood vessels	Tight fenestrations between endothelial cells around 1–10 nm	Vessel tortuosity, pericytes deficiency, large gaps between endothelial cells around 100–800 nm	39–41
Lymphatic system	Regular lymphatic network	Lack of lymphatic drainage	21,22,43
Colon	Continuous enterocyte layer; intact mucus layer	Enterocyte disruption (eEPR effect); differences in mucus thickness layer and constitution	45,107,108
pH	~7.4	~6.5 in tumor extracellular sites and inflammatory tissues	60–62
Temperature	~37 °C	~40–42 °C	64 and 65
Enzymes and other proteins	Expressed at normal values	Overexpression of enzymes such as matrix metalloproteinases; overexpression of P-gp and cytochrome P405 3A	8 and 60

affected tissues. It has been described that the fenestrations of tumor vasculature are due to abnormal levels of vascular mediators such as bradykinins, nitric oxide, vascular endothelial growth factor, fibroblast growth factor and prostaglandins, amongst others.[2,42] This unique feature of tumor vessels combined with reduced lymphatic drainage leads to the EPR effect.[22,43] This phenomenon allows macromolecules and small particulates, such as in the case of nanomedicines, to leave the systemic circulation and enter into the extravascular space to be concentrated within the tumor tissue by a mechanism called passive targeting (Figure 4.1).

Additionally, inflammatory colon tissue is characterized by enterocyte disruption, which can lead to an increase on intestinal permeability due to enterocyte barrier damage.[44] Thus, nanomedicines can accumulate at intercellular gaps by means of the epithelial enhanced permeability and retention (eEPR) effect, allowing an increase of drug concentration at the desired site. The rupture of the barrier formed by the enterocytes provides access to the entrance of luminal and mucosa bacteria, which leads to the increase of immune cell levels.[45] In this sense, the resulting immune response induces overexpression of inflammatory cytokines that disrupt intestinal homeostasis, leading to the development of IBD.[45,46]

In the case of cancer cells, nanocarriers can find membrane receptor molecules that are overexpressed as compared to normal tissues. Nanomedicines functionalized with targeting ligands can bind specifically to these receptors, thus leading to increased interaction and uptake into the cancer cells *via* receptor-mediated endocytosis (Figure 4.1).[13,47] The most common

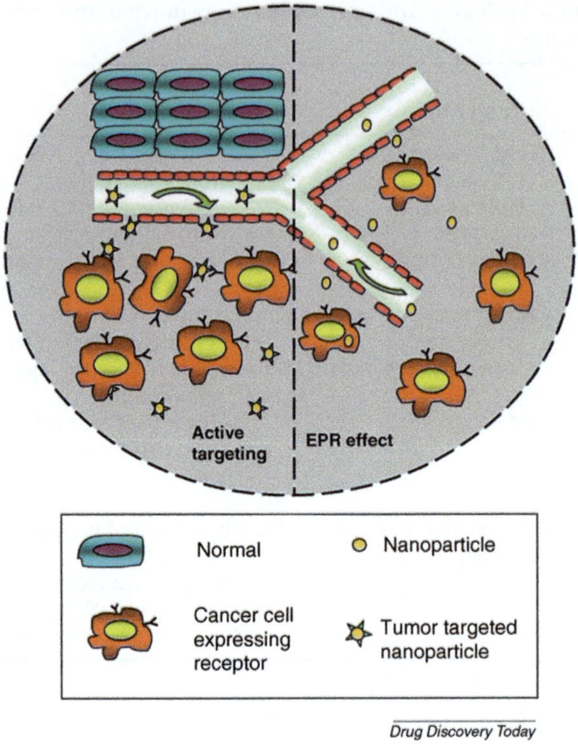

Figure 4.1 Tumor targeting. The right-hand part of the figure depicts the increased accumulation of nanoparticles in tumor owing to leaky tumor vasculature, leading to the enhanced permeability and retention effect. The left-hand part of the figure shows active targeting mediated by targeted nanoparticles. Reprinted from ref. 2, *Drug Discovery Today*, **15**, R. Misra, S. Achaerya, S. K. Sahoo, Cancer nanotechnology: application of nanotechnology in cancer therapy, 842–850, Copyright (2010), with permission from Elsevier.

surface modifiers in nanomedicines are monoclonal antibodies (and antibodies fragments),[48–50] peptides and proteins,[47] small molecules (such as folic acid),[51,52] polysaccharides (such as heparin)[53,54] and aptamers.[55,56] These and other ligands allow specific receptor binding to tumor cells. However, for successful delivery targeted receptors in the tumor cell surface must be homogeneously overexpressed in tumor cells.[2]

As in cancer, IBD has specific receptors overexpressed only at the inflamed tissue, which can lead to a more specific and effective targeting. Most of these receptor-specific ligands are used in oral colon-specific strategies such as lectin, which is a sugar binding protein with great capacity to bind specific carbohydrate residues,[57] transferrin, the receptor which is expressed at low levels in normal cells, but overexpressed in inflamed colon cells,[19] bacterial adhesins[58] or ligands for CD98, a type II membrane glycoprotein overexpressed during intestinal inflammation.[59]

Another distinction between normal and tumor tissue is the pH. It is well known that cancer cells have very rapid growth and hence a high metabolic rate. Even though there are low levels of oxygen and nutrients for tumor cells to continue proliferating, these can use glycolysis to obtain extra energy, which results in an acidic environment.[42] Thus, the extracellular environment of solid tumors present a pH around ~6.5, while healthy tissues present a pH of ~7.4.[60–62] Also, the colon has the most extensive and distinct microbial population, which is responsible for the weakly acidic pH and, together with the numerous enzymes, are responsible for drug breakdown in the colon.[63]

Finally, temperature differences between pathologic and healthy tissue are present in various diseases. In particular, inflammation and cancer tissues are characterized by having temperature values above the physiological temperature, typically between 39–41 °C.[64,65] Since cancer cells cannot dissipate heat as efficiently as normal cells, the tumor is heated and the temperature continues to increase due to the lack of vasodilatation. This may, in turn, lead to apoptosis or disruption of cell metabolism, which can make the tumor more susceptible to another type of treatment, such as chemotherapy or radiation.[66,67] On the other hand, nanomedicines may be engineered in order to take advantage of this characteristic as a therapeutic strategy in the treatment of inflammation and cancer. Such systems should be sensitive to small temperature changes in order to distinguish healthy from diseased areas. However, in most of the cases it does not happen. Thus, hyperthermia or thermotherapy are used as a complementary strategy after the administration of thermo-responsive systems in order to improve the effect of anti-cancer drugs.[17,20,68,69]

4.2.3 Inflammatory Bowel Disease

Inflammatory bowel disease is a chronic inflammatory condition of the large and/or small intestine. The most common forms of IBD are UC and CD, as mentioned previously.

UC mostly affects the innermost mucosa, never involving other layers such as the submucosa, mucularis and serosa, and it is predominant in the colon and the rectum.[63,70] CD is an inflammatory disease that can affect all the GIT (sometimes even the mouth), but it is more common in the colon and terminal ileum.[63,70] It is presents as a transmural inflammation since the disease can affect all the intestinal layers, down to the serosa layer.[71] Although immune cells act to protect our body against infections, when activated inadequately, these are the main contributors for inflammatory disease development.[72] Macrophages can release cytokines, chemokines or reactive oxygen species during the process of inflammatory diseases and these can increase damage to normal tissue.[73]

Particularly in CD, there is an activation of chronic T-helper (Th-1)-lymphocyte followed by an overexpression of tumor necrosis factor alpha (TNF-α), interferon-gamma (IFN-γ) and interleukin (IL)-17.[74] UC is already characterized by the activation of Th-2 cells followed by the overexpression of pro-inflammatory cytokines such as IL-5, IL-10 and IL-13.[75] These and other factors interrupt the intestinal homeostasis, thus leading to the development of IBD. If inflammation is not suppressed, CRC can occur. People who have chronic IBD have almost twice the risk of developing CRC as compared to otherwise healthy people.[5]

4.2.4 Gastric and Colorectal Cancer

Cancer treatment depends on the location, size, and extent of spread of cancer, as well as the patient's health.[76] However, the treatment is based on surgery in combination with radiotherapy and/or chemotherapeutic agents such as antimetabolites (5-fluorouracil, 5-FU), topoisomerase inhibitors (doxorubicin, DOX), platinum salts (cisplatin) or taxanes (paclitaxel, PTX).[42] The adjuvant or neoadjuvant therapy is given according to the cancer stage. However, this method of treatment becomes poorly effective if highly aggressive chemotherapy resistance, increased heterogeneity of the tumor and metastases have developed.[77] Thus, there is an urgent need to develop new approaches that could improve existing chemotherapeutic treatments.

The advances in nanotechnology have enabled the delivery of anticancer drugs or chemotherapeutic agents through nanoparticles, micelles or liposomes. These transport vehicles have a great potential since they can potentially overcome some barriers, as well as reduce systemic toxicity and improve biomolecules solubility, particularly hydrophobic molecules and those poorly soluble in water.[78] Additionally, these carriers have the ability to prolong the half-life of drugs in systemic circulation because of the lack of functional lymphatic drainage and due to the large abnormal tumor vasculature, preventing the clearance by MPS.[79]

4.3 Thermo-responsive Materials and Nanocarrier Systems

Stimuli-responsive materials, mostly polymers, have attracted attention for the development of potential drug delivery systems since these are able to potentially provide the ability to modulate the properties of a carrier in response to biological changes of interest. Stimuli can be classified in three different groups: chemical stimuli such as pH, ionic strength, solvent or electrochemical; physical stimuli such temperature, light, electric, magnetic, ultra-sound, mechanical; and biological stimuli such as enzymes, antibodies, receptors or proteins.[80] Stimuli can also be classified as internal (*i.e.*, pH, temperature, redox potential and enzymes) and external (*i.e.*, electromagnetic, light, radiation and ultrasound). Internal stimuli are based on the inherent differences between normal and diseased tissues, especially in tumor and inflammatory environments, where these differences are notorious. On the other hand, external stimuli can only occur through remote environmental devices, which may be disadvantageous to patients since they may not be possible to operate autonomously.[81] In any case, stimuli-sensitive polymers are able to undergo changes in conformation and/or solubility, alternations of the hydrophilic–hydrophobic state, or changes in shape due to alterations in the surrounding microenvironment, which can affect the dissolution/precipitation or drug release of engineered carriers.[25,26]

In the specific case of temperature stimuli, internal and external classification may apply. Internal stimuli occurs when the body temperature is altered from its physiological (37 °C) to pathological values (30–41 °C), which can happen in the case of inflammatory and tumorous tissues.[64,65] However, these variations in temperature may not be constant and stable for use in drug delivery and, thus, an external device may be necessary to increase the temperature in a specific location of the body. Different devices are available in the clinics in order to increase the temperature externally, namely microwaves, radiofrequency, ultrasound or hot water perfusion apparatus. This strategy is widely used in cases of cancer therapy, also referred to as hyperthermia and, it is characterized by an induced increase in the temperature (usually by six degrees Celsius above physiological temperature) of a cell or tissue in order to trigger its damage and/or destruction.[42] Furthermore, it has been used to increase and stabilize the temperature of localized tissue for improved drug delivery using nanomedicines (Figure 4.2).[16,23,82,83]

Thermo-responsive nanomedicines triggered by internal temperature change have the advantage of self-control when considering drug release,

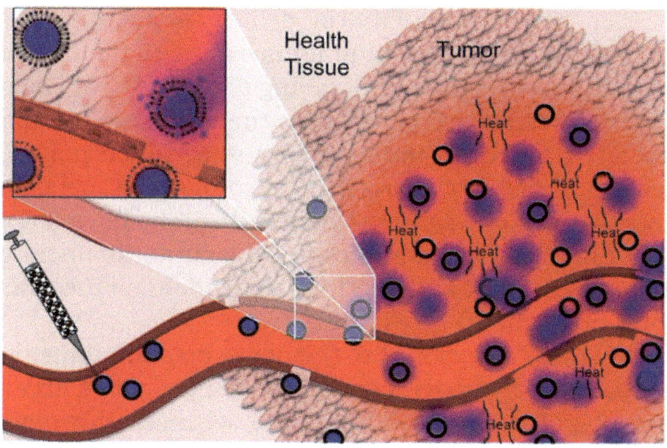

Figure 4.2 Drug release from temperature sensitive liposomes in a tumor during mild hyperthermia. Liposomes pass through the preheated tumor vasculature after intravenous administration. Subsequently, the release from the liposomes present in the vasculature and the tumor tissue is triggered due to the mild hyperthermia treatment. Reprinted from ref. 105, *International Journal of Pharmaceutics*, **515**, M. van Elk, B. P. Murphy, T. Eufrásio-da-Silva, D. P. O'Reilly, T. Vermonden, W. E. Hennink, G. P. Duffy, E. Ruiz-Hernández, 132–164, Copyright (2016), with permission from Elsevier.

since the nanocarrier will only deliver the drug when the body temperature is above or below the physiological temperature. External thermo-responsive nanomedicines are spatially and temporally more precise as the temperature is controlled by a remote device in a non-invasive manner, allowing the control of the drug released dose.[81] Although this approach may be more technically challenging, the use of an external source can guarantee constant temperatures, irrespective of patient and disease variability.

Temperature stimuli applied to nanoformulations have advantages since stimuli only exist in the pathological site, promoting high specificity, improvements in the cargo release as well as the concentration and, at same time, reduction of side effects. Also, thermo-responsive nanocarriers can be functionalized with ligands for specific cell surface receptors and, at same time, be designed to respond to the temperature differences between pathological and normal cells.[17,84–86] This type of carrier can increase the stability and solubility of the drug while the toxicity and immuno-sensitivity is decreased. On the other hand, it is known that cells are more sensitive to heat-induced damage compared to normal cells. Thus, it may be an additional advantage in the treatment of cancer by combining thermo-responsive nanomedicines with tumor hyperthermia. Actually, this is also the principle of using hyperthermia as an adjuvant in radiation therapy and chemotherapy.

Thermo-responsive polymeric materials have been largely explored in the design of drug delivery systems for the treatment of IBD and cancer due to the increased temperature in affected tissues. These systems offer great advantages over the conventional drug delivery systems, such as the possibility to control the phase transition temperature to the desired target and design versatility, besides passive targeting ability.[12,29,86,87] The main characteristic of thermo-responsive polymers is the critical solution temperature. This parameter is based on the temperature at which polymers exhibit a phase transition. It is called upper critical solution temperature (UCST) for the polymers that become soluble after heating. The polymers that become insoluble after heating have a lower critical solution temperature (LCST).[25] In most of the cases, LCST and UCST are dependent on the solution and the polymer concentration.

The value of LCST corresponds to the phase diagram transition where the enthalpy of the water hydrogen-bonded to the polymer chain turns into less than the entropy of the system. Therefore, when the hydrogen bonds are broken, the polymer precipitates (Figure 4.3).[88] Accordingly, the LCST of a polymer may be adjusted to a desired temperature by copolymerization or conjugation with other hydrophobic or hydrophilic polymers. When copolymerization with hydrophobic monomers is performed the LCST decreases, whereas the introduction of hydrophilic monomers has the opposite effect.[27] Additionally, the combination of a pH-responsive polymer with a thermo-responsive system can further change the solubility and make a system responsive to a specific temperature and/or a specific pH. This has been widely used in the biomedical field over the last few years.[25,29,87,89]

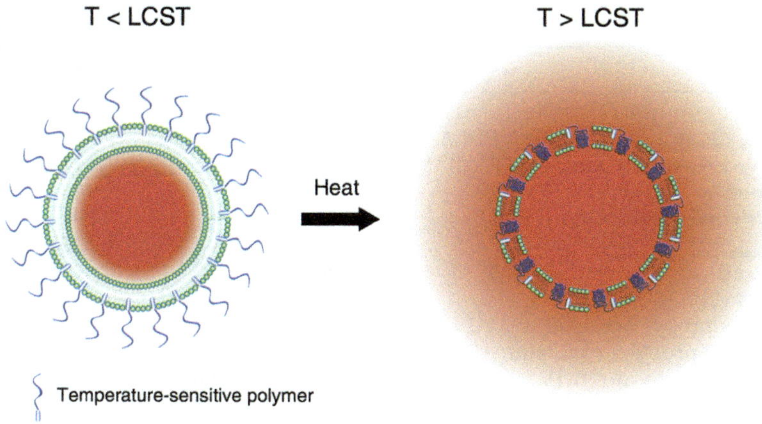

Figure 4.3 Illustration of drug release from polymer-modified thermosensitive liposomes. At temperatures greater than the lower critical solution temperature (LCST) of the polymer, polymer chains collapse, resulting in disruption of the membrane and release of the encapsulated drug. Reprinted from ref. 106, *Journal of Controlled Release*, **169**, T. Ta, T. M. Porter, Thermosensitive liposomes for localized delivery and triggered release of chemotherapy, 112–125, Copyright (2013), with permission from Elsevier.

Polymers with a LCST close to body temperature have been selected for thermo-responsive systems due to an extended conformation change below the phase transition temperature and the ability to convert from hydrophilicity to hydrophobicity when the temperature is above their LCST. These systems have been designed to be applied for both internal and external temperature stimuli. Nanocarriers are designed to only release their associated drugs at sites where the temperature is raised above normal physiological values.

The most famous thermo-responsive polymeric material is poly(*N*-isopropylacrylamide) (PNIPAM), exhibiting a LCST value of 32 °C.[90] Below the LCST there is an expansion and hydration of polymer chains, which makes the polymer hydrophilic and soluble. Also, the fact that PNIPAM is biocompatible and possesses a LCST close to body temperature makes this polymer one of the most widely explored for biomedical applications, particularly, for drug delivery. Above LCST, polymers become hydrophobic because hydrogen bounds are broken and water molecules are excluded from their structure, resulting in the collapse and aggregation of polymer chains. Consequently, there is a disassembly of nanocarriers followed by a fast release.[10] This transition is called a coil-to-globule transition and can be controlled through the monomers ratio change.[27]

Poly(*N*,*N*′-diethylacrylamide) (PDEAM) is another notorious temperature responsive polymer, with an LCST in the range of 32–34 °C.[91] Poly(methyl vinyl ether) (PMEV) has a transition temperature around 37 °C and is, for that reason, widely used for biomedical applications.[92] Conversely, poly(*N*-vinyl caprolactam) (PVCL) is less studied. However, it comprise good properties,

Table 4.2 Most common thermo-responsive polymers used in drug delivery.

Polymers	LCST	References
Poly(*N*-isopropylacrylamide) (PNIPAM)	32–33 °C	90 and 109
Poly(*N*,*N*′-diethylacrylamide) (PDEAM)	32–34 °C	91
Poly(*N*-vinylisobutylamide) (PNVIBAM)	39 °C	110
Poly(methyl vinyl ether) (PMEV)	37 °C	92
Poly(*N*-vinyl caprolactam) (PVCL)	32–34 °C	93
Pluronics	10–100 °C	96
Poly(2-ethoxy)ethoxyethyl vinyl ether (P(EOEOVE))	40 °C	101
Poly(2-isopropyl-2-oxazoline) (PiPOx) and Poly(2-*n*-propyl-2-oxazoline) (PnPOx)	36 °C	111 and 112
Poly(*N*-(1-hydroxymethyl) propylmethacrylamide) (PHMPMAAm)	30 °C (L-isomer) 34 °C (DL-mixture)	113

such as solubility in aqueous and organic solutions and biocompatibility.[93] Its LCST is between 32 and 34 °C, but an ideal temperature can be reached by copolymerization with other monomers. Table 4.2 summarizes the most widely used thermo-responsive polymers in drug delivery due to their LCST.

Poly(ethylene oxide) (PEO)-poly(propylene oxide) (PPO)-PEO triblock co-polymers, also commonly referred to Pluronics or poloxamers, are also used in thermo-responsive systems production. These polymers are not characterized by their LCST, but by their critical micellar temperature (CMT). CMT is the temperature in aqueous solution where Pluronics remain as non-associated polymers. On the other hand, above CMT the co-polymers can form micelles with the hydrophobic PPO groups forming the core and PEO hydrophilic groups forming the shell. The increasing temperature turns the polymers more hydrophobic, leading to micelle formation by self-assembly. Thus, this material can be used as thermo-responsive micelles or liposomes, where CMT controls the nanocarrier dissociation for example, below CMT Pluronics remain stable and above CMT they can cause membrane disruption releasing the associated drug.[94]

As other thermo-responsive polymers, CMT can also be controlled by varying the molecular weight, polymer concentration and the content ratio between PEO and PPO.[94,95] Thus, CMT can range from 10 °C in the case of polymers with low PEO groups, to 100 °C in the case of polymers with high PEO groups.[96]

Various nanocarriers have been described using the thermo-responsive materials described above. The following sub-sections detail the different nanosystems and explores specific examples in more detail.

4.3.1 Thermo-responsive Polymeric Nanoparticles

Polymeric nanoparticles are submicron-sized scale systems featuring a matrix composed of one or more polymers to which therapeutic agents can be associated, either inside or at the surface of the carrier *via* physical and/or chemical (covalent or non-covalent bonding) interactions.[97] Characteristics such as surface modification and drug association can be controlled during production according to specific needs.

Thermosensitive polymeric nanoparticles have been widely explored for various applications related with therapy at the GIT.

In a recent study, Rejinold *et al.*,[29] developed chitosan-graft-poly(*N*-vinyl caprolactam) nanoparticles incorporating gold nanoparticles in order to deliver curcumin (Au-CRC-TRC-NPs) using a ionic cross-linking method. Nanoparticles with diameter around 160 nm,[87] presented higher released in the presence of radiofrequency (RF) pulses that was used to increase the tissue temperature, as compared to the absence of RF pulses. The *in vivo* studies in Swiss albino mice demonstrated an increase in circulation time of Au-CRC-TRC-NPs of around 1 week compared to the free curcumin, which was eliminated in 6 h. Besides organ distribution results confirming the presence of curcumin in the liver, brain, spleen, intestine, lungs, kidney and stomach, after 7 and 30 days, it was verified there was no significant pathological changes in these organs.[29] The colon tumor model was achieved after CT-26 cell transplantation and, after reaching 4000 mm^3, Au-CRC-TRC-NPs were injected intravenously and reached the tumor on day 1, but it was on day 2 there was obvious accumulation and retention of these nanoparticles in the tumor site and it lasted about two weeks.[29] The long systemic circulation time and biodistribution can be advantageous to avoid daily injections and can be a useful approach for colon cancer treatment.

Thermo-responsive nanoparticles based on poly(*N*-isopropylacrylamide-*co*-((2-dimethylamino)ethylmethacrylate)) (poly(NIPAM-*co*-DMAEMA)) copolymers[98] were produced by free radical polymerization to 7-ethyl-10-hydroxy-camptothecin (SN-38) association, an anticancer agent used to treat colorectal and gastric cancer.[69,99] The LCST of poly(NIPAM-*co*-DMAEMA) nanoparticles was 41 °C and, when SN-38 was associated, the chosen nanoparticle formulation presented a diameter around 326 nm and association efficiency around 81%.[98] The release profile of SN-38 from these particles was the highest when heated at 42 °C. Consequently, cytotoxicity studies only demonstrated toxicity when nanoparticles were heated at 42 °C for 30 min as compared to the results performed at 37 °C, due to the preferential particle disassociation and SN-38 release at higher temperature. The anti-tumor effect of poly(NIPAM-*co*-DMAEMA) nanoparticles was assessed in a CT-26 murine colon tumor model. Results confirmed significant tumor growth suppression after intravenous administration of SN-38 nanoparticles, followed by local tumor heating using a temperature-controlled heating blanket for 30 min at 42 °C after 24 h post injection. On the other hand, SN-38 nanoparticles without hyperthermia, or only hyperthermia, did not show significant tumor suppression as compared to the control,[98] which seems to demonstrate the efficiency of the proposed nanosystem to deliver the cargo at the desired site depending on the thermal stimulus.

In another study, Bartlett *et al.*,[89] developed poly(*N*-isopropylacrylamide–2-acrylamido-2-methyl-1-propanesulfonate) [poly(NIPAM–AMPS)] nanoparticles crosslinked to a hydrolytically degradable *N*,*O*-dimethacryloyl hydroxylamine (DMHA) and/or a nondegradable *N*,*N*′-methylenebi-sacrylamide (MBA) in order to obtain a dual thermo- and pH-responsive drug delivery system. These degradable or nondegradable nanoparticles were cross-linked to prevent nanoparticle dissociation below LCST and to be able to adjust the degradation kinetics.[89] The system was also comprised of the association of a cell-penetrating peptide KAFAKLAARLYRKALARQLGVAA (KAFAK). Such a strategy allowed the protection of KAFAK from enzymatic degradation, which is responsible for the intracellular inhibition of pro-inflammatory cytokines, such as TNF-α by blocking the activity of mitogen-activated protein kinase 2 (MK2) in situations of IBD or cancer.[89] Data presented by the authors also showed that the incorporation of AMPS into the PNIPAM backbone increased the loading efficiency and stability at pH < 5 of KAFAK into PNIPAM-AMPS nanoparticles. Also, KAFAK had a faster release from the degradable systems as compared to those which were nondegradable.[89] The *in vitro* inflammatory model based on stimulated immortalized human monocytes (THP-1) was successfully used in demonstrating that KAFAK released from the degradable nanoparticles inhibit approximately 50% of TNF-α production after one day. A potentially effective release between one and four days, as correlated with the therapeutically active doses of KAFAK, was also achieved.[89] These results were only achieved because the LCST was below the physiological temperature. As these tests were performed at 37 °C, nanoparticles could dissociate effectively.

The previous group has also recently used the same type of system, but with additional modifications: PEG was added in order to minimize the interaction of nanoparticles with plasma proteins and the degradable disulfide cross-linker (*N*,*N*′-bis(acryloyl)cystamine) (NGPEGSS) was used instead of DMHA.[86] Release studies demonstrated higher KAFAK release at low pH (~4.0) and at a temperature close to LCST (~37 °C) from degradable nanoparticles as compared to nondegradable, but both systems remained stable at physiological conditions. The cellular uptake and nanoparticle localization was performed in an *in vitro* inflammatory model of a murine RAW 264.7 macrophage cell line activated with lipopolysaccharide. Results showed nanoparticle internalization and accumulation within or at the cell membrane. Also, data suggested that nanoparticles could accumulate into endosomes/lysosomes. The addition of PEG or NGPEGSS did not affect macrophage viability *in vitro.*[86] Additionally, the release of KAFAK from PEGylated and degradable nanoparticles was deemed sufficient (and higher as compared to the nondegradable nanoparticles or free KAFAK) to reduce interleukin 6 (IL-6) and TNF-α expression by approximately 67% and 64%, respectively, after 24 h.[86] The two previous studies combined[86,89] support the theory that PNIPAM-AMPS nanoparticles may be able to reduce the expression of pro-inflammatory cytokines.

In a further report, poly(*N*-isopropylacrylamide-DL-lactide) was used to produce thermo-responsive nanoparticles. A blend of PLA and the previous block copolymer was used (PLA/PNIPAM-PLA nanoparticles) to associate the anti-inflammatory drug betamethasone disodium 21-phosphate (BP) and tentatively reduce the onset of side-effects. The LCST of the co-polymer found was at 34 °C and the obtained nanoparticles presented an average diameter of 140 nm, independently of the temperature. Results demonstrated the PLA/PNIPAM-PLA nanoparticles allowed for a more extensive release of BP release (close to 90% at 37 °C in 24 h as compared to almost 40% at 30 °C). The authors also used PLA only and PLA/PEG-PLA only nanoparticles for comparison purposes and reported that BP release was considerably lower (around 20%) at 37 °C after 7 days.[10] These results showed a drug release temperature dependence only when the thermosensitive polymer is present. When PNIPAAM is not present in the nanoparticle, BP release is low and temperature independent. The intracellular uptake and BP internalization was evaluated in RAW 264.7 cells by fluorescence microscopy and time-resolved fluoroimmunoassay, respectively. Cellular uptake of PLA/PNIPAM-PLA nanoparticles was not observed below LCST (30 °C) as evidenced in Figure 4.4. Conversely, a significant difference was found at 37 °C, *i.e.*, above the LCST.[10] These results suggest that the cellular uptake of nanoparticles was

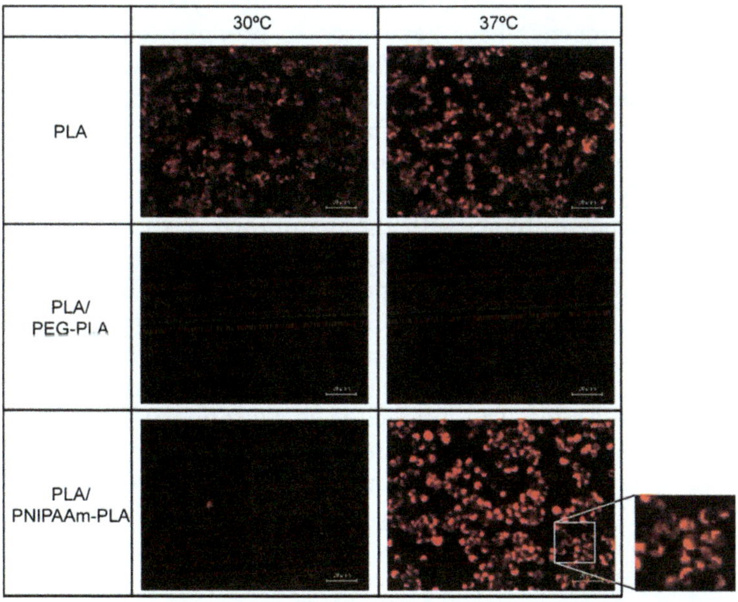

Figure 4.4 Internalization of NPs into RAW 264.7 cells. Fluorescence images after incubation with rhodamine-labeled NPs prepared from PLA, PLA/PEG–PLA and PLA/PNIPAM–PLA at 30 °C and 37 °C after 3 h. Red images show rhodamine. Scale bar, 20 μm. Reprinted from ref. 10, *Colloids and Surfaces B: Biointerfaces*, **99**, E. Ayano, M. Karaki, T. Ishihara, H. Kanazawa, T. Okano, Poly (*N*-isopropylacrylamide)–PLA and PLA blend nanoparticles for temperature-controllable drug release and intracellular uptake, 67–73, Copyright (2012), with permission from Elsevier.

truly controllable by temperature. Researchers also observed a fluorescence signal from rhodamine-conjugated PLA nanoparticles independently of the temperature due to the hydrophobic character of PLA. Additionally, the amount of BP internalized in RAW cells after 3 h was significantly higher at 37 °C as compared to 30 °C for PLA/PNIPAM-PLA nanoparticles.[10] The PLA nanoparticles did not present any significant difference of BP internalization regarding changes in temperature. In this sense, the authors concluded that PLA/PNIPAM-PLA nanoparticles may be an interesting thermo-controllable system for drug delivery applications, particularly if cellular uptake in inflammatory states is required.

4.3.2 Thermo-responsive Liposomes

Liposomes are highly versatile drug delivery systems that have been widely used in nanomedicine. These carriers are spontaneously formed by vesicles comprising one or more lipid bilayers of amphipathic phospholipids surrounding an aqueous core. Hydrophilic drugs are typically entrapped in this last, while hydrophobic compounds can be associated within the lipid bilayer.[42] This feature allows the combination of active molecules with diverse physicochemical features (namely solubility) into one single system.

Different approaches have been used to produce temperature-sensitive liposomes in order to achieve controlled release of active compounds. Two typical approaches include the modulation of the temperature-induced phase transition of the liposome membrane (*e.g.*, by simply changing the quantitative or qualitative lipid composition)[82,83] or the incorporation of thermo-responsive polymers into the liposome membrane (Figure 4.3).[18,94,100] In the last case, polymers interact with the membrane of the liposome in order to allow drug release above the LCST. As an example, Wells *et al.*,[20] produced di-oleoylphosphatidylcholine (DOPC)/cholesterol liposomes with added Pluronic® F127 (originally developed by Chandaroy *et al.*[94] with slight modifications) and tested these vesicles as potential drug delivery systems using Lucifer yellow iodoacetamide (LY) as a fluorescent marker in Colon-26 (CT-26) cells and mice. The interaction of Pluronic® F127 with liposomes at temperatures above the poloxamer CMT led to the bilayer disruption with consequent drug release.[94] Around 90% of the LY content was released *in vitro* at 42 °C as compared to only 50% at 37 °C. Higher CT-26 cell binding to LY was also observed at 42 °C. *In vivo* studies in mice with CT-26 tumors demonstrated that, after thermo-responsive liposomes injection *via* a tail vein, unheated tumors presented similar fluorescence as non-injected ones. Conversely, injection of thermo-responsive liposomes followed by local heating treatment led to significantly higher fluorescence signals as compared with the unheated group ($P = 0.00002$).[20] These results seem to back up the theory that LY release from liposomes in locally heated CT-26 tumors was considerably higher as compared with the controls without mild hyperthermia or with mild hyperthermia alone.[20]

Based on a previous work where poly[2-(2-ethoxy)ethoxyethyl vinyl ether (EOEOVE)] was incorporated in stable liposomes,[101] Kono *et al.*,[68] produced

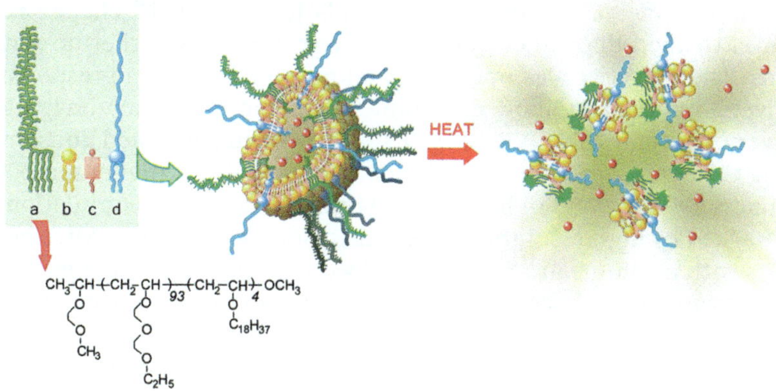

Figure 4.5 Design of temperature-sensitive liposomes composed of thermosensitive poly(EOEOVE)-OD_4 (a), membrane-forming EYPC (b), membrane-stabilizing cholesterol (c), and highly hydrophilic and non-toxic PEG-lipid (d). Heat-triggered release of DOX from liposomes is illustrated. Structure of poly(EOEOVE)-OD_4 is also shown. Reprinted from ref. 68, *Biomaterials*, **31**, K. Kono, T. Ozawa, T. Yoshida, F. Ozaki, Y. Ishizaka, K. Maruyama, C. Kojima, A. Harada, S. Aoshima, Highly temperature-sensitive liposomes based on a thermosensitive block copolymer for tumor-specific chemotherapy, 7096–7105, Copyright (2010), with permission from Elsevier.

thermo-responsive liposomes comprised of PEG-lipid (4 mol%), egg yolk phosphatidylcholine (EYPC), cholesterol and copoly(EOEOVE-*block*-octadecyl vinyl ether) to incorporate doxorubicin (DOX) (2 mol%) for tumor-specific chemotherapy. The design of this system is shown in Figure 4.5.

Increasing the content of liposomes in poly(EOEOVE)-OD_4 led to higher DOX release when tested at hyperthermia conditions, which was related to the LCST of poly(EOEOVE) (estimated to be around 40 °C). Indeed, around 85% of DOX was released from the liposomes (poly(EOEOVE)-OD_4 content at 2 mol%) over one minute after mild heating at 45 °C.[68] The *in vivo* studies were performed in mice bearing tumors with a volume of *ca.* 150 mm^3 by inoculating mouse colon carcinoma 26 cells into the left flank of Balb/c mice and liposomes were injected intravenously *via* the tail vein. Liposomes with PEG exhibited a higher percentage of distribution in blood and a lower percentage of distribution in the liver and spleen, as well as a long circulation in the blood, compared to liposomes without PEG.[68] Regarding tumor growth suppression, after 3, 6 and 12 h of liposomes injection, the tumor site was heated locally at 45 °C for 10 min using a radio frequency oscillator. It was verified that heating the tumor site 12 h after liposomes administration was when tumor suppression was more evidenced. Conversely, the injection of liposomes encapsulating DOX or saline without tumor heating did not result in tumor suppression.[68] These results indicate that liposomes were able to accumulate in tumors and DOX release is triggered by an increase in temperature. The same did not occur in the case of unheated tumor tissue.

The previous research group further developed a similar system by adding 5% polyamidoamine G3 dendron-based lipids featuring Gd^{3+} chelate residues into PEGylated liposomes composed by poly(EOEOVE) thermo-sensitive chains. The objective was not only to allow effective thermo-responsive drug delivery, but also to allow detectable magnetic resonance imaging.[17] DOX delivery by copoly(EOEOVE-*block*-ODVE)-modified PEGylated liposomes containing Gd-DOTA-dendron-based lipids was extensive at higher temperatures (around 80% at 45 °C in less than 10 min). Liposome accumulation *in vivo* was tracked by high resolution magnetic resonance imaging. Mice bearing tumors with a volume around 200 mm^3 were prepared by inoculation of mouse colon carcinoma 26 cells into the right flank of BALB/c mice and were used to test tumor growth suppression. Liposomes incorporating DOX were injected *via* the tail vein and it was verified after at 8 h post-administration that there was a higher liposome accumulation at the tumor tissue. Thus, the tumor tissue was locally heated at 44 °C for 10 min, 8 h after administration. Tumor suppression was higher with heating at 44 °C after 8 h of liposome administration as shown in Figure 4.6. The administration of the thermo-responsive liposomes without mild heating, as well as saline injection with or without heating, had no suppressive effect in tumor growth.[17]

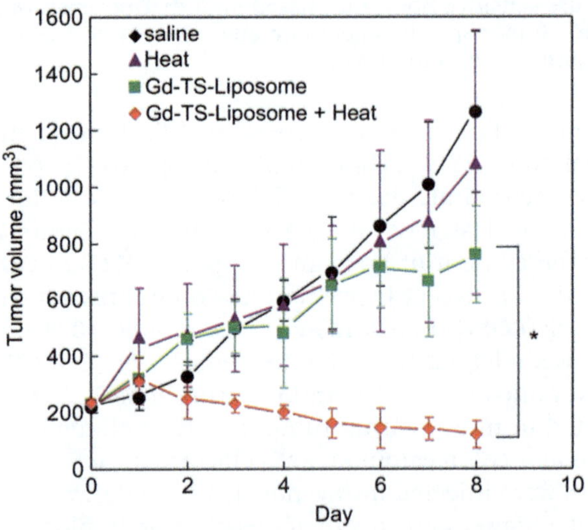

Figure 4.6 Time course of tumor growth after injection of DOX-loaded liposomes with (diamonds) or without (squares) local heating. DOX dose was 6 mg kg^{-1} body weight. The data represent mean ± S.D. ($n = 4$). *, $P < 0.01$. Injection of saline (HBS) without heating (circles) and heating without injection (triangles) were also performed as control experiments. Reprinted from ref. 17, *Biomaterials*, **32**, K. Kono, S. Nakashima, D. Kokuryo, I. Aoki, H. Shimomoto, S. Aoshima, K. Maruyama, E. Yuba, C. Kojima, A. Harada, Y. Ishizaka, Multi-functional liposomes having temperature-triggered release and magnetic resonance imaging for tumor-specific chemotherapy, 1387–1395, Copyright (2011), with permission from Elsevier.

4.3.3 Thermo-responsive Micelles

Amphiphilic molecules can form micelles spontaneously in aqueous solution by self-assembly when present in levels above the critical micelle concentration. Stimuli responsive micelles can be produced by two main methodologies: (i) the amphiphilic segments of the molecules may undergo micellization or demicellization through the alternation of the equilibrium between hydrophilicity and hydrophobicity, which can be modulated by various stimuli, namely temperature;[102] or (ii) one or both of the hydrophilic and hydrophobic segments of the amphiphilic molecule may be replaced by a stimuli-responsive component.[103]

An interesting example of stimuli-responsive micelles with potential application in gastric cancer therapy was reported by Li *et al.*[15] These researchers developed thermo- and pH-responsive polymeric micelles based on poly(*N*-isopropylacrylamide-*co*-*N*,*N*-dimethylacrylamide) (PID) thermo-responsive blocks by the reversible addition-fragmentation chain transfer (RAFT) technique. The rationale for the development of such a system took into account both the typical high temperature and low pH of tumor tissues. The hydrophobic block of PLA and poly(ε-caprolactone) (PCL) were linked to the thermo-responsive blocks by ring-open polymerization in order to produce poly(*N*-isopropylacrylamide-*co*-*N*,*N*-dimethylacrylamide-*b*-lactide) (PID_{118}-*b*-PLA_{59}) and poly(*N*-isopropylacrylamide-*co*-*N*,*N*-dimethylacrylamide-*b*-ε-caprolactone) (PID_{118}-*b*-PCL_{60}). The previous polymers were used to produce DOX-associated micelles by the dialysis method.[15] Micelles presented average size values of 170 nm and 87 nm and LCST of 39 °C and 40.5 °C for PID_{118}-*b*-PLA_{59} and PID_{118}-*b*-PCL_{60} micelles, respectively. The authors reported that such properties may be suitable for providing stability in systemic circulation and allow micelle extravasion across tumor vasculature.[15] Results from *in vitro* experiments showed that both micellar systems remain stable, allowing for very slow DOX release at physiological conditions (T = 37 °C and pH = 7.4). However, at higher temperatures (around or above) LCST and under acidic conditions (pH ~ 5.3), micelles undergo a hydrophilic to hydrophobic shift due to the presence of PID, thus releasing about 90% of DOX within 10 h. Cellular uptake was also increased.[15] Cytotoxicity studies performed in N-87 cells (stomach cancer cell) demonstrated that blank micelles presented low cytotoxicity and both micelles-DOX had a higher toxicity at 40 °C as compared to 37 °C.[15]

The previous group further used a similar system in another work, but with an increase in the hydrophobic content of PLA in order to optimize drug loading. PID_{118}-*b*-PLA_{71} micelles were further functionalized with an anti-Her2 Fab onto the micellar surface to form immunomicelles (Fab conjugated immunomicelles - FCIMs). The rationale behind the functionalization relies on the fact that overexpression of Her2 in stomach cancer cells surface may be an indication of dysregulated growth and angiogenesis.[104] Thus, the authors expected that both Fab and temperature could be useful to regulate the intracellular uptake and the accumulation of such multifunctional micelles in gastric tumors.[85] FCIMs were evaluated in terms of intracellular uptake of DOX in N87 stomach cancer cells over-expressing Her2 and shown to significantly improve intracellular uptake compared to free DOX, Doxil® (Janssen Pharmaceutica, Beerse, Belgium) or immune micelles (IMs - without Fab).[85] Moreover, this improvement was even higher when the temperature was above LCST. Additionally, Balb/c nude mice bearing gastric cancer N87 cell tumors were used to evaluate the *in vivo* tissue distribution and tumor accumulation. Imaging data collected using an IVIS® system evidenced that animals treated with FCIMs presented a higher fluorescence signal in tumors as compared to control (saline), DOX or IMs. These results seem to indicate that antibody-mediated targeting can promote higher intra-tumoral accumulation even when compared to IMs that are only responsive to temperature.[85] Regarding therapeutic efficiency, tumor volume increase in control and DOX-treated animals was higher as compared to IMs and FCIMs (Figure 4.7). Moreover,

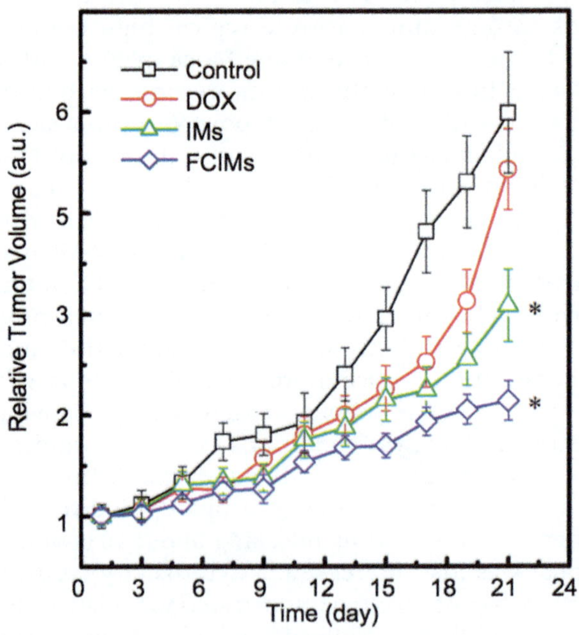

Figure 4.7 *In vivo* anticancer therapeutic effects for Balb-c/nude mice bearing N87 tumors after intravenous administration every day for three days, PBS was used as a control, free doxorubicin was used as a reference, FCIMs-DOX and IMs-DOX. Tumor volumes were measured every 2 days and normalized to the initial tumor volume (day 0). Results are presented in terms of relative tumor volumes, mean ± s.d., $N = 5$. *: $P < 0.05$ compared to control. Reprinted from ref. 85, *Biomaterials*, **33**, W. Li, H. Zhao, W. Qian, H. Li, L. Zhang, Z. Ye, G. Zhang, M. Xia, J. Li, J. Gao, B. Li, G. Kou, J. Dai, H. Wang, Y. Guo, Chemotherapy for gastric cancer by finely tailoring anti-Her2 anchored dual targeting immunomicelles, 5349–5362, Copyright (2012), with permission from Elsevier.

tumor volume of the group administered with FCIMs was around 3 times lower than the group treated with IMs.[85] These *in vivo* results appear to indicate that micelles featuring tandem targeting and temperature responsiveness may have considerable potential for anticancer therapy in gastric tumors.

4.4 Conclusions and Future Perspectives

Temperature increase to a few degrees above the physiological temperature in situations of inflammation or cancer is a well-known condition. Based on the thermally responsive properties of some polymers, researchers have taken advantage of this feature to formulate nanosystems for drug delivery, particularly those described here for GIT diseases such as cancer or IBD. In fact, there are few examples in which the increased temperature of the inflamed or tumour region is enough for the nanosystems to release the drug in sufficient amounts to have the desired effect. In most cases, it is necessary to use an external device that locally heats the damaged region in a determined time after the administration of a thermo-responsive system to reach a considerable tumour growth suppression. Additionally, the *in vivo* tumour models described here to demonstrate the effect of thermo-responsive systems with and without an external heat source are based on subcutaneous tumours, which are not the best representation of what occurs in human tumours. Although thermo-responsive systems with or without local heating exhibit considerably higher drug release profiles above LCST, and greater tumour growth suppression compared with the free drug, saline or even non-thermo-triggered systems, the transition from *in vivo* models to *in vivo* clinical applications continues to be a demanding challenge.

Acknowledgements

This chapter is a result of the project NORTE-01-0145-FEDER-000012, supported by Norte Portugal Regional Operational Programme (NORTE 2020), under the PORTUGAL 2020 Partnership Agreement, through the European Regional Development Fund (ERDF).

This work was financed by FEDER - Fundo Europeu de Desenvolvimento Regional funds through the COMPETE 2020 - Operacional Programme for Competitiveness and Internationalisation (POCI), Portugal 2020, and by Portuguese funds through FCT - Fundação para a Ciência e a Tecnologia/ Ministério da Ciência, Tecnologia e Ensino Superior in the framework of the project "Institute for Research and Innovation in Health Sciences" (POCI-01-0145-FEDER-007274). Andreia Almeida (grant SFRH/BD/118721/2016), Helena Macedo (SFRH/BD/131587/2017) and José das Neves (grant SFRH/BPD/92934/2013) would like to thank Fundação para a Ciência e a Tecnologia (FCT), Portugal for financial support.

References

1. J. M. Reichert and J. B. Wenger, *Drug Discovery Today*, 2008, **13**, 30–37.
2. R. Misra, S. Acharya and S. K. Sahoo, *Drug Discovery Today*, 2010, **15**, 842–850.
3. W. H. Organization, *Media Centre: Cancer*, 2017, http://www.who.int/en/news-room/fact-sheets/detail/cancer.
4. H. Kuper, H. O. Adami and D. Trichopoulos, *J. Intern. Med.*, 2000, **248**, 171–183.
5. A. C. Society, *Cancer Facts & Figures 2016*, 2016, https://www.cancer.org/content/dam/cancer-org/research/cancer-facts-and-statistics/annual-cancer-facts-and-figures/2016/cancer-facts-and-figures-2016.pdf.
6. S. Kapse-Mistry, T. Govender, R. Srivastava and M. Yergeri, *Front. Pharmacol.*, 2014, **5**, 159.
7. S. Parveen and S. K. Sahoo, *J. Drug Targeting*, 2008, **16**, 108–123.
8. A. Lamprecht, H. Yamamoto, H. Takeuchi and Y. Kawashima, *J. Pharmacol. Exp. Ther.*, 2005, **315**, 196–202.
9. Y. Meissner, Y. Pellequer and A. Lamprecht, *Int. J. Pharm.*, 2006, **316**, 138–143.
10. E. Ayano, M. Karaki, T. Ishihara, H. Kanazawa and T. Okano, *Colloids Surf., B*, 2012, **99**, 67–73.
11. I. Hilger, R. Hiergeist, R. Hergt, K. Winnefeld, H. Schubert and W. A. Kaiser, *Invest. Radiol.*, 2002, **37**, 580–586.
12. C. M. Abreu, H. C. Paula, V. Seabra, J. P. Feitosa, B. Sarmento and R. C. de Paula, *Carbohydr. Polym.*, 2016, **154**, 77–85.
13. W. Y. Seow, J. M. Xue and Y.-Y. Yang, *Biomaterials*, 2007, **28**, 1730–1740.
14. J. Akimoto, M. Nakayama, K. Sakai and T. Okano, *Mol. Pharmaceutics*, 2010, **7**, 926–935.
15. W. Li, J. Li, J. Gao, B. Li, Y. Xia, Y. Meng, Y. Yu, H. Chen, J. Dai and H. Wang, *Biomaterials*, 2011, **32**, 3832–3844.
16. G. Kong, G. Anyarambhatla, W. P. Petros, R. D. Braun, O. M. Colvin, D. Needham and M. W. Dewhirst, *Cancer Res.*, 2000, **60**, 6950–6957.
17. K. Kono, S. Nakashima, D. Kokuryo, I. Aoki, H. Shimomoto, S. Aoshima, K. Maruyama, E. Yuba, C. Kojima and A. Harada, *Biomaterials*, 2011, **32**, 1387–1395.
18. T. Ta, A. J. Convertine, C. R. Reyes, P. S. Stayton and T. M. Porter, *Biomacromolecules*, 2010, **11**, 1915–1920.
19. B. Tirosh, N. Khatib, Y. Barenholz, A. Nissan and A. Rubinstein, *Mol. Pharmaceutics*, 2009, **6**, 1083–1091.
20. J. Wells, A. Sen and S. W. Hui, *Int. J. Pharm.*, 2003, **261**, 105–114.
21. H. Maeda, *Adv. Enzyme Regul.*, 2001, **41**, 189–207.
22. H. Maeda, J. Wu, T. Sawa, Y. Matsumura and K. Hori, *J. Controlled Release*, 2000, **65**, 271–284.
23. J. Weinstein, R. Magin, M. Yatvin and D. Zaharko, *Science*, 1979, **204**, 188–191.
24. S. Mura, J. Nicolas and P. Couvreur, *Nat. Mater.*, 2013, **12**, 991.
25. D. Schmaljohann, *Adv. Drug Delivery Rev.*, 2006, **58**, 1655–1670.

26. B. Jeong and A. Gutowska, *Trends Biotechnol.*, 2002, **20**, 305–311.
27. P. Kujawa and F. M. Winnik, *Macromolecules*, 2001, **34**, 4130–4135.
28. N. S. Rejinold, M. Muthunarayanan, V. Divyarani, P. Sreerekha, K. Chennazhi, S. Nair, H. Tamura and R. Jayakumar, *J. Colloid Interface Sci.*, 2011, **360**, 39–51.
29. N. S. Rejinold, R. G. Thomas, M. Muthiah, K. Chennazhi, K. Manzoor, I.-K. Park, Y. Y. Jeong and R. Jayakumar, *Int. J. Biol. Macromol.*, 2015, **74**, 249–262.
30. L. Shen, *Molecular Mechanisms of Bacterial Infection via the Gut*, Springer, 2009, pp. 1–35.
31. R. N. Jaladanki and J.-Y. Wang, *Colloquium Series on Integrated Systems Physiology: From Molecule to Function*, 2011.
32. C. Pereira, F. Araújo, C. C. Barrias, P. L. Granja and B. Sarmento, *Biomaterials*, 2015, **56**, 36–45.
33. R. K. Montgomery, A. E. Mulberg and R. J. Grand, *Gastroenterology*, 1999, **116**, 702–731.
34. B. Gannon, *Gastrointestinal and Oesophageal Pathology*, Churchill Livingstone, Edinburgh, 1995, vol. 129.
35. J. Weiss, A. Sauer, N. Divac, M. Herzog, E. Schwedhelm, R. Böger, W. Haefeli and R. Benndorf, *Biopharm. Drug Dispos.*, 2010, **31**, 150–161.
36. A. W. Larhed, P. Artursson and E. Björk, *Pharm. Res.*, 1998, **15**, 66–71.
37. M. T. Abreu, *Nat. Rev. Immunol.*, 2010, **10**, 131–144.
38. A. L. Daugherty and R. J. Mrsny, *Pharm. Sci. Technol. Today*, 1999, **2**, 144–151.
39. R. K. Jain, *Adv. Drug Delivery Rev.*, 2001, **46**, 149–168.
40. T. M. Allen and P. R. Cullis, *Science*, 2004, **303**, 1818–1822.
41. P. Sapra and T. Allen, *Prog. Lipid Res.*, 2003, **42**, 439–462.
42. A. S. Thakor and S. S. Gambhir, *Ca-Cancer J. Clin.*, 2013, **63**, 395–418.
43. H. Maeda and Y. Matsumura, *Crit. Rev. Ther. Drug Carrier Syst.*, 1988, **6**, 193–210.
44. S. A. Ingersoll, S. Ayyadurai, M. A. Charania, H. Laroui, Y. Yan and D. Merlin, *Am. J. Physiol.: Gastrointest. Liver Physiol.*, 2012, **302**, G484–G492.
45. A. Lamprecht, *Nat. Rev. Gastroenterol. Hepatol.*, 2015, **12**, 195.
46. H. Ali, E.-M. Collnot, M. Windbergs and C.-M. Lehr, *Eur. J. Nanomed.*, 2013, **5**, 23–38.
47. Y.-Z. Du, L.-L. Cai, P. Liu, J. You, H. Yuan and F.-Q. Hu, *Biomaterials*, 2012, **33**, 8858–8867.
48. M. Sugano, N. K. Egilmez, S. J. Yokota, F.-A. Chen, J. Harding, S. K. Huang and R. B. Bankert, *Cancer Res.*, 2000, **60**, 6942–6949.
49. S. F. Rosebrough and M. Hashmi, *J. Pharmacol. Exp. Ther.*, 1996, **276**, 770–775.
50. H. Tada, H. Higuchi, T. M. Wanatabe and N. Ohuchi, *Cancer Res.*, 2007, **67**, 1138–1144.
51. D. Pan, J. L. Turner and K. L. Wooley, *Chem. Commun.*, 2003, 2400–2401.
52. R. Rossin, D. Pan, K. Qi, J. L. Turner, X. Sun, K. L. Wooley and M. J. Welch, *J. Nucl. Med.*, 2005, **46**, 1210–1218.

53. R. Mehvar, *Curr. Pharm. Biotechnol.*, 2003, **4**, 283–302.
54. M. Mazid, E. Moase, E. Scott, H. Hanna and F. Unger, *J. Biomed. Mater. Res., Part A*, 1991, **25**, 1169–1181.
55. O. C. Farokhzad, J. M. Karp and R. Langer, *Expert Opin. Drug Delivery*, 2006, **3**, 311–324.
56. E. N. Brody and L. Gold, *Rev. Mol. Biotechnol.*, 2000, **74**, 5–13.
57. G. P. Andrews, T. P. Laverty and D. S. Jones, *Eur. J. Pharm. Biopharm.*, 2009, **71**, 505–518.
58. B. Devriendt, B. G. De Geest, B. M. Goddeeris and E. Cox, *J. Controlled Release*, 2012, **160**, 431–439.
59. M. A. Charania, H. Laroui, H. Liu, E. Viennois, S. Ayyadurai, B. Xiao, S. A. Ingersoll, D. Kalman and D. Merlin, *Infect. Immun.*, 2013, **81**, 923–934.
60. M. T. Cook, G. Tzortzis, D. Charalampopoulos and V. V. Khutoryanskiy, *J. Controlled Release*, 2012, **162**, 56–67.
61. A. S. Ojugo, P. M. McSheehy, D. J. McIntyre, C. McCoy, M. Stubbs, M. O. Leach, I. R. Judson and J. R. Griffiths, *NMR Biomed.*, 1999, **12**, 495–504.
62. I. F. Tannock and D. Rotin, *Cancer Res.*, 1989, **49**, 4373–4384.
63. B. Xiao and D. Merlin, *Expert Opin. Drug Delivery*, 2012, **9**, 1393–1407.
64. F. Danhier, O. Feron and V. Préat, *J. Controlled Release*, 2010, **148**, 135–146.
65. T. Yahara, T. Koga, S. Yoshida, S. Nakagawa, H. Deguchi and K. Shirouzu, *Surg. Today*, 2003, **33**, 243–248.
66. B. Hildebrandt, P. Wust, O. Ahlers, A. Dieing, G. Sreenivasa, T. Kerner, R. Felix and H. Riess, *Crit. Rev. Oncol. Hematol.*, 2002, **43**, 33–56.
67. M. W. Dewhirst, Z. Vujaskovic, E. Jones and D. Thrall, *Int. J. Hyperthermia*, 2005, **21**, 779–790.
68. K. Kono, T. Ozawa, T. Yoshida, F. Ozaki, Y. Ishizaka, K. Maruyama, C. Kojima, A. Harada and S. Aoshima, *Biomaterials*, 2010, **31**, 7096–7105.
69. C.-L. Peng, P.-S. Lai, F.-H. Lin, S. Y.-H. Wu and M.-J. Shieh, *Biomaterials*, 2009, **30**, 3614–3625.
70. A. B. Pithadia and S. Jain, *Pharmacol. Rep.*, 2011, **63**, 629–642.
71. D. K. Podolsky, *N. Engl. J. Med.*, 2002, **347**, 417–429.
72. A. Vural and J. H. Kehrl, *Scientifica*, 2014, **2014**, 825463.
73. S. Gordon and F. O. Martinez, *Immunity*, 2010, **32**, 593–604.
74. K. A. Papadakis and S. R. Targan, *Annu. Rev. Med.*, 2000, **51**, 289–298.
75. S. Melgar and F. Shanahan, *Autoimmunity*, 2010, **43**, 463–477.
76. E. Fernandes, J. A. Ferreira, P. Andreia, L. Luís, S. Barroso, B. Sarmento and L. L. Santos, *J. Controlled Release*, 2015, **209**, 288–307.
77. P. Ramos and M. Bentires-Alj, *Oncogene*, 2015, **34**, 3617–3626.
78. W. Xun, H.-Y. Wang, Z.-Y. Li, S.-X. Cheng, X.-Z. Zhang and R.-X. Zhuo, *Colloids Surf., B*, 2011, **85**, 86–91.
79. G. S. Kwon and K. Kataoka, *Adv. Drug Delivery Rev.*, 1995, **16**, 295–309.
80. D. Wang, Y. Jin, X. Zhu and D. Yan, *Prog. Polym. Sci.*, 2017, **64**, 114–153.
81. A. Chan, R. P. Orme, R. A. Fricker and P. Roach, *Adv. Drug Delivery Rev.*, 2013, **65**, 497–514.
82. D. Needham, G. Anyarambhatla, G. Kong and M. W. Dewhirst, *Cancer Res.*, 2000, **60**, 1197–1201.

83. M. B. Yatvin, J. N. Weinstein, W. H. Dennis and R. Blumenthal, *Science*, 1978, **202**, 1290–1293.
84. B. Smith, I. Lyakhov, K. Loomis, D. Needle, U. Baxa, A. Yavlovich, J. Capala, R. Blumenthal and A. Puri, *J. Controlled Release*, 2011, **153**, 187–194.
85. W. Li, H. Zhao, W. Qian, H. Li, L. Zhang, Z. Ye, G. Zhang, M. Xia, J. Li and J. Gao, *Biomaterials*, 2012, **33**, 5349–5362.
86. S. Poh, J. B. Lin and A. Panitch, *Biomacromolecules*, 2015, **16**, 1191–1200.
87. N. S. Rejinold, R. G. Thomas, M. Muthiah, K. Chennazhi, I.-K. Park, Y. Y. Jeong, K. Manzoor and R. Jayakumar, *RSC Adv.*, 2014, **4**, 39408–39427.
88. C. De las Heras Alarcón, S. Pennadam and C. Alexander, *Chem. Soc. Rev.*, 2005, **34**, 276–285.
89. R. L. Bartlett and A. Panitch, *Biomacromolecules*, 2012, **13**, 2578–2584.
90. H. G. Schild, *Prog. Polym. Sci.*, 1992, **17**, 163–249.
91. I. Idziak, D. Avoce, D. Lessard, D. Gravel and X. Zhu, *Macromolecules*, 1999, **32**, 1260–1263.
92. R. Moerkerke, F. Meeussen, R. Koningsveld, H. Berghmans, W. Mondelaers, E. Schacht, K. Dušek and K. Šolc, *Macromolecules*, 1998, **31**, 2223–2229.
93. E. E. Makhaeva, H. Tenhu and A. R. Khokhlov, *Macromolecules*, 1998, **31**, 6112–6118.
94. P. Chandaroy, A. Sen and S. W. Hui, *J. Controlled Release*, 2001, **76**, 27–37.
95. P. Alexandridis and T. Hatton, *Dynamic Properties of Interfaces and Associated Structures*, AOCS Press, Champaign, 1996, pp. 231–262.
96. P. Alexandridis and T. A. Hatton, *Colloids Surf., A*, 1995, **96**, 1–46.
97. R. Misra and S. K. Sahoo, *Eur. J. Pharm. Sci.*, 2010, **39**, 152–163.
98. C.-L. Peng, H.-M. Tsai, S.-J. Yang, T.-Y. Luo, C.-F. Lin, W.-J. Lin and M.-J. Shieh, *Nanotechnology*, 2011, **22**, 265608.
99. T. E. Nakajima, K. Yanagihara, M. Takigahira, M. Yasunaga, K. Kato, T. Hamaguchi, Y. Yamada, Y. Shimada, K. Mihara and T. Ochiya, *Cancer Res.*, 2008, **68**, 9318–9322.
100. K. Yoshino, A. Kadowaki, T. Takagishi and K. Kono, *Bioconjugate Chem.*, 2004, **15**, 1102–1109.
101. K. Kono, T. Murakami, T. Yoshida, Y. Haba, S. Kanaoka, T. Takagishi and S. Aoshima, *Bioconjugate Chem.*, 2005, **16**, 1367–1374.
102. H.-H. Lin and Y.-L. Cheng, *Macromolecules*, 2001, **34**, 3710–3715.
103. F. Kohori, M. Yokoyama, K. Sakai and T. Okano, *J. Controlled Release*, 2002, **78**, 155–163.
104. M. Tanner, M. Hollmen, T. Junttila, A. Kapanen, S. Tommola, Y. Soini, H. Helin, J. Salo, H. Joensuu and E. Sihvo, *Annu. Oncol.*, 2005, **16**, 273–278.
105. M. van Elk, B. P. Murphy, T. Eufrásio-da-Silva, D. P. O'Reilly, T. Vermonden, W. E. Hennink, G. P. Duffy and E. Ruiz-Hernández, *Int. J. Pharm.*, 2016, **515**, 132–164.
106. T. Ta and T. M. Porter, *J. Controlled Release*, 2013, **169**, 112–125.
107. M. I. Filipe and A. Branfoot, *Cancer*, 1974, **34**, 282–290.

108. S. K. Lai, Y.-Y. Wang and J. Hanes, *Adv. Drug Delivery Rev.*, 2009, **61**, 158–171.
109. S. Fujishige, K. Kubota and I. Ando, *J. Phys. Chem.*, 1989, **93**, 3311–3313.
110. O. Klenina and E. Fain, *Polym. Sci. U.S.S.R.*, 1981, **23**, 1439–1446.
111. C. J. Hawker, A. W. Bosman and E. Harth, *Chem. Rev.*, 2001, **101**, 3661–3688.
112. B. M. Rosen and V. Percec, *Chem. Rev.*, 2009, **109**, 5069–5119.
113. F. A. Plamper, M. Ruppel, A. Schmalz, O. Borisov, M. Ballauff and A. H. Müller, *Macromolecules*, 2007, **40**, 8361–8366.

CHAPTER 5

Redox-responsive Drug Delivery Systems

ARCHANA BHAW-LUXIMON* AND DHANJAY JHURRY*

Biomaterials, Drug Delivery and Nanotechnology Unit, Centre for Biomedical and Biomaterials Research (CBBR), University of Mauritius, Réduit, Mauritius
*E-mail: a.luximon@uom.ac.mu, djhurry@uom.ac.mu

5.1 Redox-responsive Drug Delivery Systems

Imbalanced reactive oxygen species (ROS) and glutathione (GSH) are characteristic features of tumor cells. High intracellular GSH concentration in tumor cells is a well-documented fact that leads to a very high reducing intracellular bio-milieu. High accumulation of ROS is known to occur in almost all cancers and can act as a two-edged sword during tumor development either promoting or inhibiting growth. The dual action of ROS depends on its concentration. These two features present unique opportunities to design drug delivery systems that are responsive to reduction and/or oxidation stimuli and has attracted accrued interest from researchers. These nanocarriers change their structural integrity, either through disassembly or degradation, to deliver their payload in the presence of the trigger. The aim of this chapter is to summarize the key developments in the design of materials with redox-responsive behavior and their subsequent application in the field of

Biomaterials Science Series No. 1
Stimuli-responsive Drug Delivery Systems
Edited by Amit Singh and Mansoor M. Amiji

Published by the Royal Society of Chemistry, www.rsc.org

nanomedicine targeting cancer. Strategies into exploiting both stimuli in a single nano drug delivery system to enhance therapeutic efficacy are also addressed.

Intracellular GSH ≈ 10 mM and extracellular GSH ≈ 10 μM results in a concentration gradient across the cellular membrane and is one of the most appealing stimuli for drug delivery. GSH-responsive nanocarriers should show enhanced endocytosis, disassemble and release drugs in the cytosol where the GSH concentration is increased. Apart from their redox environment, tumor cells also have a lowered pH, which has been exploited using pH-responsive nanocarriers. However, the advantages of GSH-responsive carriers over pH-responsive carriers are their high stability against hydrolytic degradation and drug release right into the cytosol and cell nucleus, thus drug enhancing efficacy.[1]

Healthy cells have low level of ROS, while tumor cells have moderate levels of ROS which drive redox biology and activate signaling pathways for ROS-induced tumor promoting events.[2] At low to moderate ROS levels, the anti-oxidant programs of cells have the capacity to re-adapt themselves for cell survival. High or disproportional levels of ROS causes damage to DNA, proteins or lipids.[3,4] For instance, 20 nM is considered as a normal physiological hydrogen peroxide (H_2O_2) concentration, whereas 100 μM as a pathological H_2O_2 concentration. ROS-responsive nanocarriers should endocytose and disassemble at a physiological ROS concentration which sustains tumor cells.

Thus, an efficient reduction or oxidation sensitive nanocarrier should feature the following: (i) enhanced endocytosis to increase concentration of the drug in cytoplasm (ii) high drug loading (iii) prolonged circulation time and (iv) degrade intracellularly to release its cargo (Scheme 5.1). The nanocarrier material should contain redox-susceptible groups so that their drug release is triggered by the sub-cellular redox-gradient.[5] Reduction-sensitive functional

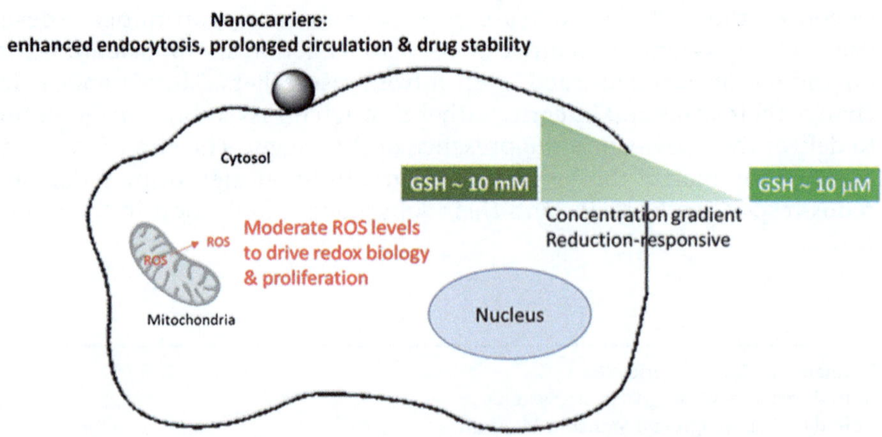

Scheme 5.1 Cellular redox tumor cells and role of the nanocarrier.

groups are mainly based on S–S bridges, which are converted to SH groups in the presence of GSH, while oxidation or ROS-susceptible groups are more diverse in chemical nature and includes the presence of thiol, selenium or boron-based linkers.[6] Dual responsive nanoparticles combine chemical segments in a single system, which can undergo reduction and oxidation in a high GSH/ROS environment.

5.1.1 Redox Profile of the Tumor Microenvironment and the Cancer Cell

The most salient features in tumors are tissue oxygen concentration, acidity, intracellular glutathione (GSH) concentration and redox status.[7] Oxygen concentration influences hypoxic conditions, whereas acidity and high ROS levels promotes cancer cell proliferation by acting on genes, signaling pathways and enzyme activity. To counter ROS-induced stress, high levels of GSH are present in tumors to prevent the balance from tipping to cell death.

5.1.1.1 Reactive Oxygen Species

ROS are by-products of aerobic metabolism. They are radicals, ions or molecules with an unpaired electron (Table 5.1).[8] In the case of cancer cells, abundant data is available on the action of superoxide, hydrogen peroxide and hydroxyl radicals on cancer pathology (Figure 5.1). The sources of higher ROS levels can be from increased metabolic and oxidases activity, mitochondrial dysfunction, peroxisome activity, increased cellular receptor signaling and oncogene activity. Cell growth/proliferation, differentiation, protein synthesis, glucose metabolism, cell survival and inflammation signaling pathways are activated by enhanced ROS in most cancers.[9] H_2O_2 can act as second messengers in cellular signaling and regulate protein activity.

Table 5.1 ROS species.

Free oxygen radical ROS	Non-radical ROS
Superoxide $O_2^{\bullet -}$	Hydrogen peroxide H_2O_2
Hydroxyl radical $^{\bullet}OH$	Singlet oxygen $^1O_2$
Nitric oxide $NO^{\bullet}$	Ozone/trioxygen O_3
Peroxyl radicals $ROO^{\bullet}$	Organic hydroperoxides ROOH
Alkoxyl radicals $RO^{\bullet}$	Hypochloride HOCl
Sulfonyl radicals $ROS^{\bullet}$	Peroxynitrite ONO^-
Thiyl peroxyl radicals $RSOO^{\bullet}$	Nitrosoperoxycarbonate anion $O{=}NOOCO_2^-$
	Nitrocarbonate anion $O_2NOCO_2^-$
	Dinitrogen dioxide N_2O_2
	Nitronium NO_2^+
	Highly reactive lipid or carbohydrate derived carbonyl compounds

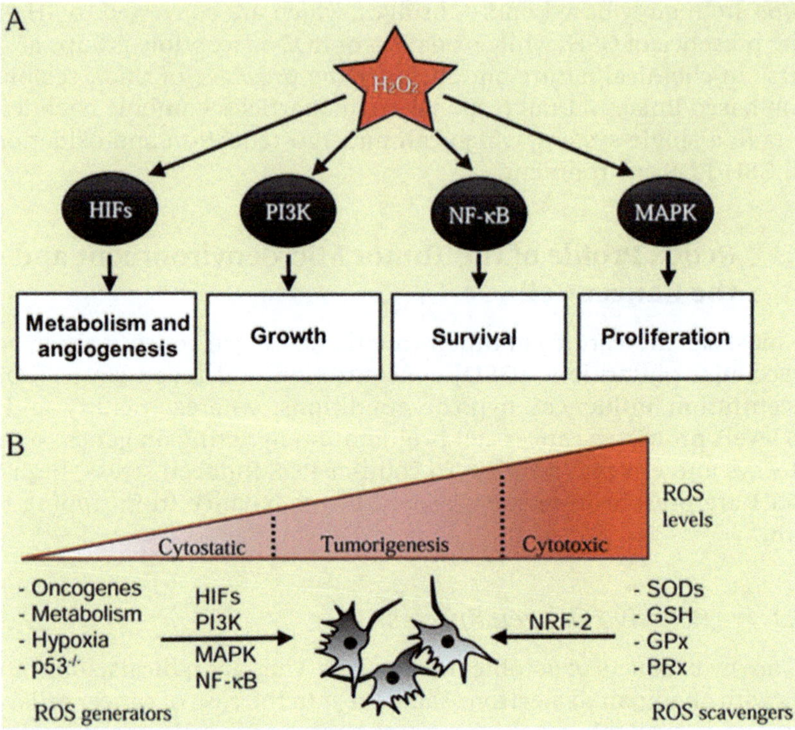

Figure 5.1 H_2O_2 (A) and ROS (B) regulation of normal and cancer cell proliferation. (Reprinted from *Current Biology*, **24**, M. Schieber and N. S. Chandel, ROS Function in Redox Signaling and Oxidative Stress, 453–462, Copyright (2014), with permission from Elsevier).[10]

Superoxide radicals are generated at the inner membrane of the mitochondria as a byproduct of the electron transport chain and by activation of NADPH oxidases (NOX). SOD enzymes in the mitochondrial matrix or Cu/ZnSOD in the cytosol reduce superoxide to H_2O_2. Both hydrogen peroxide and superoxide can form hydroxyl radicals in a cytosol Fenton reaction.

5.1.1.2 Glutathione

GSH forms part of the glutathione system, which also includes the enzymes glutathione reductase, glutathione peroxidases (GPX) and glutathione *S*-transferases (GST). In normal cells, GSH counteracts oxidative stress by reducing disulfide bonds of cytoplasmic proteins to cysteines, while it is itself oxidized to glutathione disulfide (GSSG) (Scheme 5.2). GSH acts as a radical scavenger. The reduced GSH has a higher concentration than the oxidized GSSG form by up to 100 fold.[11] The major site for biosynthesis and presence of GSH is in the cytosol and it is then distributed to different intracellular organelles such as the mitochondria. The GSSG/GSH ratio can be an indicator of the cellular redox environment.[12]

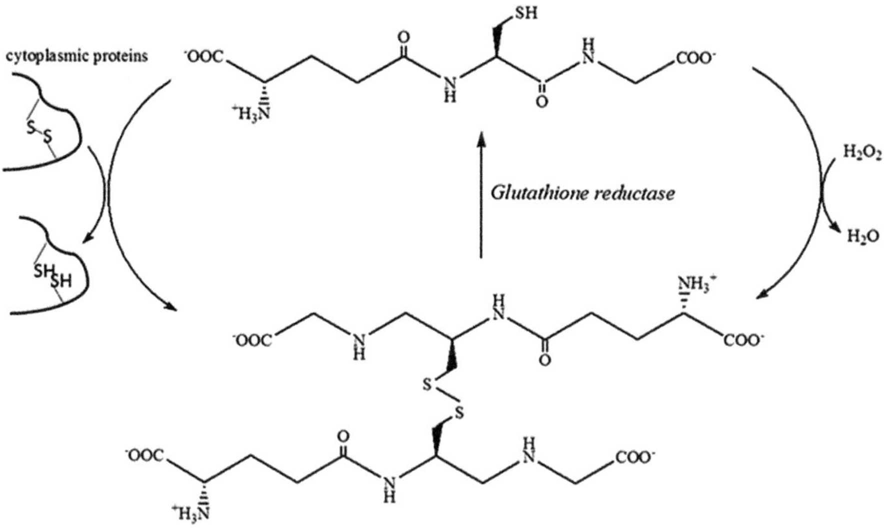

Scheme 5.2 GSH conversion to GSSG.

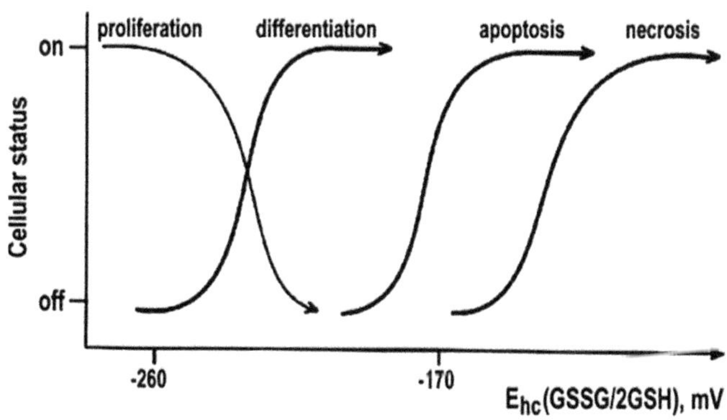

Scheme 5.3 Relationship between the GSH redox potential and biological cellular status. (Reprinted from Free Radical Biology and Medicine, 30, Schafer and Buettner, Redox environment of the cell as viewed through the redox state of the glutathione disulfide/glutathione couple, 1191–1212, Copyright (2001), with permission from Elsevier).[12]

5.1.1.3 Redox Potential in Cancer Cell

Solid tumors are characterized by low extracellular pH (pH 6.7–7.1) promoting tumor invasion, dysregulated GSH and redox potential within cancer cells and favoring proliferation.[7]

Lactic acid production and the action of carbonic anhydrase IX maintains the low extracellular pH. Inside the cell, the mitochondria reduces the superoxide at a very high rate thus promoting the production of ROS.

GSH redox potential reflects the status of cancer cells (Scheme 5.3). Proliferating cancer cells have the most negative GSH redox potential (−260 to −220 mV).

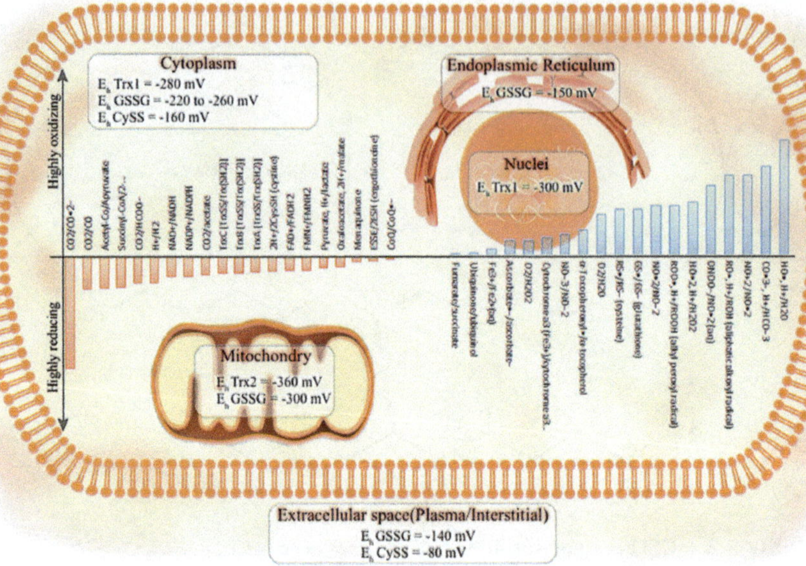

Figure 5.2 Redox profile of cultured cells (Reproduced from ref. 13 with permission from The Royal Society of Chemistry).[13]

Apoptosis is initiated if E_{hc} (GSH) becomes too positive. Severe oxidative stress increases E_{hc} (GSH) leading to necrosis. The intracellular and extracellular compartments are characterized by various redox conditions and can be exploited for drug release from nanoparticulate systems (Figure 5.2).[13]

5.1.2 Reduction (Glutathione)-responsive Systems

To mimic the action of GSH, different strategies to synthesize nanosized systems containing a S–S bridge have been designed where chemotherapeutic drugs have been covalently linked or physically entrapped.[14–17] These systems release their payload in the presence of GSH which cleaves the S–S bridge converting it to thiols (–SH HS–) while it is itself is oxidized to GSSG. Systems based on nanogels, amphiphilic micelles, modified mesoporous silica and gold nanoparticles have been reported. The ability to manipulate polymers and their combination remains an essential step to achieving redox systems. The presence of GSH-cleavable groups is not enough to qualify as a redox-sensitive system, but enhanced internalization and good drug loading ability, as well as stability, are the other parameters which come into play.

5.1.2.1 Nanogels

Nanogels combine the characteristics of hydrogels and nanoparticles (Figure 5.3).[18,19] They can be formed by either physical or chemical crosslinks between water soluble and swellable polymer chains. Their interior network

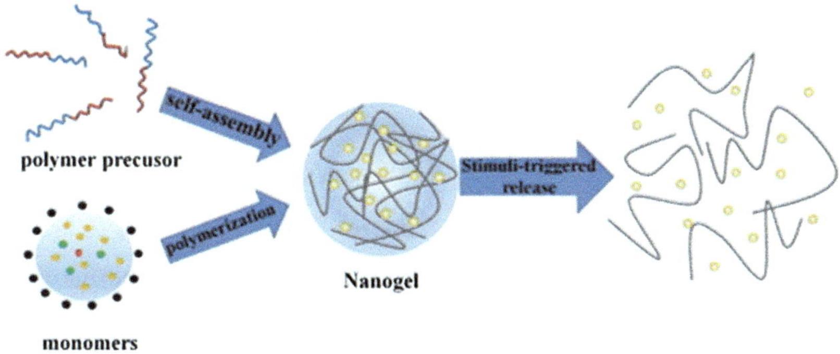

Figure 5.3 Nanogels as drug delivery devices (Reproduced from ref. 19 with permission from The Royal Society of Chemistry).[19]

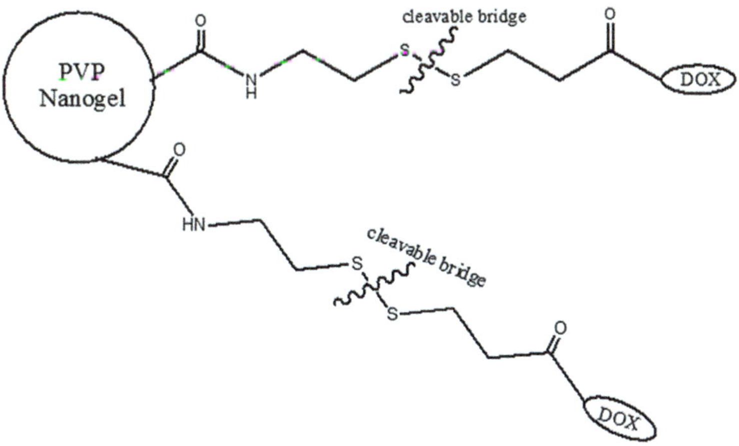

Figure 5.4 Carboxyl-functionalized PVP nanogel with S–S cleavable bridge.

can be used for encapsulation and their exterior large surface area can be used for conjugation of drug molecules.

Nanogels require the presence of a S–S bridge to become GSH responsive. A carboxyl-functionalized PVP (poly(*N*-vinylpyrrolidone)) nanogel with a linker containing a cleavable disulphide bridge (aminoethyldithiopropionic acid (AEDP)) that can be used for glutathione-mediated delivery of doxorubicin (DOX) (Figure 5.4)[20,21] was shown to be biocompatible using ECV304 cells (endothelial cell vein). DOX release from the nanogel was studied in the presence and absence of GSH. In the presence of GSH, DOX release was two-fold higher (70% *vs.* 34%) and occurred at a faster rate. Even after a prolonged time, there was no improvement of release in the absence of GSH. Only physically entrapped DOX was released.

Another system based on polyvinyl alcohol couples together the redox and thermally responsive capacity of nanogels.[22] These reversibly-crosslinked

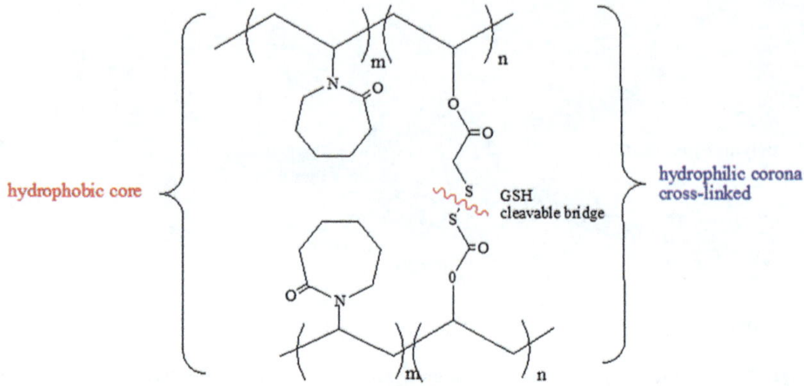

Figure 5.5 Cross-linked PVOH-*b*-PNVCL nanogels.

poly(vinyl alcohol)-*b*-poly(*N*-vinylcaprolactam) (PVOH-*b*-PNVCL, Figure 5.5) nanogels showed no toxicity in a mouse fibroblast-like L929 cell line and the Nile Red loaded nanogels had enhanced cellular uptake within human melanoma MEL-5 cells compared to free Nile Red. Triggered release was shown using a dithiothreitol (DTT) reducing agent mimicking GSH.

The presence of an S–S bridge has been identified as an important structural feature for redox responsive systems; however, cell internalization of the nanoparticle still remain a crucial requirement. Multi-responsive nanogels with the capacity to undergo acid-triggered hydrolysis and reduction-induced degradation have been used to encapsulate PTX and tested on MCF-7.[23] The nanogels have been prepared using a miniemulsion copolymerization of monomethyl oligo(ethylene glycol) acrylate (OEGA) and an ortho ester-containing acrylic monomer (DMDEA) in the presence of a disulfide-containing cross-linker (BADS) (Figure 5.6). The presence of DMDEA and OEGA provides hydrophobic microdomains in the gel network for the loading of PTX. Although in PBS the drug release was pH and reduction potential dependent (reducing agent DTT: dithiothreitol), the anticancer efficiency was lower than that of PTX alone on MCF-7. This counter result was attributed to the inefficient cellular internalization of the drug-loaded nanogels.

Nanogels fabricated from bio-derived poly(γ-glutamic acid) (PGA) nanocomplex showed GSH sensitivity.[24] The nanocomplex was formed by ionic interaction between thiolated PGA and DOX followed by H-bonding with PEG and finally crosslinking of the disulfide to yield a nanogel. On MCF-7 human breast cancer cell line, unloaded DOX was detected in the cell nuclei and the DOX-loaded γ-PGA nanogels were distributed in the cytoplasm and nuclei, that is, the cellular acidic environment. DOX showed controlled release in the presence of GSH.

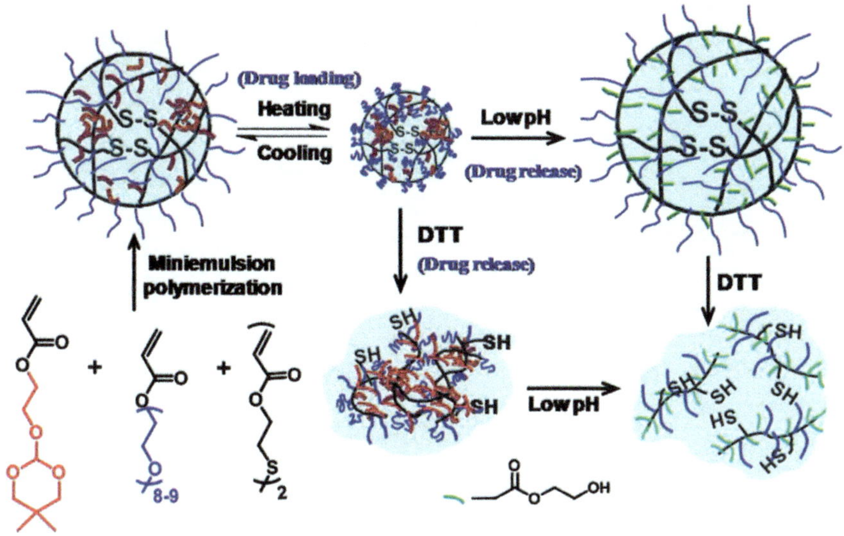

Figure 5.6 Thermo and pH-responsive nanogels (Reprinted from *Journal of Controlled Release*, **152**, Z.-Y. Qiao, R. Zhang, F.-S. Du, D.-H. Liang, Z.-C. Li, Multi-responsive nanogels containing motifs of ortho ester, oligo (ethylene glycol) and disulfide linkage as carriers of hydrophobic anticancer drugs, 57–66, Copyright (2011), with permission from Elsevier).[23]

5.1.2.2 Amphiphilic Copolymer Nanomicellar Systems

The amphiphilic block copolymer strategy rests on two main concepts, either the presence of a S–S cleavable bridge between the hydrophilic and hydrophobic segments, or a cleavable S–S bond within the hydrophilic or hydrophobic segment (Scheme 5.4). Cleaving the S–S bond by GSH within either segment causes destabilization of the micelles and drug release.

A disulfide-linked dextran-*b*-poly(ε-caprolactone) diblock copolymer (Dex–SS–PCL) has been synthesized for triggered release of doxorubicin (DOX) and tested on a mouse leukemic monocyte macrophage cell line (RAW 264.7) (Figure 5.7).[25] Cleavage of the S–S bridge in the intracellular compartment in the presence of GSH, removed the hydrophilic shell and allowed the release of DOX. DOX-loaded Dex–PCL gave a very weak fluorescence of DOX in cells whereas fluorescence images of DOX-loaded Dex–SS–PCL showed release in the cytosol and transportation in the nucleus.

The biopolymer hyaluronic acid (HA) has been linked through an S–S bridge to PLGA yielding a HA-*b*-PLGA (HASSLG) copolymer (Figure 5.8). The resulting nanoparticle has been loaded with DOX and used for targeting of tumor cells through the CD44 receptor and redox-responsive process.[26] GSH increased the release of DOX in a dose dependent manner. In an animal imaging study, CD44-positive MDA-MB231 cells and CD44-negative NIH3T3

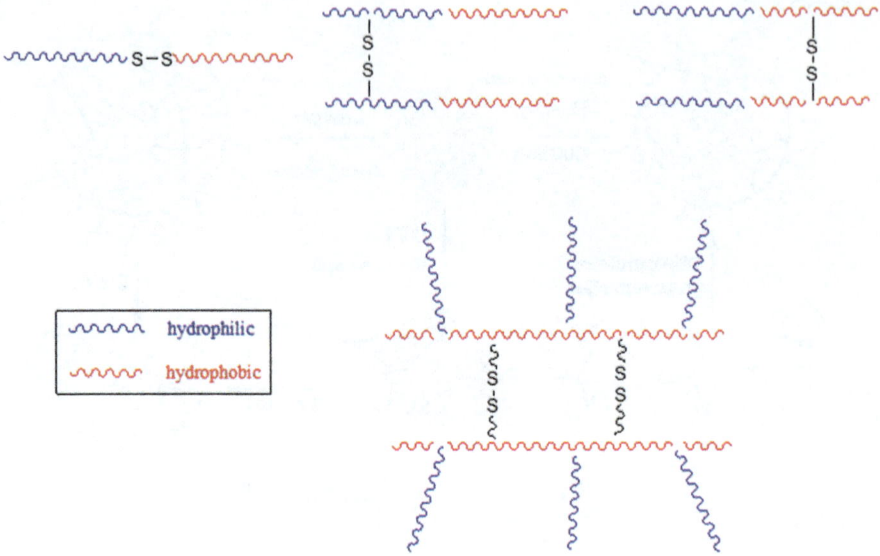

Scheme 5.4 GSH responsive segments in nanomicelles.

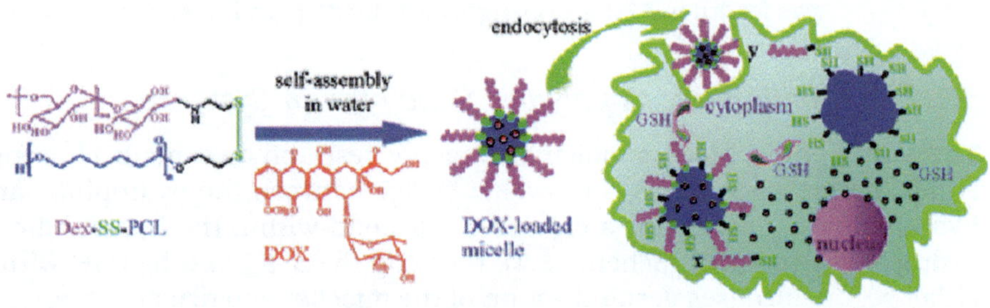

Figure 5.7 DOX-loaded Dex–SS–PCL micelles (Reprinted with permission from H. Sun, B. Guo, X. Li, R. Cheng, F. Meng, H. Liu, Z. Zhong, *Biomacromolecules*, 2010, **11**, 848. Copyright (2010) American Chemical Society).[25]

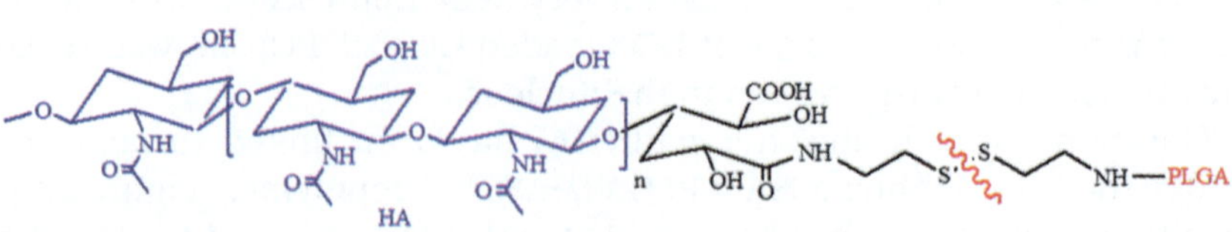

Figure 5.8 HA–SS-*b*-PLGA.

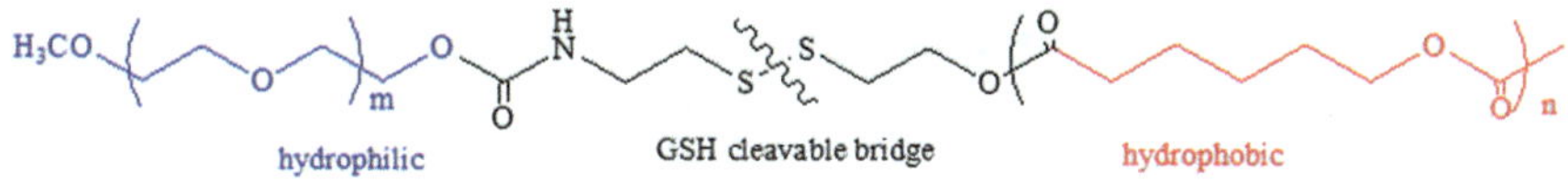

Figure 5.9 PEG–S–S–PCL amphiphile.

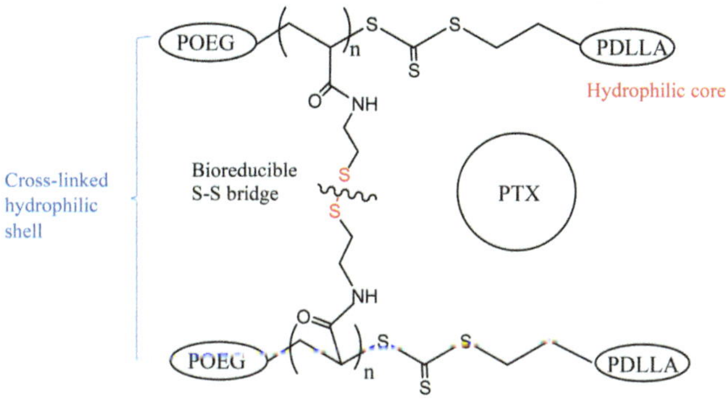

Figure 5.10 PTX-loaded hydrolytically degradable poly(lactic acid) core and a glutathione-responsive disulfide cross-linked poly(oligoethylene glycol)-containing corona.

cells implanted into the right flank and left flank of mice showed stronger fluorescence intensity at the tumor mass of MDA-MB231 cells compared toa solid mass of NIH3T3 cells. This indicated the targeted delivery aptitude of the nanoparticles.

A similar type of strategy has been adopted by linking a PEG segment with a PCL segment using an S–S bridge to give rise to an amphiphilic system for the encapsulation of DOX (Figure 5.9).[27] When compared with DOX loaded PEG-*b*-PCL, higher release of DOX and better efficacy on a mouse leukemic monocyte macrophage cell line were noted.

To enhance the release of paclitaxel (PTX) inside cancer cells, a nanomicellar system (34 ± 6 nm) consisting of a hydrophobic poly(DL-lactic acid) core and a hydrophilic shell of oligo(ethylene glycol) methyl ether acrylate-*N*-acryloyloxysuccinimide copolymer (P(OEGMEA-*co*-NAS)) cross-linked with a bioreducible disulfide cross-linker was engineered (Figure 5.10).[28] In the presence of GSH, the S–S bridge is cleaved, enhancing the cumulative release of PTX at both pH 5.5 and pH 7.4, reaching 65% after 8 days compared to 50% (pH 5.5) and 35% (pH 7.4) in the absence of glutathione. This system was tested on human ovarian adenocarcinoma cells (OVCAR-3) and showed an 11-fold lower IC_{50} than a Taxol-mimicking formulation with similar PTX content after 72 h of incubation. The higher efficacy of the nanomicellar system compared to the free drug was attributed to the coupling effect of the nanoparticle and pH responsive release triggered by the intracellular reductive medium.

The S–S functionality can also be introduced as a pendant group on either the hydrophobic or hydrophilic segment. For instance, DOX-loaded

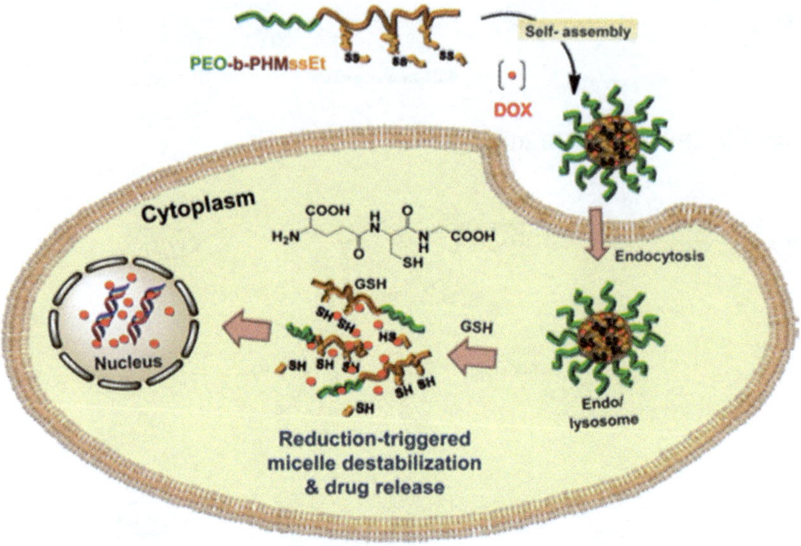

Figure 5.11 Thiol-responsive degradable micelles consisting of a pendant disulfide-labeled methacrylate polymer block (PHMssEt) and a hydrophilic poly(ethylene oxide) (PEO) block (Adapted with permission from B. Khorsand, G. Lapointe, C. Brett, J. K. Oh, *Biomacromolecules*, 2013, **14**, 2103. Copyright (2013) American Chemical Society).[29]

disulfide-labeled PEO-*b*-polymethacrylate (PHMssEt) showed both enhanced cellular uptake and intracellular release of GSH triggered DOX for HeLa cancer cells (Figure 5.11).[29]

Polysaccharides can be interesting segments to construct S–S bridges due to their numerous hydroxyl groups. Such a strategy has been applied on starch-*g*-PEG micelles with the introduction of a S–S bridge from pendant groups attached to the starch segments, leading to cross-linking of the hydrophobic core (Figure 5.12).[30] The DOX-loaded micelles lend themselves to GSH-triggered drug release through the reduction of the S–S bridges in the core. HeLa cells pretreated with GSH-Oet (glutathione monoester) showed the highest DOX released with subsequent localization of DOX in the cell nucleus, while cells pretreated with BSO (buthionine sulfoximine) showed the lowest DOX release.

Carboxymethyl dextran (CMD)-*g*-lithocholic acid (LCA) with a disulfide linkage released DOX in the presence of 10 mM GSH mimicking the intracellular environment, whereas release is slowed down at pH 7.4 (Figure 5.13).[31] CMD is the hydrophilic shell, whereas LCA is the hydrophobic core. They showed higher toxicity with SCC7 cancer cells than DOX-loaded nanoparticles without the disulfide bond. This is an indication of the higher endocytosis of the nanoparticle with the drug released in the cytoplasm and nuclei *via* S–S bond cleavage. In tumor-bearing mice, the Cy5.5-labeled CMD–SS–LCAs accumulated at tumor sites.

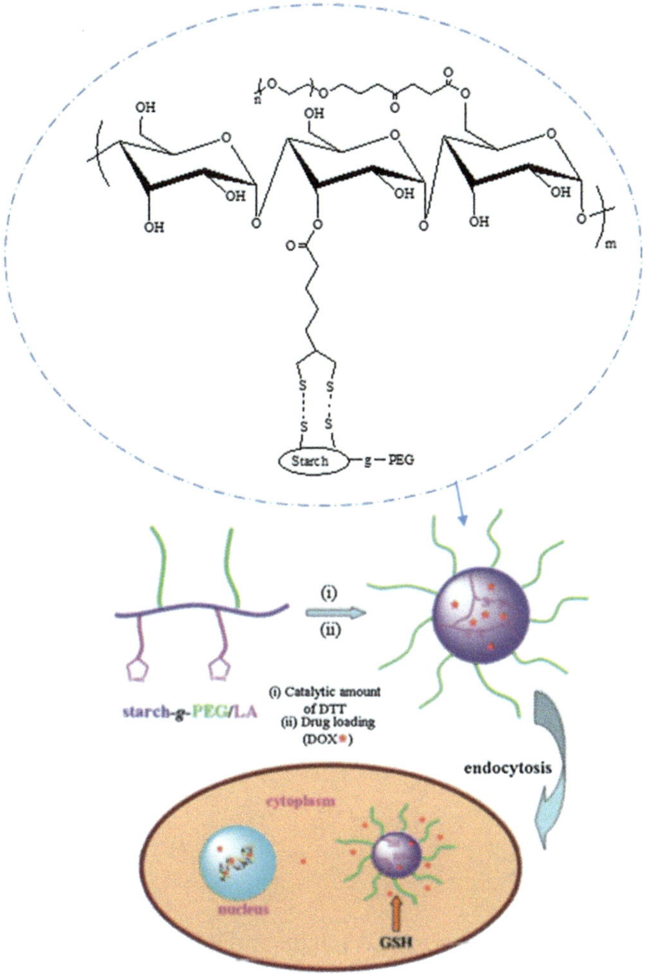

Figure 5.12 S–S crosslinked Starch-*g*-PEG (Modified and Adapted from ref. 30 with permission from The Royal Society of Chemistry).[30]

The hydrophilicity and stealth effect of PEG has been exploited to devise a PEGylated nano-graphene oxide encapsulating DOX·HCl (NGO–SS–mPEG) (Figure 5.14).[32] At the tumor intracellular level of GSH in HeLa cells, the S–S link is cleaved and the PEG shell removed and then the drug released.

Liu *et al.*[33] (2011) designed amphiphilic hyperbranched polyphosphate (HPHDP) (30–78 nm) with alternative hydrophobic disulfide and hydrophilic polyphosphate segments showing redox responsive properties in living cells (Figure 5.15). To show that GSH can effectively cleave the S-S bridge and release the drug, DOX loaded HPHDP micelles were used on HeLA cells in the presence of buthionine sulfoximine (BSO), an inhibitor of intracellular synthesis of GSH and GSH-Oet (glutathione monoester). The IC_{50} of the control,

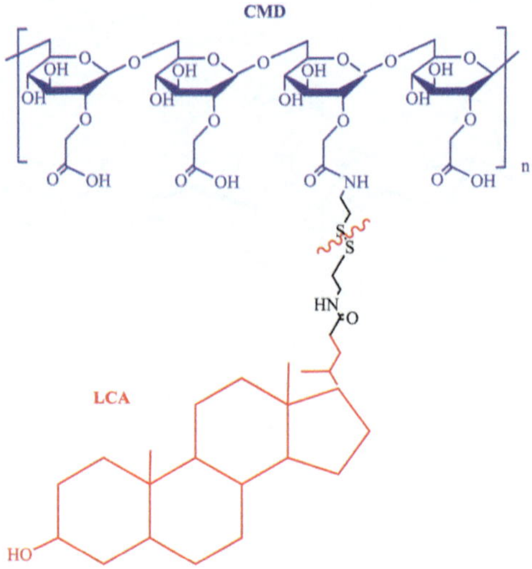

Figure 5.13 Carboxymethyl dextran (CMD)-*g*-lithocholic acid (LCA).

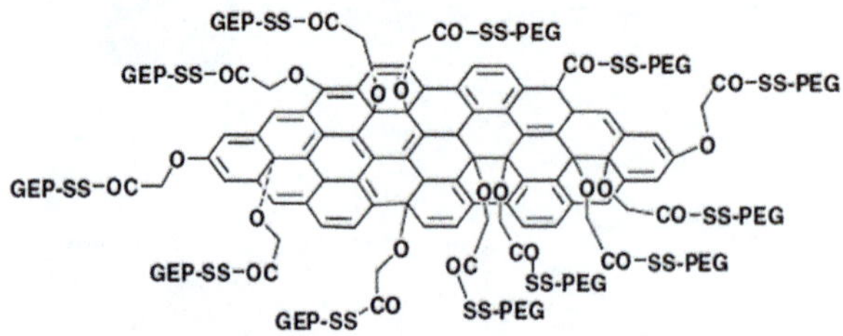

Figure 5.14 NGO–SS–mPEG.

the GSH-Oet and the BSO were 0.35, 0.08 and 0.64 mg mL^{-1}, respectively. BSO prevented the disaggregation of DOX-loaded micelles while GSH-Oet triggered rapid destabilization of the micelles leading to increased DOX release and thus reducing proliferation of HeLA cells.

A polyamidoamine (PAMAM) dendrimer with crosslinked disulphide encapsulating an all-*trans* retinoic acid has been reported.[34] This system influenced the cellular redox status of HT1080 (human fibrosarcoma cells) and MCF-7 (human breast cancer cells) by decreasing the GSH level by 54% in HT1080 and 41% in MCF-7 cells. Cleavage of the S–S link by GSH decreased its intracellular concentration and at the same time released the encapsulated drug. Apoptosis of cells markedly increased and reached 59.8% and 40.6% in HT1080 and MCF-7 cells.

5.1.2.3 Lipid-based Nanocarriers

To enhance the cellular uptake of nanoparticle, as well as having redox-sensitivity, lipids which are compatible with the cell membrane have proved to be useful. One such example is a chitosan based glycolipid-like nanocarrier linked to stearic acid (CSO–SS–SA) (Figure 5.16).[35] 3,3′-Dithiodipropionic

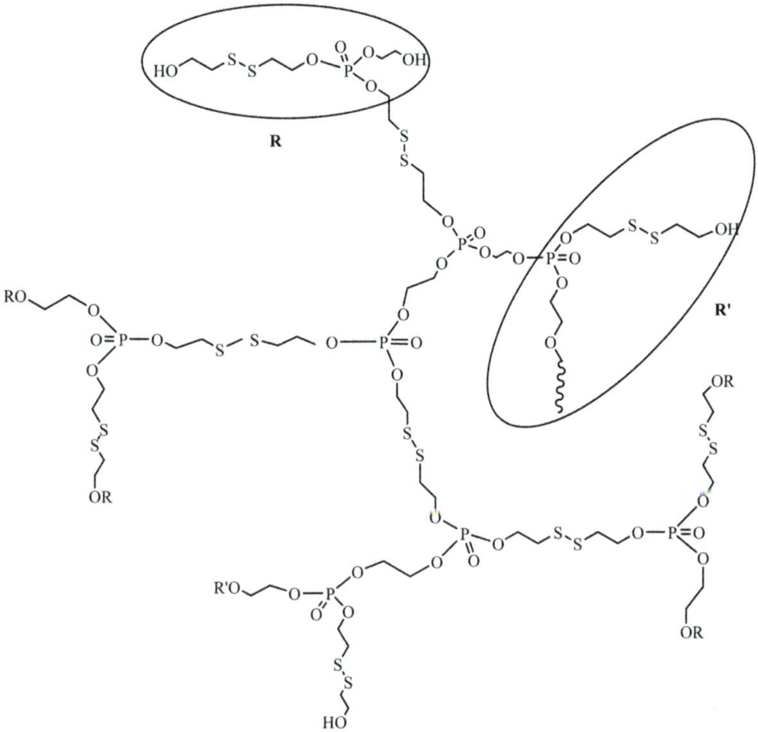

Figure 5.15 HPHDP self-assembled micelles.

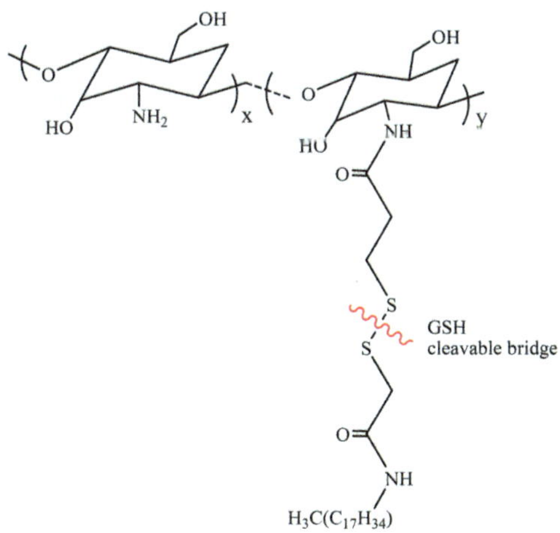

Figure 5.16 Glycolipid nanocarrier.

was used as a linker between chitosan and octadecylamine through a condensation reaction between the amino and carboxylic groups. This nanocarrier exhibited higher tumor treating capacity than a CSO–SA nanoparticle. Fast release of PTX from CSO–ss–SA was observed in human ovarian cancer SKOV-3 cells compared to human normal liver L-02 cells. Furthermore, with

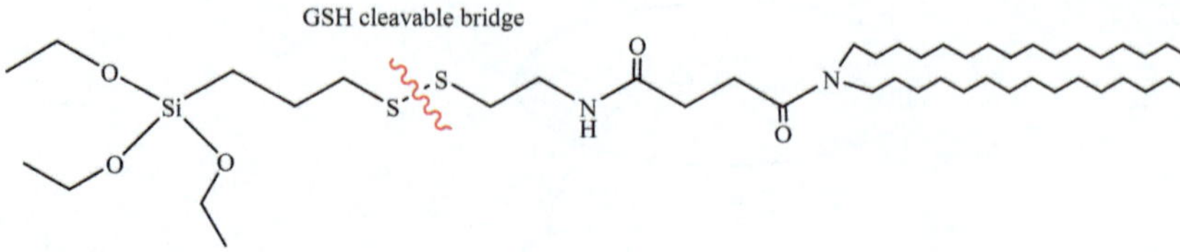

Figure 5.17 Redox responsive cerasome.

SKOV-3 cells, cellular inhibition showed a positive correlation with GSH concentration. In tumor bearing nude mice, PTX–CSO–SS–SA, showed an inhibition of tumor growth of 60.5% (12.5 mg kg^{-1}) and 88.4% (25.0 mg kg^{-1}), compared to Taxol (59.4%) at the same doses.

Redox responsive cerasome has a liposome-resembling lipid bilayer structure, with a cleavable disulfide bond as a connector unit of a triethoxysilyl head and a hydrophobic alkyl double chain, showed accelerated release of DOX in the presence of GSH (Figure 5.17).[36]

5.1.2.4 *Vesicles: Liposomes and Polymersomes*

Natural proteins with their numerous functionality can trap a variety of active substances and interact with receptors at cell membranes. However, to be used as drug delivery nanocarriers similar to polysaccharides, chemical modifications of the –OH or $–NH_2$ groups are required which adds to the chemical synthetic steps. The possibility of using native proteins, which can in addition respond to redox stimuli, could open up new perspectives for these macromolecules. Keratin has abundant cysteine residues and has been used to obtain redox responsive drug delivery systems without chemical modification.[37] The cysteine residues are oxidizable by air to give inter- and intra-molecular disulfide bonds. Polyethyleneglycol-40 stearate (PEG40ST) was attached to keratin using cysteine residues which subsequently led to the formation of S–S bonds. This system was loaded with hydrophilic methotrexate (MTX) and hydrophobic curcumin (CUR). These polymersomes showed slow release of drugs in a phosphate buffer without GSH or with 10 μM GSH (extracellular GSH), but at 10 mM GSH (intracellular GSH) a faster release was observed. On HeLa (human cervical cancer cell line) and CHO-K1 (a cell line derived from the ovary cancer of the Chinese hamster) higher toxicity was observed compared to the free drug. This result indicated higher endocytosis of the polymersomes with release in the cytoplasm.

Hybrid hydrogels were prepared *via* the use of a maleimide-functionalized liposome cross-linkers and thiolated polyethylene glycol polymers (Figure 5.18).[38] DOX and Cytochrome *C* were trapped in the network. GSH degrades the network by cleaving the thiol links and thus triggered the release of encapsulated molecules at physiological and pathological concentrations.

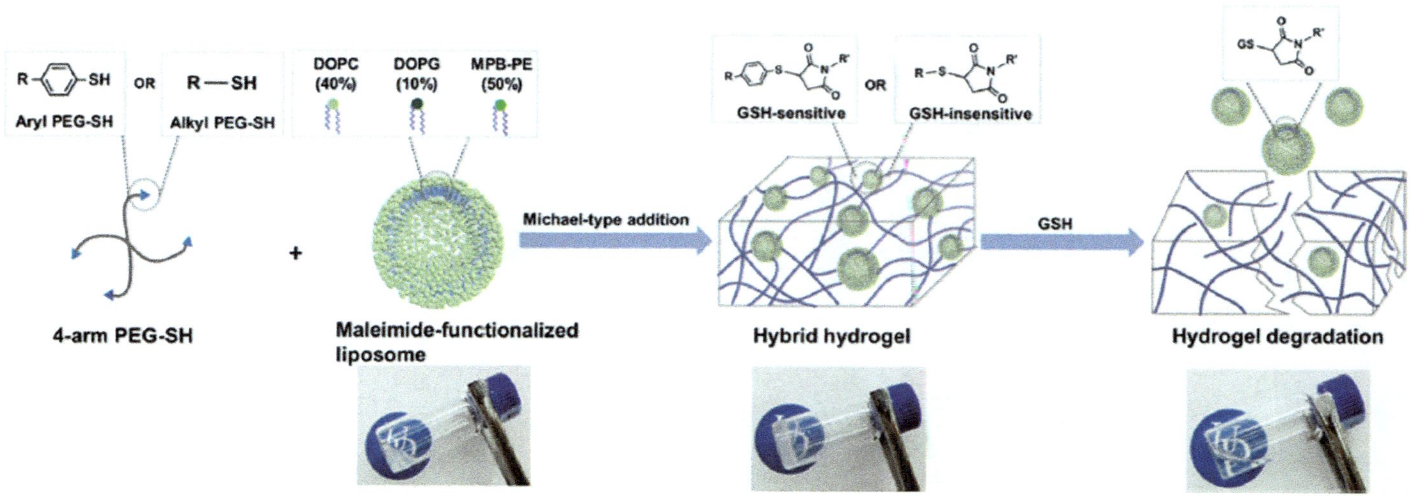

Figure 5.18 Schematic representation of the formation and degradation of the liposome cross-linked hybrid hydrogels[38] (Adapted with permission from Y. Liang, K. L. Kiick, *Biomacromolecules*, 2016, **17**(2), 601. Copyright (2016) American Chemical Society).

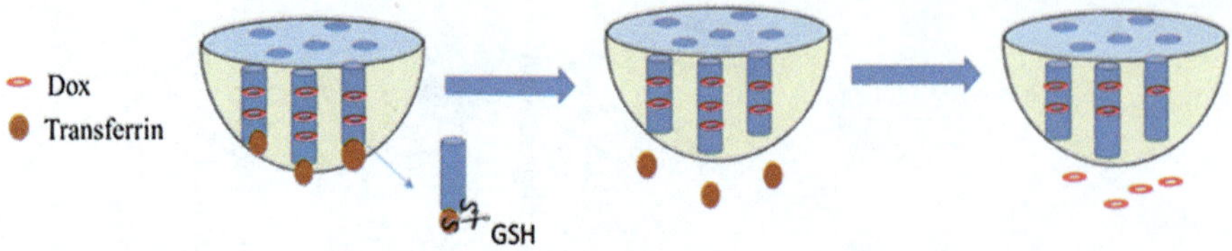

Scheme 5.5 Release of DOX after cleavage of S–S bridge by GSH.

For combined prolonged circulation and enhanced endocytosis, PTX-loaded liposomes functionalized with TAT–S–S–PEG (TAT: cell penetrating peptide) have been developed.[39] The liposome core was made from DSPE (1,2-distearoyl-*sn-glycero*-3-phosphoethanolamine) and PTX was physically entrapped in the liposome. TAT was shielded by the PEG layer until it reached the tumor site where exogenous GSH cleaved the S–S linker and exposed it for enhanced cell internalization. This TAT decorated liposome inhibited the proliferation of murine melanoma B16F1 tumor cells *in vitro* and *in vivo* on B16F1-bearing mice.

5.1.2.5 *Mesoporous Silica Nanoparticles*

Apart from organic polymeric systems, mesoporous silica nanoparticles (MSNs) also present certain versatility which have been explored for the triggered release of drugs.[40] MSNs MCM-41 (hexagonal arrangement of the mesopores), MCM-48 (with a cubic arrangement of the mesopores) and MCM-50 (with a laminar structure)[41] are used in redox systems as they have sufficient surface silanol groups (Si–OH) for further modifications.[42] Surface modification of mesoporous silica allows for attaching redox sensitive groups as molecular gates.

Gate keepers such as rotaxanes and β-cyclodextrin have been attached onto mesoporous nanosilica (MSN) using a GSH-cleavable disulfide bond as linker. Transferrin (Tf), as a gate keeper in a DOX loaded-hollow MSN with the dual property of tumor-targeting and GSH-responsive release has shown good promise (Scheme 5.5).[43] The particle was capped with transferrin using an S–S bridge. The S–S bridge is cleaved by GSH allowing the release of DOX (78%). In the absence of GSH negligible DOX release was observed (5.2%). *In vivo* experiments on A549 tumor-bearing mice using free DOX, DOX–HMSN–SH and DOXHMSN–*s*–*s*–Tf, showed that the tumor growth was effectively inhibited by DOXHMSN–*s*–*s*–Tf after 14 days.

The use of mesoporous silica–S–S–PEG, where PEG acts as a gatekeeper also showed cleavage of the S–S bond in the presence of GSH with release of the encapsulated model drug.[44] The unloaded prepared silica carriers were non-cytotoxic towards MCF-7 cells.

A double pH–GSH responsive DOX-loaded hollow mesoporous silica nanoparticle was blocked with cytochrome *C* conjugated lactobionic acid (CytC–LA) *via* redox-cleavable disulfide bonds and pH-disassociation boronate ester bonds as intermediate linkers (HMSNs–S–S–CPA–CytC–LA@DOX where

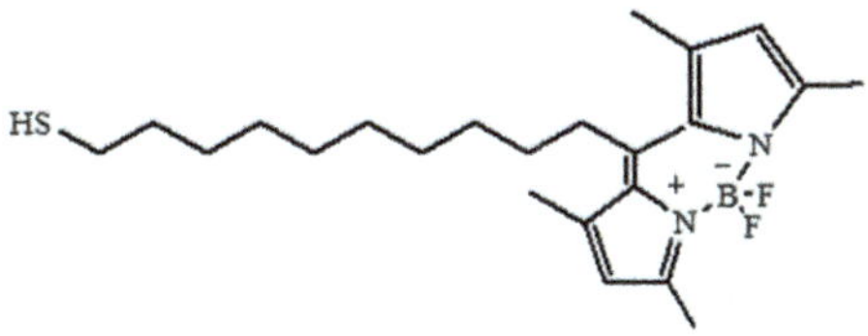

Figure 5.19 Thiolated Bodipy dye.

Table 5.2 The use of gold nanoparticles in GSH-responsive systems.

Diameter of Au core (nm)	Agent	Effect in presence of GSH
27.0 ± 3–27.6 ± 3	β-lapachone[52]	Enhanced apoptosis than free β-lapachone (A549 cells)
20–40	Paclitaxel[53]	Enhanced cellular uptake on HeLa, A549, and MG63 cell lines
3	Thiolated chemoagent: captopril and 6-mercaptopurine[54]	Improved cytotoxicity on HeLa cells

CPA is 4-carboxyphenylboronic acid).[45] These particles were endocytosed by HepG2 cells and DOX release was triggered by both GSH and low pH. Male nude mice injected with HepG2 cells showed the highest tumor inhibition with the HMSNs–S–S–CPA–CytC–LA@DOX than with free DOX and simple HMSN–DOX.

CeO_2 nanoparticles (nanoceria) with their anti-oxidant properties detoxify superoxide and hydrogen peroxide species as they can reversibly switch oxidation states (+3 and +4). They have excellent biocompatibility and biocatalytic activity. Nanoceria have been applied as capping agents for MSNs through immobilization. They are rapidly eroded by GSH, liberate the nanochannels and thus allow controlled release of camptothecin from the mesopores.[46]

5.1.2.6 *Gold Nanoparticles*

Gold nanoparticles have low toxicity and a small size with long circulation time, low cytotoxicity, and high payload-to-carrier ratios.[47] The surface of gold nanoparticles can be modified for tumor specific cell membrane penetration.[48] Monolayer protected gold nanoparticles can be applied for GSH-responsive drug delivery. A 2 nm gold nanoparticle covered with a mixed monolayer of tetra(ethylene glycol)-lyated cationic ligand and a thiolated Bodipy dye (HSBDP, mimicking hydrophobic drug) (Figure 5.19) showed an eight-fold increase in release of HSBDP in the presence of GSH in human liver cells (HepG2).[49,50] The tetra(ethylene glycol)-lyated cationic ligand generated a cationic surface for enhanced cellular uptake, whereas the HSBDP–Au thiol link is cleaved by GSH.

Similarly, other agents have been attached to gold nanoparticle surface *via* a thiol link (Table 5.2).[51]

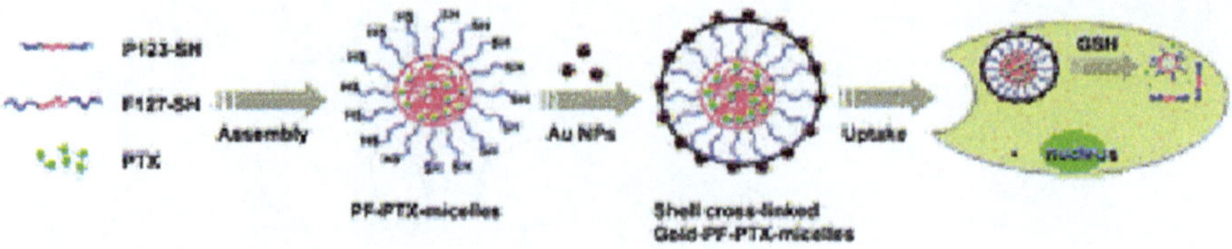

Figure 5.20 Gold–Pluronic–Paclitaxel nanomicelles[52] (Modified and Adapted from ref. 52 with permission from The Royal Society of Chemistry).

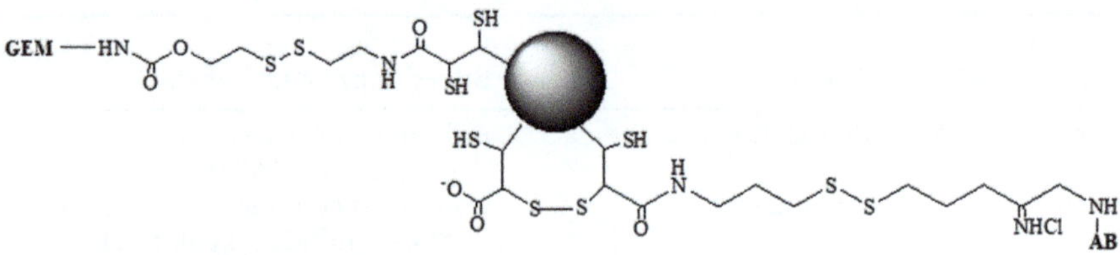

Figure 5.21 Multifunctionalized Fe MNP with attached gemcitabine (GEM) and anti CD44 antibody (AB).

A somewhat different strategy has been reported by Tao *et al.* (2013) where a gold-nanoparticle-crosslinked pluronic micelle was used as a carrier for paclitaxel (69 nm in diameter) (Figure 5.20).[55] Tested on human glioma U87 cells under a glutathione stimulus, the disulphide linkages were cleaved disrupting the micelles and inducing rapid intracellular drug release. In addition, the nanomicelles had a longer systemic circulation time than and were cleared more slowly from plasma than pluronic–paclitaxel micelles.

5.1.2.7 Magnetic Iron Oxide Nanoparticles

Magnetic iron oxide (Fe_3O_4, γ-Fe_2O_3) nanoparticles (Fe MNPs) can be used both for anti-tumor therapy and imaging. They are biocompatible with low toxicity and allow the tracking of drug delivery, and monitoring of therapeutic response *in vivo*.[56] Similar to gold nanoparticles, surface modification allows the development of multifunctional iron oxide nanoparticles. Drug molecules can be either incorporated within the particle or loaded onto the surface. Multifunctionalized iron MNPs with antiCD44 antibody and gemcitabine have been devised using this strategy for the treatment of CD44-positive cancer cells.[57] A S–S containing linker was used to covalently attach gemcitabine onto the surface (Figure 5.21). Studies indicated that the release of GEM from MNP–GEM–antiCD44 is intracellular and strongly dependent on the reducing environment. Up to 96–98% of GEM was released when treated with 1 mM GSH (intracellular GSH concentration) and 3–5% released with 1 μM GSH (extracellular GSH). On CD44-positive cancer cells (Panc-1 and MDA-MB231) greater antiproliferative effect was observed for MNP–GEM–antiCD44 compared to the control MNP–GEM.

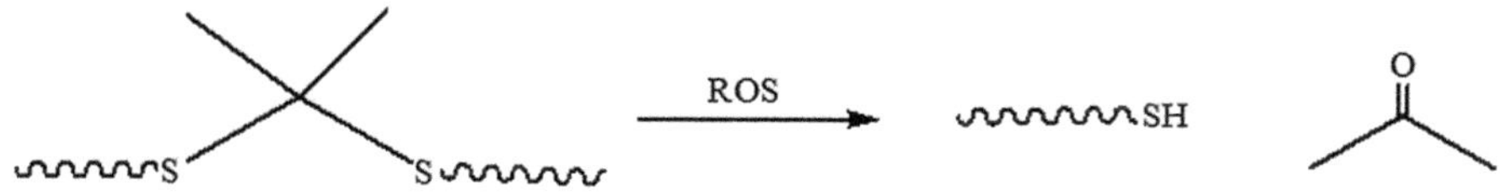

Scheme 5.6 Oxidative cleavage of thioketal.

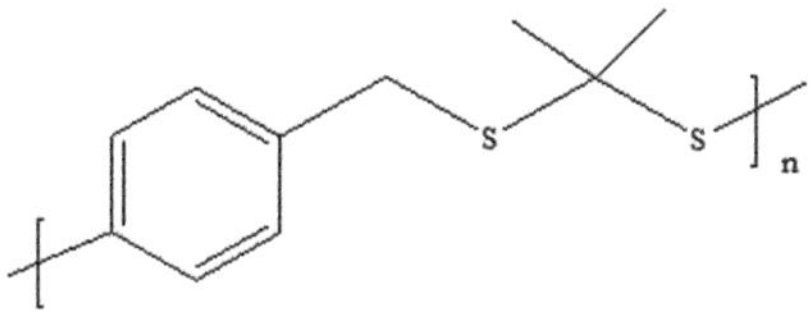

Figure 5.22 Poly(1,4-phenyleneacetone dimethylene thioketal) (PPADT).

5.1.3 ROS-responsive Systems

ROS-responsive materials or linkers for nano drug delivery systems exploit the high level ROS weakness of tumor cells to enhance drug delivery and trigger targeted release.[6,58,59] A range of ROS-responsive drug delivery systems have been synthesized and investigated. They are mainly dependent on two basic chemical principles which are induced solubility switch or degradation.[60] These systems mostly function through the oxidation of a particular atom (*e.g.* S, Se) or oxidative cleavage of carbon-heteroatom bonds (*e.g.* C–S) in the presence of H_2O_2 resulting in destabilization of the nano systems and release of drugs. Thus drug release is effected through carrier disassembly, cleavage induced carrier degradation and carrier-drug linker cleavage. One major requirement remains the capability of the ROS-responsive material to react to the physiological ROS level.[61]

This section will focus on polymeric nanoparticles which have shown encouraging results *in vitro*/*in vivo*.

5.1.3.1 Sulphur as Oxidative Stress Responsive Agent for Nanoparticles

The rationale behind the use of thioketal linkers is that they readily undergo oxidative cleavage by ROS, resulting in the formation of ketones and thiols (Scheme 5.6). Thioketals usually link the hydrophobic with the hydrophilic segment of block copolymer nanoparticles. Their oxidative cleavage by ROS leads to disassembly of nanoparticles resulting in drug release.

Simple systems such as paclitaxel loaded poly(1,4-phenyleneacetone dimethylene thioketal) (PPADT) nanoparticles (Figure 5.22) reduced cell viability of human prostate cancer cells (PC-3) known to have a ROS level as high as 34%.[62] ROS-triggered release was demonstrated using Nile Red encapsulated nanoparticles in PBS solutions with (i) 10 mM H_2O_2 solution with ferrous ions containing hydroxyl radicals and (ii) 10 mM KO_2 solution

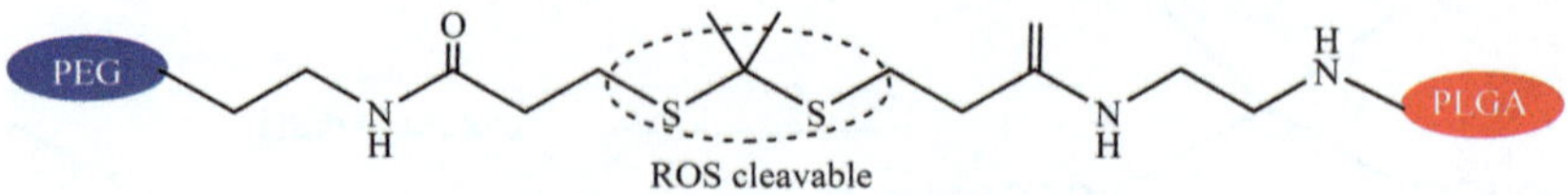

Figure 5.23 ROS-responsive amphiphilic PEG–TK–PLGA.

containing superoxide anion radicals. In the absence of ROS, degradation of the nanoparticles was delayed due to stability of the thioketal bond and almost no drug release observed up to 60 h, compared to 80% release with KO_2.

Classical PEG-based amphiphilic copolymers with a ROS-cleavable thioketal linker between the hydrophilic and hydrophobic segments also falls within oxidative stress responsive systems for cancer nanotherapeutics.[63] DOX-loaded amphiphilic PEG–TK–PLGA (TK: thioketal linker) (Figure 5.23) resulted in less than 30% viable human oral cancer Cal27 cells (IC50 ~ 4.67 ± 0.21 mg L^{-1}) compared to 90% for the unloaded micelles. Upon incubation with a ROS inhibitor (YCG063), the anticancer efficacy of the nanomicelles was reduced with a two-fold increase in the IC_{50} (~8.21 ± 0.10 mg L^{-1}). In the absence of ROS, the DOX-loaded nanoparticles showed around 25% DOX release within 6 h and the cumulative release reached approximately 52% after 72 h. Drug release showed a dependency on ROS concentration with 35% release at 6 h and approximately 90% after 72 h at a KO_2 concentration of 50 mM, 60% release at 6 h and more than 90% after 72 h at a KO_2 concentration of 100 mM.

More sophisticated systems have included other functions on the nanoparticles, such as the GRGDS–PEG–TK–PLGA nanoparticle with GRGDS peptides on the nanoparticle surface to facilitate endocytosis for $\alpha_v\beta_3$ specific expression tumor cells and the thioketal (TK) linker for ROS-triggered release.[64] This system was used to encapsulate DOX, and in Balb/c mice bearing tumors it showed better anti-tumor efficacy than free DOX. *In vitro*, DOX was released in a KO_2 concentration dependent manner. After 48 h, cumulative release reached 31% in PBS, 76% in PBS containing 50 μM KO_2 and more than 80% in PBS containing 50 μM KO_2.

The thioketal can also be used as a cross-linker to improve nanoparticle stability for prolonged circulation and to release the drug in response to ROS from the tumor. Cleavage of the cross-links leads to disassembly and drug release. In addition these nanoparticles can be dually responsive to oxidative stress and reduced pH. One such system based on a copolymer of PEG and poly(amino acids), poly(ethyleneglycol)–poly(N^6-carbobenzyloxy-L-lysine)–poly(β-benzyl-L-aspartate) (PEG–PLys–PBla), used the PLys block for grafting the ROS cleavable cross-linker and hydrophobic PBla block as a pH responsive moiety *via* an acid-labile covalent conjugation with DOX (Figure 5.24).[65] DOX release from the nanoparticles in the presence of H_2O_2 at pH 5.0 was over 60% of the drug after 48 h. In the absence of H_2O_2 and at pH 7.4 very little drug was released.

Dual reduced pH- and ROS-responsive nanoparticles can be therapeutically efficient on inflamed tissues. Both inflammation and reduced pH are

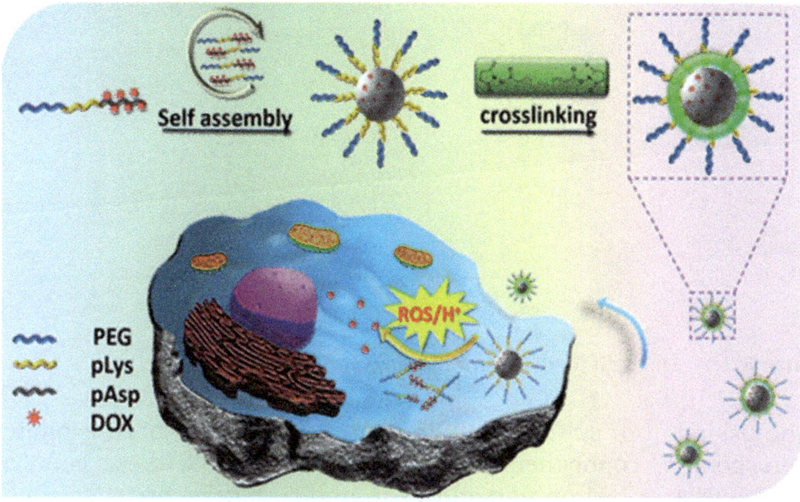

Figure 5.24 Thioketal cross-linker to improve the formulation's stability during circulation and to foster quick intracellular drug release in response to tumor's ROS potential and drug conjugation strategy with an acid-labile chemical bond.[65] (Reproduced with permission from Zhang Y., Guo Q., An S., Lu Y., Li J., He X., Liu L., Zhang Y., Sun T., Jiang C., *ACS Appl Mater Interfaces*, 2017, **9**(14), 12227. Copyright (2017) American Chemical Society).

associated with cancer and bioactive molecules such as curcumin, which is an immune-modulatory agent with potent anti-inflammatory and antioxidant capabilities, have shown very good therapeutic potential. The thioketal linkage plays a major role in these systems. They are cleaved in the presence of ROS and undergo hydrophobic to hydrophilic switch but remains unchanged in response to pH fluctuations. Another segment of the nanoparticle then responds to the reduced pH. For instance, a nanoparticle made of pH responsive *N*-palmitoyl chitosan (NPCS) with a Cy3 (cystamine, fluorescent molecule) moiety and ROS responsive poly-(1,4-phenyleneactone dimethylene thioketal) (PPADT) has been explored for the delivery of curcumin.[66]

In vivo biodistribution of the nanoparticles loaded with curcumin was studied in the ankles of healthy and acutely inflamed mice. No change in fluorescence intensity of curcumin was observed in healthy mice, but in the inflamed ankles, the fluorescence intensity was very high showing triggered release of encapsulated curcumin *via* oxidative stress and reduced pH.

Another pH–ROS responsive nanocarrier contained polycaprolactone, having acid-labile β-carboxylic amide segments which can reverse the charge at reduced pH and mPEG with a ROS-cleavable thioether linker.[67] The action of pH and ROS brought about disassembly of the nanoparticles. DOX was encapsulated in the nanoparticles and the release was studied at pH 7.4 and 6.0 with or without 10 mM H_2O_2. Release was faster in the presence of H_2O_2 (about 62%) and was further increased at pH 6.0 (80%).

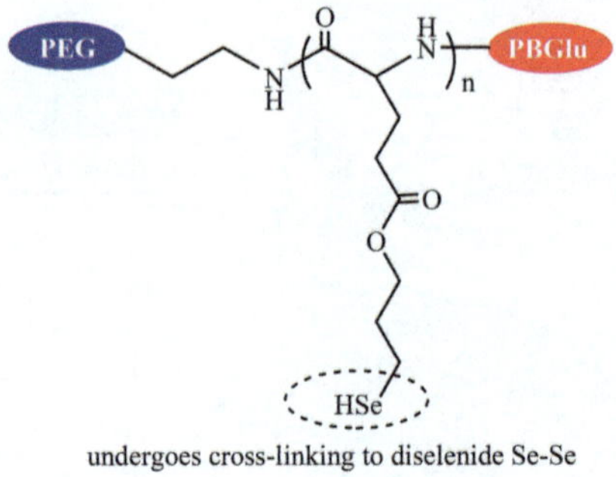

Figure 5.25 Diselenide ROS responsive nanoparticle.

The oxidation of sulfur to sulfone and sulfoxide has been exploited for ROS-responsive copolymers. After oxidation by ROS, these copolymers, which consist of hydrophilic and hydrophobic blocks, are thus converted to double hydrophilic blocks leading to drug release. For instance, amphiphilic mPEG-*b*-polydiethylsulfide undergoes cancer-specific oxidation of the polysulfide *via* high levels of H_2O_2.[68] Thus, the hydrophobic polysulfide core is converted to hydrophilic sulfone/sulfoxide and releases its drug cargo. In addition, the oxidized copolymers can destabilize the endosomal membranes to effect specific cytosolic drug delivery. α-tocopheryl succinate release rates from the micelles was almost 100 wt% at high H_2O_2 (133 μM) but only 40 wt% at low H_2O_2 (1.33 μM). The release was pH-independent.

5.1.3.2 *Selenium Linker/Crosslink as a ROS Responsive Agent for Amphiphilic Nanoparticles*

The use of selenium crosslinks has been exploited as it can be oxidized from its divalent to tetravalent state by ROS agents such as H_2O_2.

The DOX-loaded nanocarrier with PEG as the outer shell and an inner shell consisting of Se–Se cross-linked poly (glutamate) coupled with a hydrophobic core of poly (γ-benzyl glutamate) undergoes Se–Se bond cleavage in the presence of H_2O_2 resulting in a H_2O_2-dependent DOX release (Figure 5.25).[69] After 3 days, the drug release was approximately 17.3% in the absence of H_2O_2 and at normal physiological H_2O_2 concentration (20 nM), but reached a value of 65% at pathological H_2O_2 concentration (100 μM). In PC3 tumor-bearing nude mice treated for a period of 29 days, a significant reduction in tumor volume was obtained compared to nanoparticles without a diselenide bond and free DOX (210.10, 490.52, 680.86 mm^3).

5.1.3.3 *Boronic Ester Oxidation by ROS Triggered Drug Release*

Pendant aryl boronic ester units linked to copolymers or polymers forming nanoparticles undergo oxidative cleavage in the presence of ROS, resulting in either disassembly of nanoparticles[70] and drug release or direct drug release[71] (Figure 5.26).

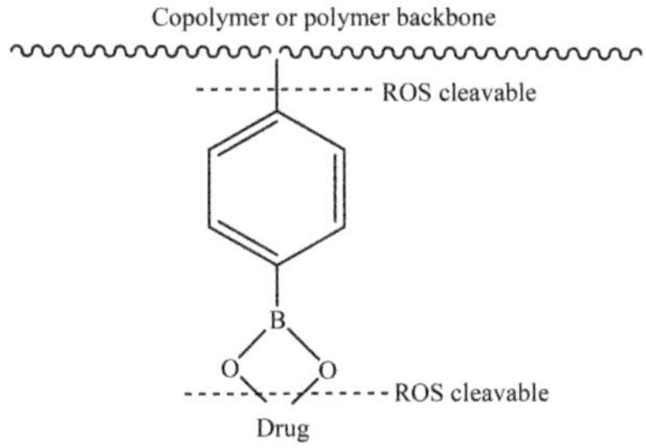

Figure 5.26 Aryl boronic ester oxidation.

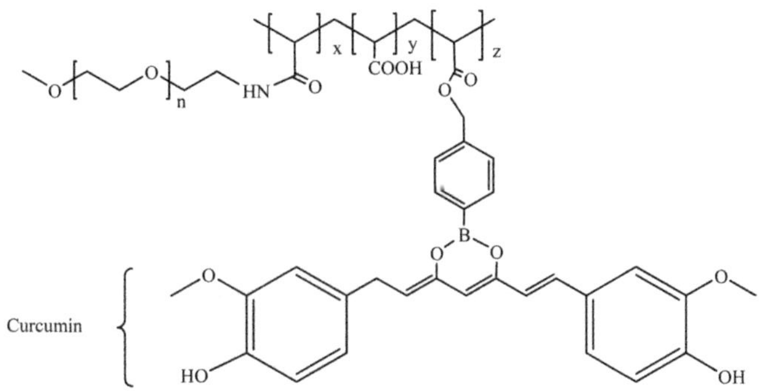

Figure 5.27 PPH curcumin nanoparticle.

4-(Hydroxymethyl) phenylboronic acid (HPBA)-modified PEG-grafted poly-(acrylic acid) polymer (PPH) with curcumin coordinated to the boronic acid, improved the stability of curcumin and released it in response to H_2O_2 (Figure 5.27).[71] In the presence of NAC, an antioxidant reported to strongly decrease the cellular level of H_2O_2, the cytotoxicity of the curcumin-loaded nanoparticles decreased, showing its dependency on the ROS-trigger. In the presence of 2 mM H_2O_2, an increase in particle size occurred (150 to ~475 nm) with degradation of boronic acid destroying the nanoparticle structure with accelerated drug release. After endocytosis by A549 cells, curcumin-loaded nanoparticles were mainly localized in lysosomes after 4 h incubation and were still there after the next 8 h incubation, whereas curcumin alone was mainly localized on the cell membrane and nucleus.

5.1.3.4 Double Stimuli: Intrinsic ROS and External Light

Photodynamic therapy is based on the use of an external source of light to activate a photosensitizer which results in the formation of ROS. Combining these two phenomena results in ROS responsive nanoparticles which can be activated by an external light stimulus. ROS/light responsive drug delivery systems are thus disassembled to release their cargo drug both in a time and space controlled manner. The irradiation time and amount determine the rate of drug release (Figure 5.28).[72] Some of these systems can function independently of

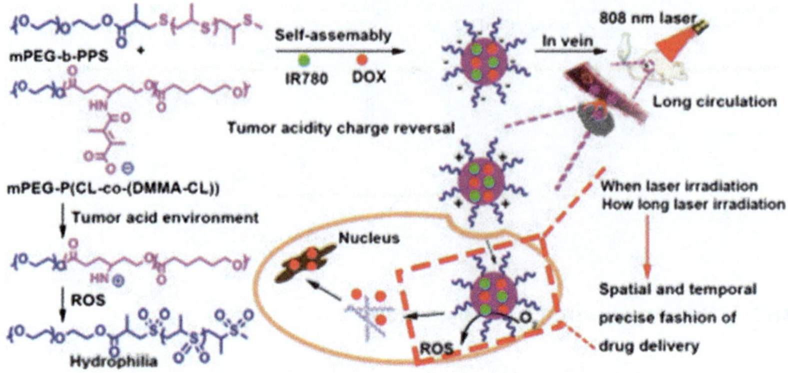

Figure 5.28 Light triggered ROS generation and subsequent ROS activated DOX release temporally and spatially (Reprinted from *Journal of Controlled Release*, H. Deng, X. Zhao, L. Deng, J. Liu, A. Dong, Reactive oxygen species activated nanoparticles with tumor acidity internalization for precise anticancer therapy, Copyright 2017, with permission from Elsevier).

each stimulus. For instance, PEG-bilirubin nanoparticles undergo oxidation in the presence of ROS to water soluble biliverdin while in the presence of light ($\lambda = 450$ or 650 nm) water soluble photoisomers are formed leading to both disassembly and drug release.[73] Interestingly, this DOX-loaded bilirubin nanoparticle showed a 55% tumor reduction in xenografted mice bearing human lung adenocarcinoma cell (A549) tumors which further increased to 71.9% upon irradiation compared to 27.8% with free DOX. These nanoparticles are also easily prepared and can be applied to large scale production.

All these systems are PEGylated to promote cell internalization and long circulation time.

Mitochondria are the main source of ROS in cells and using them as a primary target of photodynamic therapy is another strategy explored under dual stimuli. The thioketal linker still plays a major role in such systems. Polyethylene glycol functionalized with thioketal linker-modified camptothecin (TL–CPT) and triphenylphosphonium, TL–CPT–PEG1K–TPP and Zinc phthalocyanine (ZnPc, photosensitizer)/CPT–TPP nanoparticles can accumulate selectively several hundred-fold within mitochondria.[74] (Yue *et al.* 2016) CPT is released when the thioketal linker is ROS cleaved. 633 nm laser is absorbed by ZnPc to produce ROS which acts through both photodynamic therapy and thioketal linker cleavage, releasing camptothecin reducing tumor growth in NCI-H460 tumor bearing mice (Figure 5.29).

5.1.3.5 Immunotherapy

T-cell-based therapy is being used for cancer treatment. Efficient T cell activation is required for an effective antitumor response.[75] ROS participate in T cells activation, apoptosis, and hyporesponsiveness and it has been reported

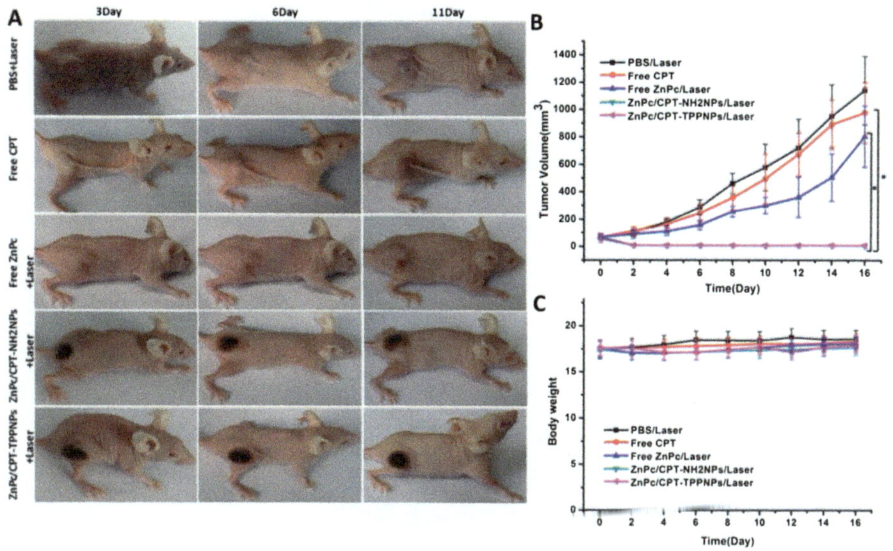

Figure 5.29 *In vivo* chemo- and photodynamic therapy of the ZnPc/CPT-TPPNPs: (A) Images of mice bearing NCI-H460 tumors after treatment. The mice were irradiated by 633 nm laser with the power density of 50 mW/cm^2 for 30 min. (B) Growth curves of NCI-H460 tumors in different groups after treatments. (C) Body weight curves after various treatments as indicated. The data are shown as mean ± SD ($n = 6$), * indicates $P < 0.05$.[74] (Reproduced from Yue *et al.* 2016, http://dx.doi.org/10.7150/thno.15433. Published under the terms of the CC BY 4.0 licence, https://creativecommons.org/licenses/by/4.0/).

that ROS produced in the TME can also suppress immune responses (Figure 5.30).[76] T-cell response is influenced by the redox bio-milieu created by the surrounding cells.[77] For instance, in inflamed tissues, a large amount of ROS, which can suppress/kill T cells, is produced by neutrophils. Mitochondrial ROS regulates IL-2 and IL-4 expression and can thus control T-cell activation[78] and also enhance immunity.[79] For T cells survival, low concentration of ROS is required, whereas high concentration leads to apoptosis/necrosis.[80]

ROS-responsive nanoparticles encapsulating agents used for immunotherapy could be very interesting but no such studies have been reported yet. A number of different types of nanoparticles have been reported for cancer immune therapy including pH-responsive nanoparticles.[81,82] Iron oxide-zinc oxide nanoparticles (15 nm) have been used for the delivery of tumor-associated antigens to antigen-presenting cells, liposomes (150 nm) for adjuvant delivery to tumor-draining lymph nodes, pH-responsive polymeric micelles (23 nm) for co-delivery of antigen and adjuvant, and iron-dextran particles (50–100 nm)/quantum dot nanocrystals (30 nm) for direct activation of tumor-associated antigen specific T-cells. Immunotherapy also faces the challenge of efficient and targeted delivery of molecules (antigen/adjuvant)

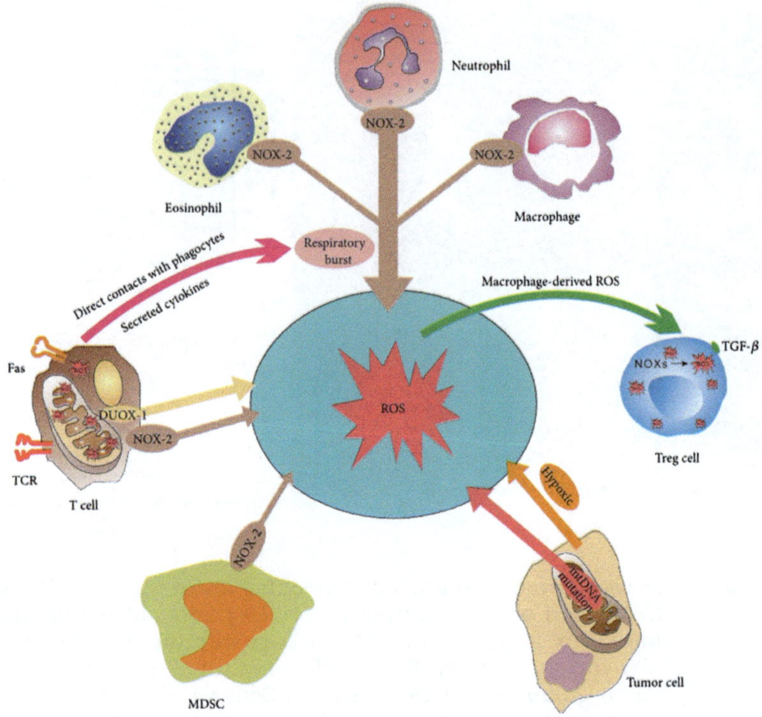

Figure 5.30 ROS produced in the tumor microenvironment[76] (Reproduced from ref. 76, http://dx.doi.org/10.1155/2016/1580967. Copyright © 2016 The Authors. Published under the terms of the CC BY 4.0 licence, https://creativecommons.org/licenses/by/4.0/).

to appropriate cells. Nanoparticle bound antigens and adjuvants can better trigger T- and B-cell responses as compared to soluble antigens given with various adjuvant types. ZnO nanoparticles have been shown to exhibit a strong preferential ability to kill cancerous T cells (~28–35×) compared to normal cells.[83]

The ability to trigger delivery of antigen/adjuvant in a high ROS milieu would enhance the immunotherapy efficacy.

5.1.4 Dual ROS–GSH Redox Responsive Systems

The trend is towards the fabrication of multi-stimuli responsive carriers for anti-cancer chemotherapy.[84]

Both ROS and GSH are characteristic features of tumor cells. Due to the heterogeneous nature of these cells, the possibility to exploit both in a single nano drug delivery system would enhance the therapeutic efficacy. One such system is based on PEGylated PTX–S–oleic acid conjugate (100 nm) achieving a very high drug loading of 57.4% (Figure 5.31). The PEGylation was performed using tocopheryl polyethylene glycol 2000 succinate ($TPGS_{2k}$).[85] Drug release

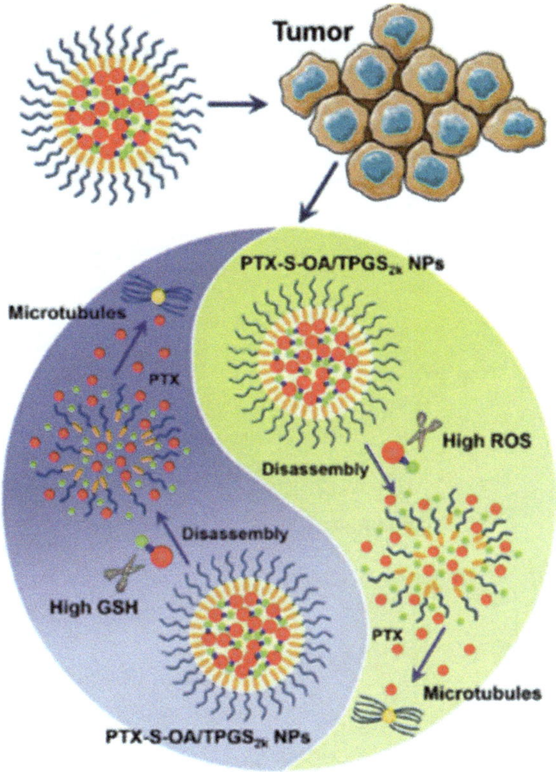

Figure 5.31 Dual ROS–GSH responsive nanosystem[85] (Adapted with permission from C. Luo, J. Sun, D. Liu, B. Sun, L. Miao, S. Musetti, J. Li, X. Han, Y. Du, L. Li, L. Huang, Z. He, Nano Lett, 2016, **16**(9), 5401. Copyright (2016) American Chemical Society.)

in PBS in the presence of the ROS simulatant H_2O_2 and GSH simulatant DTT (dithiothreitol) showed no release for the control PTX–OA/$TPGS_{2k}$, whereas PTX–S–OA/$TPGS_{2k}$ NPs exhibited a redox-responsive drug release. The thiolysis process for GSH stimulus and oxidation of the thioether to hydrophilic sulfone, and hydrolysis of oleic acid 2-hydroxyethyl ester for ROS stimulus were responsible for the release of PTX. The pharmacokinetic profiles of the nanoparticles in Sprague Dawley (SD) rats showed enhanced half-life and mean residence time compared to taxol and the non-PEGylated systems.

The capacity of being dually responsive towards cancer cells, as well as non-harming towards normal cells and organs, would further add to the efficacy. Only high concentrations of ROS or GSH can trigger the release of the drug and not the balanced ROS/GSH as found in healthy cells. One such system based on amphiphilic mPEG-*b*-P(Des-*alt*-Cys) (Des: diethyl sulfide, Cys: cystamine) has been used to encapsulate camptothecin (CPT).[86] Cys with its S–S bridge are cleaved in the presence of GSH leading to micelle fragmentation. In the presence of ROS (H_2O_2), Des undergoes oxidation and micelles

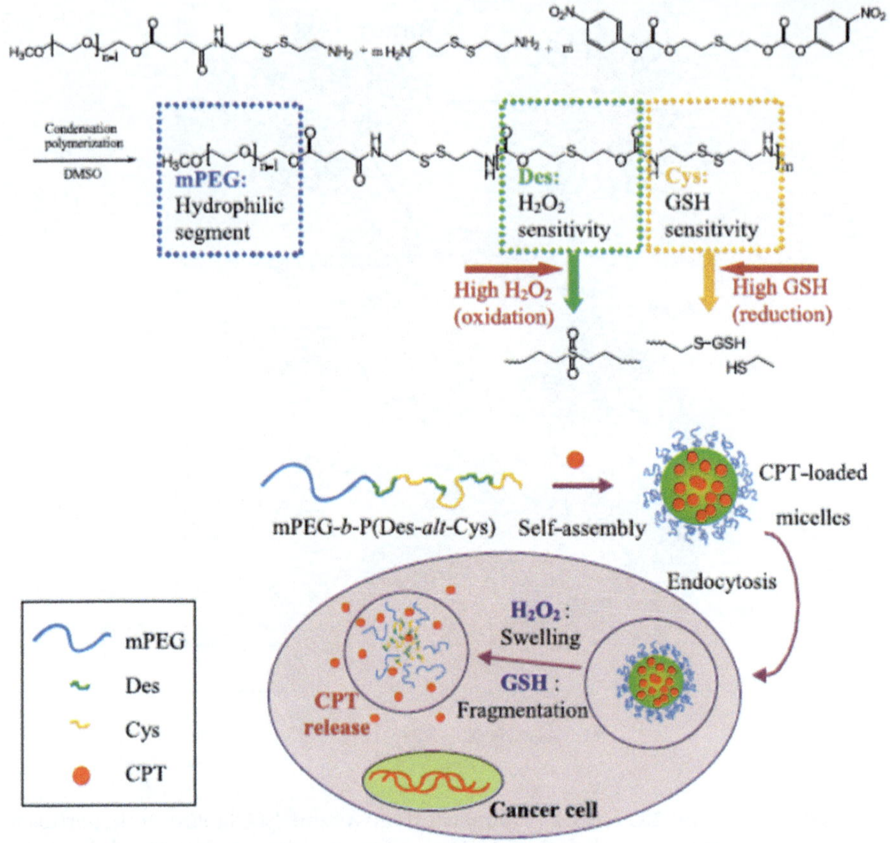

Figure 5.32 Dual ROS/GSH responsive mPEG-*b*-P(Des-*alt*-Cys) micelles[86] (Reprinted from *Biomaterials*, **61**, Y.-T. Chiang, Y.-W. Yen, C.-L. Lo, Reactive oxygen species and glutathione dual redox-responsive micelles for selective cytotoxicity of cancer, 150–161, Copyright (2015), with permission from Elsevier).

are swollen and ultimately disassemble (Figure 5.32). This system was able to maintain the stability of CPT and trigger its selective release as shown in mice bearing HCT 116 tumors. Free CPT showed a 4.7 ± 0.69 fold increase in size, whereas CPT-loaded micelles showed a 3.4% increase over 27 days.

5.1.5 New Avenues in Redox Responsive Drug Release

5.1.5.1 The Increase ROS and Decrease GSH Strategy

The threshold concept of ROS for both tumor and healthy cells translates into a balance of intracellular ROS with antioxidant programs (including GSH) for cell survival.[87] When the concentration of ROS in the tumor micro-environment increases, antioxidants have the capacity to re-adapt, thus low-to-moderate levels of ROS production may enhance the proliferation of

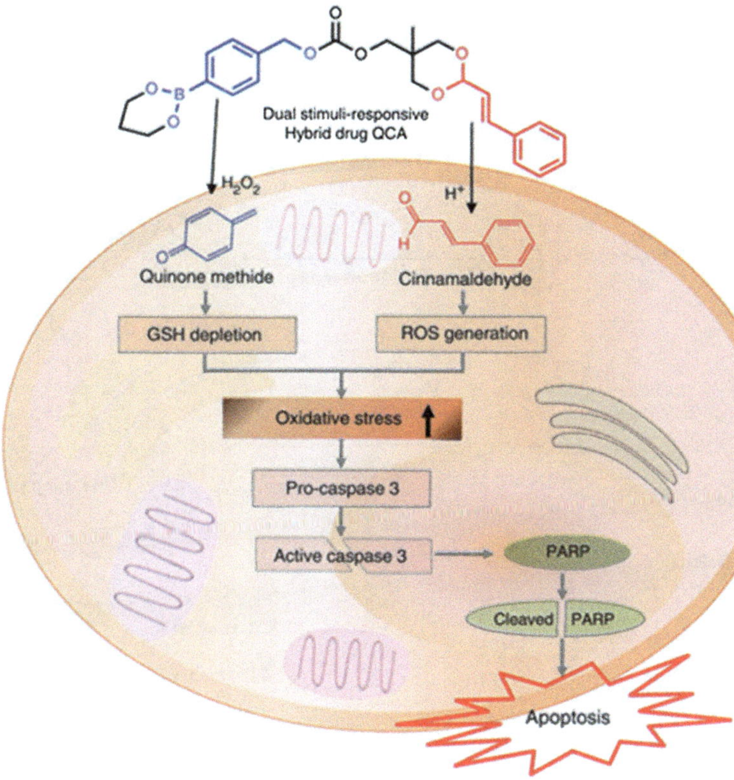

Figure 5.33 A dual stimuli-responsive hybrid anticancer drug QCA. QCA is activated by H_2O_2 and acidic pH and amplifies oxidative stress in cancer cells, leading to apoptotic cell death[90] (Reproduced from Noh *et al.* 2015, Copyright (2015) with permission from Nature Publishing Group).

cancer cells.[88,89] When ROS levels exceed the optimum concentration, the antioxidant programs forgo their re-adaptation capacity. As a result, irreversible oxidative damage occurs and induces cell death *via* apoptosis, necrosis, and autophagy.

Thus, cancer cells produce ROS, but not enough to kill them due to the presence of the counter-balance GSH. Increasing further ROS while scavenging GSH can effectively amplify the oxidative stress and kill the cells. The dual stimuli-responsive hybrid anticancer drug QCA (2-dioxaborinan-2-yl) benzyl ((5-methyl-2-styryl-1,3-dioxan-5-yl) methyl) carbonate can be activated by H_2O_2 to produce glutathione-scavenging quinone methide and activated by acidic pH to form ROS-generating cinnamaldehyde (Figure 5.33).[90] These two molecules thus enhance oxidative stress leading to cell death. The effect of GSH was studied in a prostate cancer DU145 cell line, a colon cancer SW620 cell line and a fibroblast NIH3T3 cell line. More GSH depletion was observed in DU145 than SW620, while no effect on GSH was seen in non-malignant NIH3T3 cells. In contrast, induced significant ROS

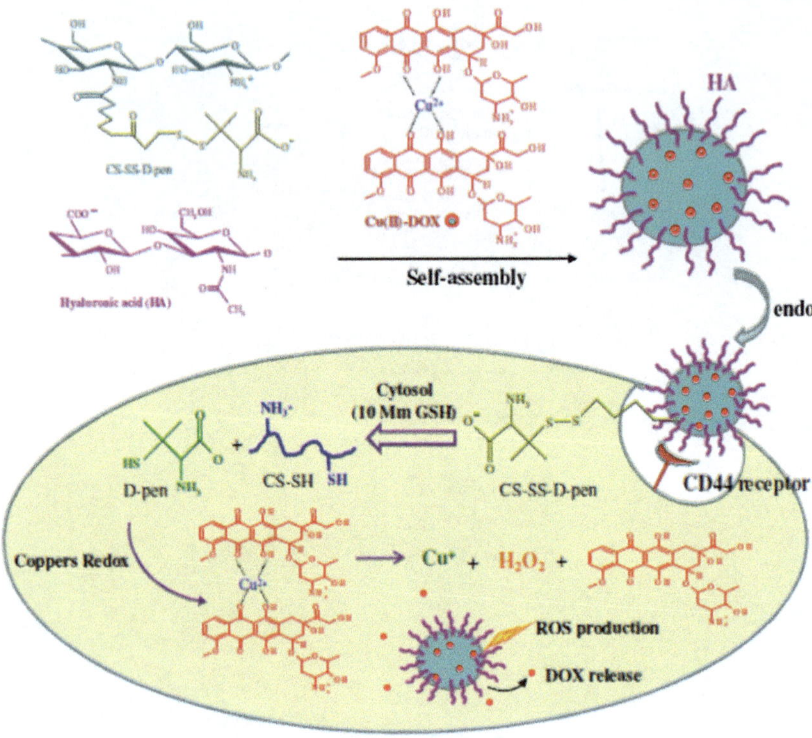

Figure 5.34 CD44 targeting and redox-triggered ROS production and DOX release[91] (Reprinted from *Acta Biomaterialia*, **35**, C.-W. Lin, K.-Y. Lu, S.-Y. Wang, H.-W. Sung, F.-L. Mi, CD44-specific nanoparticles for redox-triggered reactive oxygen species production and doxorubicin release, 280–292, Copyright (2016) Acta Materialia Inc., with permission from Elsevier).

generation was observed in SW620. Xenograft mouse models of SW620 and DU145 cells showed marked anti-tumor effects with the synergistic action of QCA. It would be interesting to encapsulate such a system in a nanocarrier and enhance its cellular uptake.

5.1.5.2 Intracellular GSH-triggered ROS Production

Hyaluronic acid (HA) has a high binding affinity for CD44, which is overexpressed in some types of cancers. HA was used to coat a chitosan–S–S–D-penicillamine (Cu(II) chelator) (CS–S–S–D-pen) nanoparticle.[91] Intracellularly, D-pen is released through GSH cleavage of the S–S bond. DOX and Cu(II) ions were incorporated in the nanoparticle in the form of a Cu(II)–DOX. Inside the colon cancer cells (HT 29), released D-pen initialized ROS production *via* a series of reactions that involved Cu(II)-catalyzed D-pen oxidation. ROS resulted in depolymerization of HA which increased the DOX release (Figure 5.34).

References

1. R. Cheng, F. Meng, C. Deng and Z. Zhong, *Nano Today*, 2015, **10**, 656.
2. R. Burdon, *Free Radical Biol. Med.*, 1995, **18**, 775.
3. W. Fiers, R. Beyaert, W. Declercq and P. Vandenabeele, *Oncogene*, 1999, **18**, 7719.
4. J. Matés, J. Segura, F. Alonso and J. Márquez, *Arch. Toxicol.*, 2008, **82**, 273.
5. D. J. Phillips and M. I. Gibson, *Antioxid. Redox Signaling*, 2014, **21**(5), 786.
6. G. Saravanakumar, J. Kim and W. J. Kim, *Adv. Sci.*, 2017, **4**, 1600124.
7. D. Chen, A. A. Bobko, A. C. Gross, R. Evans and C. B. Marsh, *et al.*, *PLoS One*, 2014, **9**(10), e107511.
8. G.-Y. Liou and P. Storz, *Free Radical Res.*, 2010, **44**(5), 479.
9. P. Storz, *Front. Biosci.*, 2005, **10**, 1881.
10. M. Schieber and N. S. Chandel, *Curr. Biol.*, 2014, **24**, R453.
11. K. Aquilano, S. Baldelli and M. R. Ciriolo, *Front. Pharmacol.*, 2014, **5**, 196.
12. F. Q. Schafer and G. R. Buettner, *Free Radical Biol. Med.*, 2001, **30**, 1191.
13. M. Karimi, A. Ghasemi, P. Sahandi Zangabad, R. Rahighi, S. Moosavi Basri, H. Mirshekari, M. Amiri, Z. Shafaei Pishabad, A. Aslani, M. Bozorgomid, D. Ghosh, A. Beyzavi, A. Vaseghi, A. R. Aref, L. Haghani, S. Bahrami and M. R. Hamblin, *Chem. Soc. Rev.*, 2016, **45**, 1457.
14. R. Cheng, F. Feng, F. Meng, C. Deng, J. Feijen and Z. Zhong, *J. Controlled Release*, 2011, **152**, 2.
15. R. L. McCarley, *Annu. Rev. Anal. Chem.*, 2012, **5**, 391.
16. H. Wen and Y. Li, *Med. Chem.*, 2014, **4**(11), 748.
17. R. Mo and G. Zhen, *Mater. Today*, 2016, **19**(5), 274.
18. J. K. Oha, D. I. Leea and J. M. Park, *Prog. Polym. Sci.*, 2009, **34**(12), 1261.
19. H. Zhang, Y. Zhai, J. Wang and G. Zhai, *Mater. Sci. Eng., C*, 2016, **60**, 560.
20. N. Grimaldi, M.-A. Sabatino, D. Bulone, G. Adamo, S. Rigogliuso, G. Ghersi and C. Dispenza, *NSTI-Nanotech.*, 2013, **3**, 155, ISBN 978-1-4822-0586-2.
21. G. A. Adamo, N. Grimaldi, S. Camporaa, M. A. Sabatino, C. Dispenza and G. Ghersi, *Chem. Eng. Trans.*, 2014, **38**, 457.
22. J. Liu, C. Detrembleur, M. Hurtgen, A. Debuigne and M.-C. De Pauw-Gillet, *et al.*, *Polym. Chem.*, 2014, **5**(1), 77.
23. Z.-Y. Qiao, R. Zhang, F.-S. Du, D.-H. Liang and Z.-C. Li, *J. Controlled Release*, 2011, **152**, 57.
24. H. H. Bae, M. Y. Cho, J. H. Hong, H. Poo, M.-H. Sung and Y. T. Lim, *J. Microbiol. Biotechnol.*, 2012, **22**(12), 1782.
25. H. Sun, B. Guo, X. Li, R. Cheng, F. Meng, H. Liu and Z. Zhong, *Biomacromolecules*, 2010, **11**, 848.
26. H.-K. Park, S. J. Lee, J.-S. Oh, S.-G. Lee, Y. IL Jeong and H. C. Lee, *Nanoscale Res. Lett.*, 2015, **10**, 288.
27. H. Sun, B. Guo, R. Cheng, F. Meng, H. Liu and Z. Zhong, *Biomaterials*, 2009, **30**, 6358.
28. S. Samarajeewa, R. Shrestha, M. Elsabahy, A. Karwa, A. Li, R. P. Zentay, J. G. Kostelc, R. B. Dorshow and K. L. Wooley, *Mol. Pharmaceutics*, 2013, **10**, 1092.

29. B. Khorsand, G. Lapointe, C. Brett and J. K. Oh, *Biomacromolecules*, 2013, **14**, 2103.
30. A. Zhang, Z. Zhang, F. Shi, J. Ding, C. Xiao, X. Zhuang, C. He, L. Chen and X. Chen, *Soft Matter*, 2013, **9**, 2224.
31. T. Thambi, D. G. You, H. S. Ha, V. G. Deepagan, S. M. Jeon, Y. D. Suh, K. Y. Choi, K. Kim, I. C. Kwon, G.-R. Yi, J. Y. Lee, D. S. Lee and J. H. Park, *Adv. Healthcare Mater.*, 2014, **3**, 1829–1838.
32. H. Wen, C. Dong, H. Dong, A. Shen, W. Xia, X. Cai, Y. Song, X. Li, Y. Li and D. Shi, *Small*, 2012, **8**(5), 760.
33. J. Liu, W. Huang, Y. Pang, P. Huang, X. Zhu, Y. Zhou and D. Yan, *Angew. Chem., Int. Ed.*, 2011, **50**, 9162.
34. J. Yang, Y. Duan, X. Zhang, Y. Wang and A. Yu, *J. Mater. Chem. B*, 2016, **4**, 3868.
35. Y.-W. Hu, Y.-Z. Du, N. Liu, X. Liu, T.-T. Meng, B.-L. Cheng, J.-B. He, J. You, H. Yuan and F.-Q. Hu, *J. Controlled Release*, 2015, **206**, 91.
36. G. Zhou, L. Li, J. Xing, Y. Li, J. Cai, J. Chen, P. Liu, N. Gu and M. Ji, *Colloids Surf., B*, 2016, **148**, 518.
37. M. Curcio, B. Blanco-Fernandez, L. Diaz-Gomez, A. Concheiro and C. Alvarez-Lorenzo, *Bioconjugate Chem.*, 2015, **26**(9), 1900–1907.
38. Y. Liang and K. L. Kiick, *Biomacromolecules*, 2016, **17**(2), 601.
39. H. Fu, K. Shi, G. Hu, Y. Yang, Q. Kuang, L. Lu, L. Zhang, W. Chen, M. Dong, Y. Chen and Q. He, *J. Pharm. Sci.*, 2015, **103**, 1160.
40. K. Möller and T. Bein, *Chem. Mater.*, 2017, **29**(1), 371.
41. L. Wei, N. Hu and Y. Zhang, *Materials*, 2010, **3**, 4066.
42. L. Han, X.-Y. Zhang, Y.-L. Wang, X. Li, X.-H. Yang, M. Huang, K. Hu, L.-H. Li and Y. Wei, *J. Controlled Release*, 2017, **259**, 40–52.
43. Y. Tian, R. Guo, Y. Jiao, Y. Sun, S. Shen, Y. Wang, D. Lu, X. Jiang and W. Yanga, *Nanoscale Horiz.*, 2016, **1**, 480–487.
44. Y. Wang, N. Han, Q. Zhao, L. Bai, J. Li, T. Jiang and S. Wang, *Eur. J. Pharm. Sci.*, 2015, **72**, 12.
45. L. Huang, Q. Zhang, L. Dai, X. Shen, W. Chen and K. Cai, *Regener. Biomater.*, 2017, 111–124.
46. F. Muhammad, A. Wang, W. Qi, S. Zhang and G. Zhu, *ACS Appl. Mater. Interfaces*, 2014, **6**(21), 19424.
47. E. C. Dreaden, L. A. Austin, M. A. Mackey and M. A. El-Sayed, *Ther. Delivery*, 2012, **3**(4), 457.
48. G. Han, P. Ghosh and V. M. Rotello, *Nanomedicine*, 2007, **2**(1), 113.
49. R. Hong, G. Han, J. M. Fernandez, B.-J. Kim, N. S. Forbes and V. M. Rotello, *J. Am. Chem. Soc.*, 2006, **128**, 1078.
50. P. Ghosh, G. Han, M. De, C. K. Kim and V. M. Rotello, *Adv. Drug Delivery Rev.*, 2008, **60**, 1307.
51. L. Tian, L. Lu, Y. Qiao, S. Ravi, F. Salatan and M. P. Melancon, *J. Funct. Biomater.*, 2016, **7**, 19.
52. Y. HeeáKook, E. TaxáOh and H. JooáPark, *J. Mater. Chem.*, 2009, **19**, 2310.
53. D. N. Heo, D. H. Yang, H.-J. Moon, J. B. Lee, M. S. Bae, S. C. Lee, W. J. Lee, L.-C. Sun and I. K. Kwon, *Biomaterials*, 2012, **33**, 856.

54. X. Wang, X. Cai, J. Hu, N. Shao, F. Wang, Q. Zhang, J. Xiao and Y. Cheng, *J. Am. Chem. Soc.*, 2013, **135**, 9805.
55. Y. Tao, J. Han, C. Ye, T. Thomas and H. Dou, *J. Mater. Chem.*, 2012, **22**, 18864.
56. X. Peng, H. Chen, J. Huang, H. Mao and D. M. Shin, Biomedical engineering – from theory to applications, in *Targeted Magnetic Iron Oxide Nanoparticles for Tumor Imaging and Therapy*, ed. R. Fazel-Rezai, August 29 2011, ch. 9, ISBN 978-953-307-637-9.
57. A. Aires, S. M. Ocampo, B. M. Simões, M. J. Rodríguez, J. F. Cadenas, P. Couleaud, K. Spence, A. Latorre, R. Miranda, Á. Somoza, R. B. Clarke, J. L. Carrascosa and A. L. Cortajarena, *Nanotechnology*, 2016, **27**, 065103 (10pp).
58. Q. Xu, C. He, C. Xiao and X. Chen, *Macromol. Biosci.*, 2016, **16**, 635.
59. J. Liang and B. Liu, *Bioeng. Transl. Med.*, 2016, **1**(3), 239.
60. S. H. Lee, M. K. Gupta, J. B. Bang, H. Bae and H.-J. Sung, *Adv. Healthcare Mater.*, 2013, **2**(6), 908.
61. S. Joshi-Barr, C. de Gracia Lux, E. Mahmoud and A. Almutairi, *Antioxid. Redox Signaling*, 2014, **21**(5), 730.
62. J. S. Kim, S. D. Jo, G. L. Seah, I. Kim and Y. S. Nam, *J. Ind. Eng. Chem.*, 2015, **21**, 1137.
63. Q. Li, Y. Wen, J. Wen, Y.-P. Zhang, X.-D. Xu, A. Victorious, R. Zavitz and X. Xu, *RSC Adv.*, 2016, **6**, 38984.
64. Q. Li, Y. Wen, X. You, F. Zhang, V. Shah, X. Chen, D. Tong, X. Wei, L. Yin, J. Wu and X. Xu, *J. Mater. Chem. B*, 2016, **4**, 4675.
65. Y. Zhang, Q. Guo, S. An, Y. Lu, J. Li, X. He, L. Liu, Y. Zhang, T. Sun and C. Jiang, *ACS Appl. Mater. Interfaces*, 2017, **9**(14), 12227.
66. H.-L. Pu, W.-L. Chiang, B. Maiti, Z.-X. Liao, Y.-C. Ho, M. S. Shim, E.-Y. Chuang, Y. Xia and H.-W. Sung, *ACS Nano*, 2014, **8**(2), 1213.
67. H. Deng, X. Zhao, J. Liu, L. Deng, J. Zhang, J. Liu and A. Dong, *J. Mater. Chem. B*, 2015, **3**, 9397–9408.
68. L.-Y. Yu, G.-M. Su, C.-K. Chen, Y.-T. Chiang and C.-L. Lo, *Biomacromolecules*, 2016, **17**(9), 3040.
69. V. G. Deepagan, S. Kwon, D. G. You, V. Q. Nguyen, W. Um, H. Ko, H. Lee, D.-G. Jo, Y. M. Kang and J. H. Park, *Biomaterials*, 2016, **103**, 56.
70. K. E. Broaders, S. Grandhe and J. M. J. Frechet, *J. Am. Chem. Soc.*, 2011, **133**, 756.
71. C.-Q. Luo, L. Xing, P.-F. Cui, J.-B. Qiao, H.-J. He, B.-A. Chen, L. Jin and H.-L. Jiang, *Int. J. Nanomed.*, 2017, **12**, 855.
72. H. Deng, X. Zhao, L. Deng, J. Liu and A. Dong, *J. Controlled Release*, 2017, **S0168-3659**(17), 30513.
73. Y. Lee, S. Lee, D. Y. Lee, B. Yu, W. Miao and S. Jon, *Angew. Chem., Int. Ed. Engl.*, 2016, **55**(36), 10676.
74. C. Yue, Y. Yang, C. Zhang, G. Alfranca, S. Cheng, L. Ma, Y. Liu, X. Zhi, J. Ni, W. Jiang, J. Song, M. J. de la Fuente and D. Cui, *Theranostics*, 2016, **6**(13), 2352.
75. B. Molon, B. Calì and A. T. Viola, *Front. Immunol.*, 2016, **7**(20), DOI: 10.3389/fimmu.2016.00020.

76. X. Chen, M. Song, B. Zhang and Y. Zhang, *Oxid. Med. Cell. Longevity*, 2016, 1580967.
77. A. V. Belikov, B. Schraven and L. Simeoni, *J. Biomed. Sci.*, 2015, **22**, 85.
78. M. M. Kaminski, S. W. Sauer, C. D. Klemke, J. G. Suss, P. H. Okun and K. Krammer, *J. Immunol.*, 2010, **184**, 4827.
79. M. P. Murphy and R. M. Siegel, *Immunity*, 2013, **38**, 201.
80. P. Kesarwani, A. K. Murali, A. A. Al-Khami and S. Mehrotra, *Antioxid. Redox Signaling*, 2013, **18**(12), 1497.
81. K. Shao, S. Singha, X. Clemente-Casares, S. Tsai, Y. Yang and P. Santamaria, *ACS Nano*, 2015, **9**(1), 16.
82. R. Tong and D. S. Kohane, *Annu. Rev. Pharmacol. Toxicol.*, 2016, **56**, 18.
83. C. Hanley, J. Layne, A. Punnoose, K. M. Reddy, I. Coombs, A. Coombs, K. Feris and D. Wingett, *Nanotechnology*, 2008, **19**(29), 295103.
84. J. Wen and S. Sun, *Curr. Trends Biomed. Eng. Biosci.*, 2017, **3**(2), 555607.
85. C. Luo, J. Sun, D. Liu, B. Sun, L. Miao, S. Musetti, J. Li, X. Han, Y. Du, L. Li, L. Huang and Z. He, *Nano Lett.*, 2016, **16**(9), 5401.
86. Y.-T. Chiang, Y.-W. Yen and C.-L. Lo, *Biomaterials*, 2015, **61**, 150.
87. L. Zhang, J. Li, L. Zong, X. Chen, K. Chen, Z. Jiang, L. Nan, X. Li, W. Li, T. Shan, Q. Ma and Z. Ma, *Oxid. Med. Cell. Longevity*, 2016, 161678, 9 pages.
88. Q. Kong, J. A. Beel and K. O. Lillehei, *Med. Hypotheses*, 2000, **55**(1), 29.
89. J. E. Klaunig and L. M. Kamendulis, *Annu. Rev. Pharmacol. Toxicol.*, 2004, **44**, 239.
90. J. Noh, B. Kwon, E. Han, M. Park, W. Yang, W. Cho, W. Yoo, G. Khang and D. Lee, *Nat. Commun.*, 2015, **6**, 6907.
91. C.-W. Lin, K.-Y. Lu, S.-Y. Wang, H.-W. Sung and F.-L. Mi, *Acta Biomater.*, 2016, **35**, 280.

CHAPTER 6

Magnetically-responsive DDS

E. GUISASOLA, A. BAEZA AND M. VALLET*

Dpto. Química en Ciencias Farmacéuticas, Instituto de Investigación Sanitaria Hospital 12 de Octubre i+12, Universidad Complutense de Madrid, Plaza Ramon y Cajal s/n and Centro de Investigación Biomédica en Red de Bioingeniería, Biomateriales y Nanomedicina (CIBER-BBN), Av. Monforte de Lemos, 3-5, Pabellón 11, Planta 0 28029, Madrid, Spain
*E-mail: vallet@ucm.es

6.1 Introduction

The direct administration of drugs either *via* intravenous injection or orally, usually induces some toxicity or low efficiency due to the unbalance of the drug concentration within the blood stream, which is higher than the proper dosage immediately after the administration and lower after a few hours as a consequence of the clearance of these therapeutic agents from the host. Therefore, the efficient administration of therapeutic agents requires the development of systems able to release the convenient drug dosage at the proper time. In recent years, a vast number of nanometric stimuli-responsive drug-delivery systems (DDS) have been described for releasing the required drugs in a controlled manner, only in the presence of certain stimuli.[1] These stimuli can be divided into external stimuli, in which the drug release process is triggered by the application of an externally applied stimulus such as light, magnetic field, temperature or ultrasound; and internal stimuli, in which the trigger event is provoked by some internal characteristic of the treated disease such as lower pH, presence of specific enzymes or biomolecules.[2]

Biomaterials Science Series No. 1
Stimuli-responsive Drug Delivery Systems
Edited by Amit Singh and Mansoor M. Amiji

Published by the Royal Society of Chemistry, www.rsc.org

The latter ones have the advantage of not requiring any external equipment but as main drawback, these nanodevices present a reduced control in the drug release than the externally triggered systems. Thus, DDS controlled by external stimuli are specially recommended when the administered drug presents high toxicity and requires a more precise control in the drug release, as in the case of chemotherapeutic agents. Among the different external stimulus which can be employed, magnetic field is one of the most employed trigger stimuli due to its interesting properties of having excellent biocompatibility and higher penetration capacity in living tissues. These particular features give rise to a great versatility in the treatment of diseases through nanotechnology. Permanent magnets present a static magnetic field that interact with magnetic nanoparticles (MNPs) by attracting them to its proximity. This simple fact allows researchers to develop a localization method that can be used from cell magnetic separation[3] or to trigger the drug release from the DDS,[4,5] to the induction of targeted particle accumulation in tumors.[6,7]

In order to create magnetic-responsive drug delivery systems (MR-DDS) it is compulsory to introduce magnetic nanocores within the formulation. Thus, these nanocarriers are usually based on superparamagnetic iron oxide nanoparticles (SPION) coated with an organic or hybrid organic/inorganic temperature-responsive shell which act as a drug reservoir. The application of an alternating magnetic field (AMF) induces a temperature increase as a consequence of its interaction with the SPION and the achieved higher temperature provokes a change of the temperature-responsive moiety which triggers the drug departure. SPION are used as magnetically-sensitive devices because, thanks to their superparamagnetic behaviour, they can dissipate heat under alternating magnetic field exposition without suffering aggregation due to their lack of magnetization once the AMF application is stopped. Moreover, this drug release mechanism allows a synergic improvement of the therapeutic efficacy of the released drugs, especially in oncology, because many cytotoxic compounds are more effective in combination with the hyperthermia achieved in the tissue exposed to AMF.[8] Finally, the presence of magnetic cores within these nanocarriers can also be exploited for imaging of the particle fate by magnetic resonance imaging (MRI).[9]

6.2 Synthesis and Heating Mechanism

The synthesis of magnetic nanoparticles is based on two main routes, chemical wet synthesis and physical techniques.[10] Physical methods such as vapour deposition or electron beam lithography have a lack of control over the nanoparticle size, therefore other techniques such as laser ablation are preferable to obtain optimally sized MNPs for biomedical applications.[11,12] Within the wet synthetic methods the co-precipitation of metal ions is probably the most studied approximation. As well as other chemical techniques, the formation of the nanoparticles is described by the LaMer model. This theory differentiates three different consecutive steps in the nanoparticle growth: nucleation, crystal growth and Ostwald ripening. The first step consists of the ion precipitation that leads to the formation of several atom clusters within the solvent. Tuning the reaction conditions is possible to control the rate of the nucleation and therefore the nanoparticle size, this is, since higher amounts of nuclei are formed less dissolved ions are available to grow the nanoparticles, so smaller nanoparticles are obtained. The crystal growth occurs by ion diffusion from the solution to the formed clusters. The Oswald ripening refers to the smaller nanoparticle dissolution to deposit on larger particles by means of a better thermodynamic stabilization through the surface to volume ratio minimization. The nanoparticle formation is optimized by the time differentiation between the two first steps.[13] Researchers have developed several wet methods to obtain nanoparticles for biomedical uses as hydrothermal synthesis at high pressure or thermal decomposition of organic precursors between some other techniques.[14,15] The microbial synthesis through magnetotactic bacteria should also be mentioned due to its good reproducibility and control over the size and composition of the MNPs that are produced by biomineralization.[16] This bacterial route can afford high yields of single or multidomain magnetic nanoparticles between 20–45 nm.

The magnetic properties arise from the atomic structure and composition of this sort of material. This way, different magnetic behaviours are observed towards a magnetic field depending on the electronic configuration (Figure 6.1). For example, in a pure ferromagnetic material all the magnetic moments

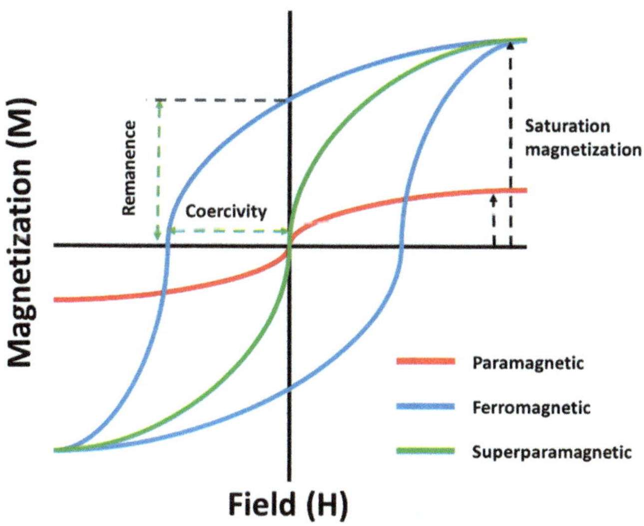

Figure 6.1 Different magnetic behaviours for magnetic materials obtained by a vibrating sample magnetometer (VSM). The magnetization curve of superparamagnetic materials shows no hysteresis loop, no remanence (residual magnetization when the magnetic field is ceased) and high saturation magnetization values. As a consequence of the absence of remanence the nanoparticles do not trend to aggregate, this way ultra-stable colloids can be achieved.

are aligned to the same direction, which creates a permanent magnetization even in absence of a magnetic field. Diamagnetic materials do not have unpaired electrons, which implicate that it is repelled by a magnetic field. In the case of paramagnetic behaviour, the existence of unpaired electrons in the material shows magnetic dipole moments randomly oriented that do not interact with one another. Once the paramagnetic material is exposed to a magnetic field, the magnetic moments align parallel to the applied field causing a small attraction to the magnet, but no magnetization is observed when the magnetic field application is ceased.

As a consequence of the size reduction of a bulk magnetic material to the nanoscale another important magnetic behaviour is observed, superparamagnetism. Here the formation of single domain nanoparticles is thermodynamically favourable when the system has enough thermal energy to overcome the magnetic anisotropy barrier of the nanoparticle. This fact means that every nanoparticle would act as a single magnet with its magnetic moment pointing randomly. In the superparamagnetic state the magnetic moment is oriented easily in the direction of an applied magnetic field, having a high saturation magnetization that leads to a strong magnetic response.[17]

The properties of superparamagnetic nanoparticles allow them to develop a heating effect under an AMF that increase the temperature of the surroundings. The heat produced by the MNPs exposed to an AMF is being used for hyperthermia therapies, since cancer tissues are more sensitive to high temperatures than normal tissues.[18] The relaxation process is the time period while the magnetic moment is aligned in a given orientation and is responsible for the heating mechanism. Two different relaxation processes contribute to the temperature rise, the Brown and Néel relaxations (Figure 6.2).

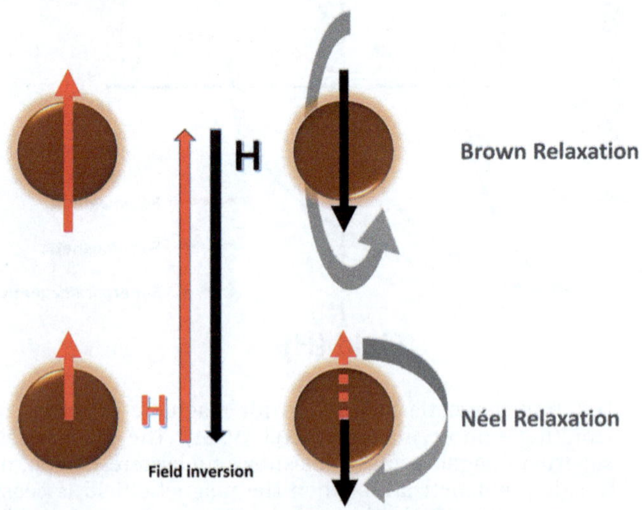

Figure 6.2 Brown and Néel relaxations. AMF application provokes the particle physical spin (Brown) and the inversion of the magnetic moment (Néel) of the superparamagnetic nanoparticles.

The first one produces heat due to the friction with the dispersant while the particle is spinning upon itself. The heat dissipation caused by Néel relaxation occurs because the rotation of the magnetic moments within the particle. These relaxation mechanisms take place when the magnetic relaxation of the MNPs have a lower frequency than the applied AMF.[19] The inversion speed of magnetic moments depends mainly on the size and saturation magnetization of the nanoparticles and the applied field characteristics of amplitude and frequency, besides the viscosity of the surroundings where the nanocrystal is placed. When the nanoparticles are embedded in a large matrix, in a biological fluid or tissue the spinning is blocked because of the high viscosity of the medium, therefore the only contribution to the heating effect is due to the Néel relaxation process.[20]

Parameters such as magnetic field amplitude and frequency, size distribution and composition, participate in the heating capacity of the MNPs. The heating efficiency of MNPs is described by the specific absorption rate (SAR) or specific loss power (SLP) under certain magnetic field frequency (f) and dispersant medium (eqn (6.1)). It rates the energy absorbed per unit of mass of nanoparticles and depends on the specific heat capacity (C) of the dispersant and the macroscopic temperature change of the sample ($\Delta T/\Delta t$) over the time of the AMF application.[21] To report the heating efficiency independently from the frequency and intensity of the applied AMF the "intrinsic loss power" (ILP) is also described.[22]

$$\text{SAR (or SLP)}, = C\cdot(\Delta T/\Delta t)\ [\text{W g}^{-1}] = (\text{Hysteresis loop area}) \times f \qquad (6.1)$$

The attention attracted by MNPs in the last decades is owing to not only the heating capacity, but also because of their good biocompatibility, biodistribution and biodegradability.[23] In order to develop smart nanocarriers that respond to an external magnetic field, the MNPs surface must be decorated with one or more moieties that enhances the stability of the nanodevices and therefore raises the half-live despite the decrease from its capture by the reticuloendothelial system (RES) and macrophages in living tissues.[24]

6.3 Applications of Magnetically-sensitive Nanoparticles as Drug Delivery Systems

As it has been mentioned in the introduction, the use of nanometric carriers for the selective and controlled transport of therapeutic agents has received huge attention in the recent decades. In the case of magnetically-sensitive nanodevices, their application in oncology has concentrated most of the efforts due to the possibility of combining magnetically-triggered release with the generation of a local hyperthermia in the target tissue, which provides a great advantage in attempts to eradicate the tumoural cells. The use of nanoparticles as drug carriers in oncology was proposed for the first time by Maeda and Matsumura in 1986.[25] These Japanese researchers reported that the intravenous administration of macromolecules with a molecular weight higher than 40 KDa have a tendency to be passively accumulated in tumoural tissues. The reason for this phenomenon lies in the particular characteristics of tumoural blood vessels which are really irregular and poorly built, presenting fenestrations of a few hundreds of nanometres as a consequence of their rapid growing rate. Therefore, when a nanometric entity is injected into the blood stream and reaches the tumoural tissue, it is able to escape from the blood vessel passing through these pores resulting in accumulation within the malignant tissue. Additionally, the rapid expansion rate of the tumoural cells in the surrounding tissue compromises the normal function of the lymphatic system in the zone due to the compression of the lymphatic vessels. Therefore, the elimination of these extravasated particles is slower than in normal tissues because there is not an efficient drainage system in the diseased tissue. These two characteristics, higher permeation and enhanced retention of nanometric objects, explain the passive accumulation of nanoparticles in tumoural tissues and is termed the enhanced permeation and retention (EPR) effect. This unique characteristic has been extensively exploited by researchers in order to design novel nanocarriers able to deliver cytotoxic drugs specifically to the tumoural tissues reducing the effect on the rest of the healthy organs.[26] In the case of magnetic nanoparticles, this accumulation can be improved using the magnetic properties of the nanocarrier by placing a potent permanent magnet in the target tissue.[27]

One of the most widely employed strategies for the development of magnetic-sensitive nanocarriers consists of the use of thermosensitive polymeric coatings around magnetic cores, which are usually SPION. Thus, the organic shell acts as a sponge able to retain the therapeutic species within its branches due to intramolecular interactions and, when AMF is applied, the generated heat induces the collapse of the polymer layer pushing out the trapped drugs. Poly-(*N*-isopropylamide) (PNIPAM) is the most representative thermosensitive polymer and one of the most widely employed by far. This polymer suffers linear-to-globular transition when the temperature exceeds 32 °C and this transition has been employed for drug delivery applications in a vast number of nanocarriers.[28] The transition temperature of the polymer layer can be increased above the physiological condition adding more hydrophilic monomers to the polymer composition.[29] For instance, Guisasola *et al.* have reported the use of mesoporous silica nanoparticles which contains SPION trapped within the silica matrix and coated with a shell composed of a thermosensitive polymer formed by *N*-isopropylacrylamide and *N*-hydroxymethyl acrylamide (NHMA) (90 : 10).[30] Poly(NIPAM/NHMA) with this particular composition presents a transition temperature of 42–43 °C, which fits in the hyperthermia range. This system is capable of retaining large amounts of drugs inside the highly porous mesoporous silica matrix and releasing them once AMF is applied, even when the external temperature is maintained below the transition temperature of the polymer. The reason is that the SPION trapped inside acts as hot spots under AMF exposition, as it will be discussed below, and the temperature in the area surrounding the particle is much higher than the global temperature, being enough to provoke the transition of the polymer shell (Figure 6.3).

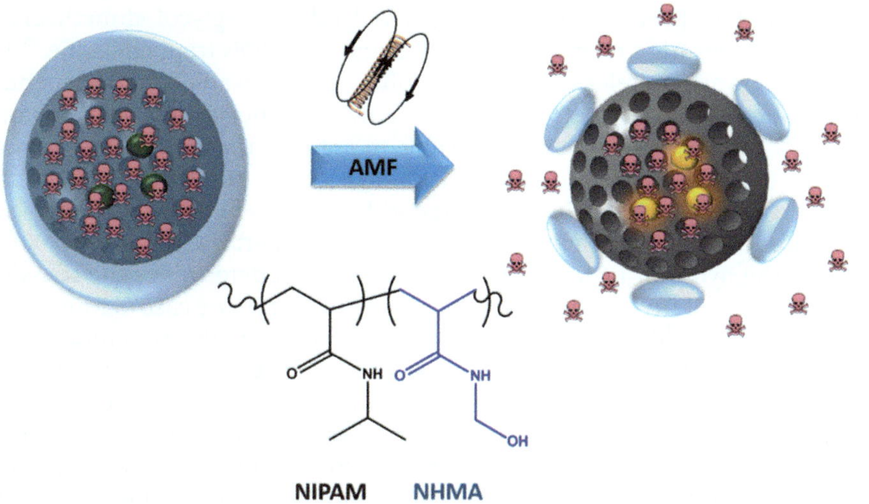

Figure 6.3 Magnetic-sensitive nanodevice based on magnetic MSN coated with PNIPAM/NHMA as a thermosensitive polymer.

Poloxamers are another thermosensitive polymer family composed of poly(ethylene oxide)-*b*-poly(propylene oxide)-*b*-poly(ethylene oxide) (PEO–PPO–PEO) triblock sequences which have received high attention in the recent years. In this case, the transition temperature of the polymer can be controlled by increasing the amount of the hydrophilic block (PEO) and can be tuned from 19 °C up to 100 °C. Chen *et al.* have reported the use of this type of polymer in the formation of thermosensitive micelles with iron oxide nanocrystals trapped inside in order to release vitamin B12 only when AMF is applied.[31]

Biopolymers have been employed as coatings of magnetic particles in order to develop more biocompatible magnetically-sensitive nanodevices. Alginate is a natural polymer which can be easily and inexpensively extracted from brown algae. Le Visage *et al.* have reported the synthesis of alginate-based microbeads which contains SPION of 8 nm diameter encapsulated inside.[32] These microbeads were loaded with doxorubicin (Dox), a potent antitumoral drug, and were able to release their payload under AMF exposition. More recent, Sotiriou *et al.* have described a similar system composed of silica-coated SPION embedded within an alginate sub-micron bead which was capable of increasing the temperature in the media up to 42 °C and releasing a biomolecule model (dextran) once the system was exposed to the magnetic field.[33] Alginate has also been employed in the formulation of dextran-coated iron oxide nanoparticles functionalized with carbonyl–ruthenium complexes providing materials able to release controlled amounts of CO by magnetic heating.[34] Chitosan, another polysaccharide widely employed in nanomedicine due to its excellent biocompatibility, has also been employed in the formation of magnetically-responsive systems. Thus, in one recent example, chitosan has been cross-linked with polyethylene glycol dimethacrylate (PEGDMA) forming microbeads which contains SPION trapped inside.[35] This system was able to release antibiotics following a tuneable kinetic depending on the cross-linker percentage. Additionally, the exposition to AMF enhances the release rate providing a system able to regulate the administered dosage at will. There are certain polypeptides which exhibit temperature-responsive properties such as elastin-like polipeptides.[36] These biopolymers are characterized by the repetition of this pentapeptide pattern (Val-Pro-Gly-X-Gly), where the hydrophilic nature of the amino acid X determines the transition temperature of the resulting polymer, being higher in the case of more hydrophilic amino acids. Another natural biopolymer sensitive to temperature is DNA. DNA in the natural form is a double helix in which each strand is composed by the repetition of four different nucleotides. These nucleotides present in common a sugar (deoxyribose) and one phosphate group and differ from each other in the nitrogenous base, which can be adenine (A), thymine (T), guanine (G) and cytosine (C). The double-helix are formed by the formation of Watson–Crick bonds between the complementary bases A-T and G-C. This supramolecular structure is sensitive to the temperature, because the double-helix can be disassembled when the temperature exceeds a certain temperature, called the melting temperature (Tm), as a consequence of the rupture of the Watson–Crick bonds. It is possible to engineered specific DNA double strands with a desired Tm simply by varying the GC content in the strand and therefore, it is relatively easy to obtain sequences with the specific Tm required for each use. Baglioni *et al.* have reported the use of single DNA strands anchored on core–shell magnetic nanoparticles formed by a Cobalt/Ferrite core coated with a gold layer for the controlled delivery of the complementary DNA single strands in response to AMF (Figure 6.4).[37]

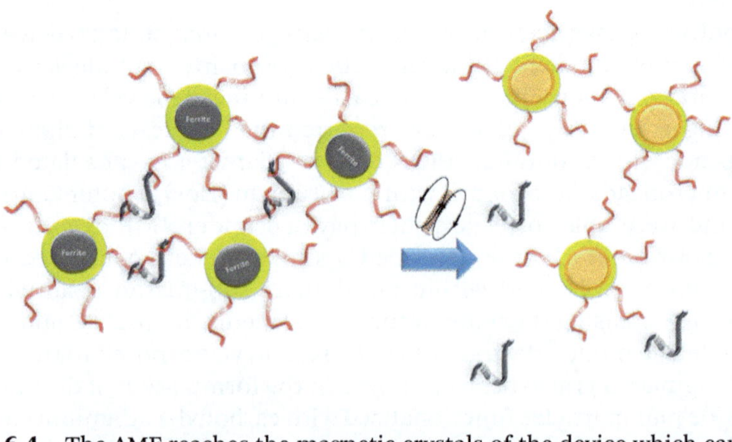

Figure 6.4 The AMF reaches the magnetic crystals of the device which causes the elevation of the temperature at the cellular level and activates the apoptotic cascade in diseased cells, as well as triggering the drug departure, and enhancing the toxicity through the synergic effect of the treatment combination.

In this example, one DNA strand act as a linker between the magnetic particles because part of its sequence is complementary with the DNA strand attached to the particle. Thus, a clustered structure is formed until the AMF application breaks the DNA hybridization and disassembles the cluster releasing the DNA single strand.

In another example, DNA single strands with 15 base pairs with a Tm of 47 °C have been anchored on the surface of mesoporous silica nanoparticles and the complementary DNA strand was attached on the surface of SPION.[38] The hybridization between both strands produced a closure in the pore channels keeping the drug trapped inside. When an AMF was applied, the generated heat induces the dehybridization between the DNA strands opening the system and releasing the drug. The pores of mesoporous silica nanoparticles can also be closed using organic–inorganic moieties on the pore outlets which are attached through temperature responsive linkers. Cucurbit[6]uril have been employed as reversible pore-blockers in mesoporous silica nanoparticles which contains zinc-doped iron oxide nanocrystals inside and *N*-(6-*N*-Aminohexyl)aminomethyl groups placed on the particle surface.[39] This moiety is electrostatically bound to the functional groups present on the surface and it closes the pores avoiding premature drug release. The exposition to AMF increases the temperature in the area surrounding the particle producing pore uncapping and drug release.

Another possibility for the synthesis of magnetically sensitive nanocarriers consists of the encapsulation of magnetic cores within liposomes. Liposomes were the first nanometric systems used in clinical medicine for the transportation of cytotoxic drugs for antitumoral applications, specifically DOXIL® which was approved by the Food and Drug Administration (FDA) in 1995, which is a liposome-based formulation of doxorubicin. Liposomes are vesicles formed by a lipids bilayer composed of different phospholipids and they are capable of transporting cytotoxic drugs dissolved in the central hydrophilic core and even hydrophobic drugs, which can be transported inside the lipophilic part of the bilayer. Inside these liposomes different magnetic nanoparticles such as iron oxide[40] or mangansese-ferrites[41] have been encapsulated in order to provide magnetic-responsiveness to these formulations because the temperature increase achieved under AMF destabilizes the lipid bilayer, releasing the drugs trapped inside. These magnetic nanocrystals can be located both in suspension in the aqueous compartment when the magnetic particles exhibit a hydrophilic surface,[42] as well as inserted within the lipid bilayer, as is the case of systems with a hydrophobic nature.[43] Plank *et al.* have reported the use of liposomes functionalized with folate on the external surface and loaded with SPION and Dox for the eradication of HeLa tumoural cells.[44] The presence of folic acid on the liposome surface enhances the internalization within the diseased cells, which usually overexpress folate receptors due to the high demand of this biomolecule by the tumoural cells. Additionally, this liposome uptake is improved by the application of a permanent magnetic field which pushes the liposomes towards the cell surface. Finally, AMF application triggers the drug release producing significantly higher cell destruction.

When a nanocarrier is injected within the blood stream, it is immediately covered by the proteins present in the blood stream. Opsonins are a family of proteins in charge of the labelling of pathogen bodies which circulate in the blood in order to facilitate their capture by macrophages. Thus, if a long circulation time of the nanocarrier is presented, it is necessary to avoid opsonin adsorption. One of the most widely employed approaches consists of coating the carrier with polyethylene glycol (PEG) chains. The hydrophilic nature of this polymer creates a hydration barrier around the particle which hampers protein adsorption. Thus, Bothun *et al.* have synthesized PEG-stabilized magnetoliposomes for delivering Dox to hepatocellular carcinoma cells obtaining around 90% cell destruction under AMF exposition in only 24 hours.[45] Other alternatives such as carboxymethyl dextran[46] have been proposed to PEG coatings. DNA and RNA strands have been transported to the inner cellular space using magnetic liposomes and liposome-like structures due to the higher transfection capacity obtained when a permanent magnetic field is present. As an example, SPION have been coated with cationic lipids in order to adsorb oligonucleotides by electrostatic interactions creating nanometric structures which protect these sensitive agents against degradation until the system reaches the target cell.[47] Wilhelm *et al.* have reported very interesting systems able to exploit other external stimuli besides the magnetic field for increasing the therapeutic efficacy of liposomes.[48] In this system, SPION were loaded in the aqueous core and a photosensitizer, *m*-Tetrahydroxyphenylchlorin (*m*-THPC) which presents high lipophilicity, was trapped within the lipid bilayer. The administration of this multifunctional liposome allows the synergism between two effects. On the one hand, the induction of magnetic hyperthermia in the tumoural zone that enhances the blood flow and oxygenation in the tissue, which produces a significant enhancement in the production of reactive oxygen species (ROS) when the photosensitizers are exposed to near infrared (NIR) radiation. On other hand, ROS generation compromises the cellular response against the temperature increase, improving the hyperthermia effect. The result of the application of both stimuli led to the complete eradication of the tumour in mice models, whereas the use of only one stimulus only inhibited the tumour growth. In another article published by the same research group, SPION and photosensitizers were loaded in a similar way, but in this case within cell vesicles, which are vesicles naturally released from cells.[49] These systems present superior properties to liposomes such as higher immunotolerance, targeting properties due to the presence of cell receptors on their surface and improved cell entry properties, being able to reach the tumoural area and once there to be captured in a selective manner by tumoural cells causing their destruction.

Finally, another alternative consists of the direct grafting of therapeutic compounds on the surface of magnetic nanoparticles employing temperature-responsive linkers. Cycloadditions are a reaction family characterized by their thermal reversibility. Thus, different drugs can be anchored through copper-catalysed azide–alkyne cycloaddition and can be released as a consequence of the high temperature present in the area surrounding the particle when AMF is applied.[50] Despite the fact that the cycloadduct reversion usually requires high temperatures, which can be up to 80 °C in some cases, the temperature reached in the intimate particle proximity under high frequency AMF can be high enough to achieve the rupture of these adducts. Zink *et al.* have synthesized mesoporous silica nanoparticle which contain one SPION and one $NaYF_4:Yb^{3+}$, Er^{3+} nanocrystal per particle.[51] This last inorganic crystal presents two green emission bands by upconversion, and the ratio is dependent on the temperature. Therefore, it is possible to measure the temperature increment close to this particle under AMF by monitoring the variation in this ratio. This experiment has shown temperature increments of up to 20 °C depending on the intensity and frequency of the applied AMF and the exposition time. Taking into account these findings, new approaches to enhance the cytotoxic effect by the combination of magnetic hyperthermia with drug release have been explored.

6.4 Hot Spot Effect Adds to Chemotherapy

Since the 1980s researchers have been evaluating the effect of temperature elevation plus chemotherapeutic treatment in malignant tumours. Despite that, the sensitivity between healthy and cancerous cells towards a temperature rise hardly presents a difference *in vitro*, effective treatments have been developed *in vivo* through mild hyperthermia at temperatures between 40 °C and 43 °C. The higher sensitivity of cancer tissues comes from the disordered tumour vasculature that leads to hypoxic and low pH areas, unlike normal tissues.[52] Several studies have demonstrated that many chemotherapeutic agents enhance its cytotoxicity when the temperature is raised to hyperthermia levels in different cancer tissues.[53,54] The macroscopic temperature elevation causes an increase in the permeation through the vascular cells and blood flow that facilitates the drug diffusion to the tumour mass. Besides this, the heat shock produced by hyperthermia triggers numerous mechanisms for cytotoxicity maximization as increased chemotherapeutic uptake, higher alkylation constants and inhibition of the cell auto-repairing pathways, provoking a lethal or sub-lethal damage inside the cancer cells.[55]

Classical hyperthermia treatments are based on simple technologies such as hot wax baths and saunas, but after the irruption of MNPs in the medical field a novel methodology was developed, called magnetic fluid hyperthermia (MFH).[56] The first works on MFH were conducted by Chan *et al.* and Jordan *et al.* in 1993, achieving a temperature rise and cancer cell death at tolerable nanoparticle concentrations.[57,58] Later, tumour shrinkage and tissue necrosis was addressed by injecting high doses of superparamagnetic nanoparticle suspensions. The limitation of this treatment is that high concentrations of nanocrystals in the cancer tissues are required to give enough heat transfer and therefore the temperature increase needed, especially *in vivo*. This is normally achieved by the direct injection of the nanoparticles into the tumoural masses.[59,60] Hayashi *et al.* have described an AMF-responsive system coating for SPIONs by the polymerization of pyrrole-3-carboxylic acid in the presence of Dox. This polymer acts as a container of the drug and has a glass-transition temperature (T_g) of 44 °C. At the T_g the polymer phase become softer allowing the release of the chemotherapeutic. When the nanocarrier was exposed to an AMF (H = 8 kA m^{-1}, f = 230 kHz, H_f = 1.8 × 109 A m^{-1} s^{-1}) the Dox was expelled out from the device by diffusion. In this way the researchers injected the system intratumorally in mice bearing subcutaneous xenograft tumours derived from the RPMI8226 multiple myeloma cell line followed by the AMF application above 42.5 °C. Thus, they observed that the mice treated only with chemotherapy or MHT had inhibition of the cancer growth, but they could not remove the tumour completely. Nevertheless, the combination of both treatments achieve a tumour volume reduction to zero within 12 days.[61] Zhang *et al.* developed a device formed by coating SPIONs with a poly(organophosphazene) (PPZ) in the presence of a tumour necrosis factor related apoptosis-inducing ligand (TRAIL) which has a specific binding to the death receptors DR4 and DR5 over-expressed in most cancer cells. The PPZ hydrogel has a temperature-responsive phase transition and gelifies at body temperature, but dissolves in the hyperthermia range, releasing the drugs trapped inside its matrix. *In vivo* experiments exhibited a high tumour killing effect by submitting the mice treated with the nanocarrier to two magnetic hyperthermia cycles, keeping the principal organs undamaged.[62]

Recently, researchers have focused on a different approximation that consists of nanoscale heating. The nanoparticle accumulation in diseased tissues can trigger several intracellular responses under AMF application without the need for increasing the macroscopic temperature conditions.[63–65] The mechanisms of cell death induction are still under investigation, but several attempts shed light on this matter. As an example, Huang *et al.* synthesized $MnFe_2O_4$ nanoparticles decorated with streptavidin (Strep) and exposed them to a genetically modified cell line that expresses a thermo-sensitive TRPV1 protein (HEK 293) and a membrane protein that conjugates the Strep. Before the attachment of the nanoparticles to the cell membrane the AMF was applied. The results show that the temperature rise at the beginning of the treatment is restricted to the nanoparticle surroundings and to the plasma membrane a few seconds later, but the system accomplished the activation of TRPV1 channels inducing a response in neuronal cells without cellular damage. Moreover, they were able to remotely induce a behavioural response in *C. elegans* worms loaded with nanoparticles, changing their locomotion by heat induction.[66] To obtain high nanoparticle accumulation internalization is a requisite for this sort of treatment, which can be done by the functionalization of the nanoparticles with targeting ligands such as antibodies, carbohydrates, small molecules or peptides. This way, the temperature rise at the nanoscale can induce toxicity in the treated cells without an observable macroscopic heating.[67]

The modification of the cytotoxicity of chemotherapeutics in combination with heating treatments is called 'thermal chemosensitisation' (Figure 6.5) and can be quantified by means of cancer cell growth comparison

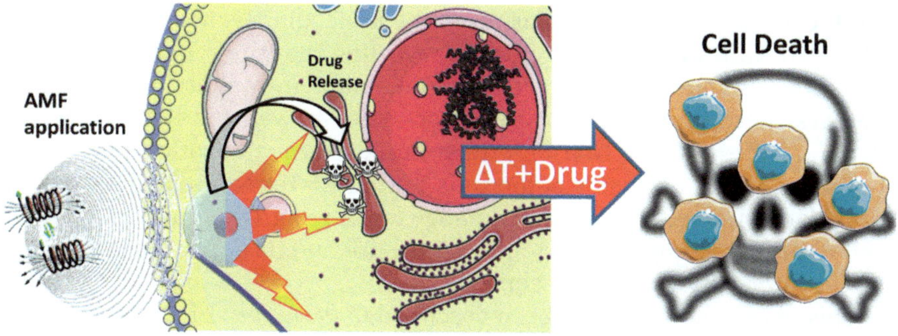

Figure 6.5 Once the MNPs are internalized by cancer cells, an AMF is applied, provoking the temperature rise and chemotherapeutics release obtaining a synergic effect that kills the tumoural tissues without a macroscopic temperature elevation.

between the treatment of the diseased cells with the drug alone and the same drug at hyperthermia temperatures. The combined therapies show an increasing rate of drug toxicity with the temperature rise. In general, cyclophosphamide, ifosfamide, platinum compounds and nitrosoureas have a linear cytotoxic effect with temperatures from 37–40 °C, while compounds such as doxorubicin or bleomycin have threshold temperatures around 42.5 °C in their interactions with temperature. Generally, the chemotherapeutics administered immediately before hyperthermia application show the best effectiveness, but the hyperthermia conditions induce complex changes in drug pharmacokinetics and should be taking into account.[52] An example of the synergic effect is that a variety of chemotherapeutic drugs are DNA-intercalating agents which kill cells during the S-phase replication, as well as hyperthermia treatment, resulting in a lower risk of secondary cancers induced by the cancer treatment.[68–70] Another hypothesis suggests that hyperthermia combined chemotherapy could help the immune system by developing an adaptive and innate immunity.[71]

Some theoretical work suggested that intracellular hyperthermia might not be possible while other authors maintain the opposite stance. Therefore, Creixell *et al.* decided to study this controversy by conjugation of epithelial grow factor (EGF) to SPIONs in order to target the epithelial grow factor receptor (EGFR) overexpressed in adenocarcinoma cell lines. The targeting moiety accomplished better accumulation of MNPs within the tumour site than non-targeted nanoparticles before the AMF application. The results showed that a 99.9% reduction in cell viability was achieved and that the clonogenic survival depended on the thermal heat dose, without any appreciable macroscopic temperature rise.[72] A striking work was performed by Connord *et al.* that permitted the analysis of hyperthermia treatment in real-time on living cells. The authors built a miniaturized electromagnet with similar parameters to the standard coils used in magnetic hyperthermia and fitted it under a confocal microscope. With this setup the molecular events occurring during AMF exposition could be monitored revealing that the cell damage rate was the same as that achieved with the ones used in magnetic hyperthermia. Besides these findings, the researchers also showed the ROS formation, a dynamic study of the lysosome membrane permeabilization and even the lysosome lining up inside some cells during AMF treatment. This data also provided evidence that the cell damage event was taking place at the single-cell level with no significant temperature rise and its dependency was on the MNPs internalization in lysosomes.[73]

This phenomena has also been described in *in vivo* experiments. Moros *et al.* confirmed this theory *in vitro* and *in vivo* by measuring the heat shock protein expression (HSP70) during AMF application. They found that the over-expressed HSP70 levels in the experiments carried out with hyperthermia without a macroscopic temperature change were comparable to the controls by external heating.[63] Other interesting responses were reported by Calatayud *et al.* showing that the heat shock can even affect the MNPs-free cells if they are in contact with MNPs-loaded cells, by a sort of bystander effect.[74] The intracellular hyperthermia would negate the issue of high concentrations of nanoparticles to achieve the cellular response because it could circumvent the potential thermal barrier that represents the cell membrane.[75] The temperature increment by itself is capable of inducing a potent cytotoxic response in many tumoural cell lines, as has been reported by different groups.[72]

6.5 Conclusion

Great advances have been made in the development of MR-DDSs in recent years. The non-invasive nature of the magnetic fields, the capability of these systems to apply two different treatments at once and their ability to reach the tumour tissues through the EPR effect and active vectorization makes them a promising tool to treat cancer. Major challenges in hyperthermia cancer therapy by MR-DDSs includes higher tumour accumulation in systemic delivery, the control of temperature elevation and the enhancement of energy dissipation rates. Nevertheless, nanoscale hyperthermia could help to avoid certain limitations of traditional hyperthermia such as the possibly unachievable doses of MR-DDSs in the tumour tissue. The cell death induced by the temperature rise inside the cells should be studied in depth to untangle the mechanisms of the intracellular responses and know the best chemotherapeutic agents to use in combination with the magnetic heating. Also, a huge effort should be made for the sake of designing better treatments, and achieving reproducible chemotherapeutic and heat doses. Despite the technology being well developed, tumour growth prevention by intracellular heating without a temperature rise of the treated tissues has not been demonstrated yet. Another unexplored potential of MR-DDSs could be to develop treatments that can use the heating effect and drugs to positively induce a behavioural change in the cells.

Acknowledgements

This work was supported by the European Research Council (Advanced Grant VERDI; ERC-2015-AdG Proposal No. 694160) and the project MAT2015-64831-R.

References

1. M. W. Tibbitt, J. E. Dahlman and R. Langer, *J. Am. Chem. Soc.*, 2016, **138**, 704–717.
2. S. Mura, J. Nicolas and P. Couvreur, *Nat. Mater.*, 2013, **12**, 991–1003.
3. H. Xu, Z. P. Aguilar, L. Yang, M. Kuang, H. Duan, Y. Xiong, H. Wei and A. Wang, *Biomaterials*, 2011, **32**, 9758–9765.
4. A. Baeza, M. Colilla and M. Vallet-Regí, *Expert Opin. Drug Delivery*, 2015, **12**, 319–337.
5. J. Kost, R. Noecker, E. Kunica and R. Langer, *J. Biomed. Mater. Res.*, 1985, **19**, 935–940.
6. Y. Namiki, T. Namiki, H. Yoshida, Y. Ishii, A. Tsubota, S. Koido, K. Nariai, M. Mitsunaga, S. Yanagisawa, H. Kashiwagi, Y. Mabashi, Y. Yumoto, S. Hoshina, K. Fujise and N. Tada, *Nat. Nanotechnol.*, 2009, **4**, 598–606.
7. M. Chorny, I. Fishbein, J. E. Tengood, R. F. Adamo, I. S. Alferiev and R. J. Levy, *FASEB J.*, 2013, **27**, 2198–2206.
8. M. Torres-Lugo and C. Rinaldi, *Nanomedicine*, 2013, **8**, 1689–1707.
9. L. K. Bogart, G. Pourroy, C. J. Murphy, V. Puntes, T. Pellegrino, D. Rosenblum, D. Peer and R. Lévy, *ACS Nano*, 2014, **8**, 3107–3122.
10. L. Mohammed, H. G. Gomaa, D. Ragab and J. Zhu, *Particuology*, 2017, **30**, 1–14.
11. R. Arbain, M. Othman and S. Palaniandy, *Miner. Eng.*, 2011, **24**, 1–9.
12. J. G. King, W. Williams, C. D. W. Wilkinson, S. McVitie and J. N. Chapman, *Geophys. Res. Lett.*, 1996, **23**, 2847–2850.
13. F. A. Périgo, G. Hemery, O. Sandre, D. Ortega, E. Garaio, F. Plazaola and F. J. Teran, *Appl. Phys. Rev.*, 2015, **2**, 41302.
14. A.-H. Lu, E. L. Salabas and F. Schüth, *Angew. Chem., Int. Ed.*, 2007, **46**, 1222–1244.
15. M. Hammad, V. Nica and R. Hempelmann, *IEEE Trans. Magn.*, 2017, **53**, 1–6.
16. J. Baumgartner, L. Bertinetti, M. Widdrat, A. M. Hirt and D. Faivre, *PLoS One*, 2013, **8**, 1–6.
17. Y. A. Koksharov, *Magnetic Nanoparticles*, 2009, pp. 197–254.
18. S. Dutz and R. Hergt, *Int. J. Hyperthermia*, 2013, **29**, 790–800.
19. S. Laurent, S. Dutz, U. O. Häfeli and M. Mahmoudi, *Adv. Colloid Interface Sci.*, 2011, **166**, 8–23.
20. Suriyanto, E. Y. K. Ng and S. D. Kumar, *Biomed. Eng. Online*, 2017, **16**, 36.
21. A. Kolhatkar, A. Jamison, D. Litvinov, R. Willson and T. Lee, *Int. J. Mol. Sci.*, 2013, **14**, 15977–16009.

22. R. R. Wildeboer, P. Southern and Q. A. Pankhurst, *J. Phys. D: Appl. Phys.*, 2014, **47**, 495003.
23. M. Arruebo, R. Fernández-Pacheco, M. R. Ibarra and J. Santamaría, *Nano Today*, 2007, **2**, 22–32.
24. L. H. Reddy, J. L. Arias, J. Nicolas and P. Couvreur, *Chem. Rev.*, 2012, **112**, 5818–5878.
25. Y. Matsumura and H. Maeda, *Cancer Res.*, 1986, **46**, 6387–6392.
26. A. Wicki, D. Witzigmann, V. Balasubramanian and J. Huwyler, *J. Controlled Release*, 2015, **200**, 138–157.
27. M. Wahajuddin and S. Arora, *Int. J. Nanomed.*, 2012, **7**, 3445–3471.
28. M. Pernia Leal, A. Torti, A. Riedinger, R. La Fleur, D. Petti, R. Cingolani, R. Bertacco and T. Pellegrino, *ACS Nano*, 2012, **6**, 10535–10545.
29. H. Wei, S.-X. Cheng, X.-Z. Zhang and R.-X. Zhuo, *Prog. Polym. Sci.*, 2009, **34**, 893–910.
30. E. Guisasola, A. Baeza, M. Talelli, D. Arcos, M. Moros, J. M. de la Fuente and M. Vallet-Regí, *Langmuir*, 2015, **31**, 12777–12782.
31. T. Y. Liu, K. H. Liu, D. M. Liu, S. Y. Chen and I. W. Chen, *Adv. Funct. Mater.*, 2009, **19**, 616–623.
32. S. Brulé, M. Levy, C. Wilhelm, D. Letourneur, F. Gazeau, C. Ménager and C. Le Visage, *Adv. Mater.*, 2011, **23**, 787–790.
33. A. Teleki, F. L. Haufe, A. M. Hirt, S. E. Pratsinis and G. A. Sotiriou, *RSC Adv.*, 2016, **6**, 21503–21510.
34. H. Meyer, F. Winkler, P. Kunz, A. M. Schmidt, A. Hamacher, M. U. Kassack and C. Janiak, *Inorg. Chem.*, 2015, **54**, 11236–11246.
35. M. Harris, H. Ahmed, B. Barr, D. LeVine, L. Pace, A. Mohapatra, B. Morshed, J. D. Bumgardner and J. A. Jennings, *Int. J. Biol. Macromol.*, 2017, **104**, 1407–1414.
36. W. Hassouneh, S. R. MacEwan and A. Chilkoti, *Fusions of Elastin-like Polypeptides to Pharmaceutical Proteins*, Elsevier Inc., 1st edn, 2012, vol. 502.
37. M. Banchelli, S. Nappini, C. Montis, M. Bonini, P. Canton, D. Berti and P. Baglioni, *Phys. Chem. Chem. Phys.*, 2014, **16**, 10023.
38. E. Ruiz-Hernández, A. Baeza and M. Vallet-Regí, *ACS Nano*, 2011, **5**, 1259–1266.
39. C. R. Thomas, D. P. Ferris, J.-H. Lee, E. Choi, M. H. Cho, E. S. Kim, J. F. Stoddart, J. Shin, J. Cheon and J. I. Zink, *J. Am. Chem. Soc.*, 2010, **132**, 10623–10625.
40. M. Gonzales and K. M. Krishnan, *J. Magn. Magn. Mater.*, 2005, **293**, 265–270.
41. P. Pradhan, J. Giri, R. Banerjee, J. Bellare and D. Bahadur, *J. Magn. Magn. Mater.*, 2007, **311**, 208–215.
42. M. R. Faria, M. M. Cruz, M. C. Gonçalves, A. Carvalho, G. Feio and M. B. F. Martins, *Int. J. Pharm.*, 2013, **446**, 183–190.
43. E. Amstad, J. Kohlbrecher, E. Müller and T. Schweizer, *Nano Lett.*, 2011, 1664–1670.
44. P. Pradhan, J. Giri, F. Rieken, C. Koch, O. Mykhaylyk, M. Döblinger, R. Banerjee, D. Bahadur and C. Plank, *J. Controlled Release*, 2010, **142**, 108–121.

45. Y. Chen, Y. Chen, D. Xiao, A. Bose, R. Deng and G. D. Bothun, *Colloids Surf., B*, 2014, **116**, 452–458.
46. H. Guo, W. Chen, X. Sun, Y. N. Liu, J. Li and J. Wang, *Carbohydr. Polym.*, 2015, **118**, 209–217.
47. S. Jiang, A. Eltoukhy, K. Love, R. Langer and D. Anderson, *Nano Lett.*, 2013, 1–6.
48. R. Di Corato, G. Béalle, J. Kolosnjaj-Tabi, A. Espinosa, O. Clément, A. K. A. Silva, C. Ménager and C. Wilhelm, *ACS Nano*, 2015, **9**, 2904–2916.
49. A. K. A. Silva, J. Kolosnjaj-Tabi, S. Bonneau, I. Marangon, N. Boggetto, K. Aubertin, O. Clément, M. F. Bureau, N. Luciani, F. Gazeau and C. Wilhelm, *ACS Nano*, 2013, **7**, 4954–4966.
50. T. T. T. N'Guyen, H. T. T. Duong, J. Basuki, V. Montembault, S. Pascual, C. Guibert, J. Fresnais, C. Boyer, M. R. Whittaker, T. P. Davis and L. Fontaine, *Angew. Chem., Int. Ed.*, 2013, **52**, 14152–14156.
51. J. Dong and J. I. Zink, *ACS Nano*, 2014, **8**, 5199–5207.
52. R. D. Issels, *Eur. J. Cancer*, 2008, **44**, 2546–2554.
53. M. Urano, *Int. J. Hyperthermia*, 1999, **15**, 79–107.
54. M. Takemoto, M. Kuroda, M. Urano, Y. Nishimura, S. Kawasaki, H. Kato, Y. Okumura, S. Akaki, S. Kanazawa, J. Asaumi, I. Joja and Y. Hiraki, *Int. J. Hyperthermia*, 2003, **19**, 193–203.
55. R. K. Jain, *Science*, 2005, **307**, 58–62.
56. N. M. Bleehen, *Br. J. Cancer*, 1982, **45**, 96.
57. D. C. F. Chan, D. B. Kirpotin and P. A. Bunn, *J. Magn. Magn. Mater.*, 1993, **122**, 374–378.
58. A. Jordan, P. Wust, H. Fählin, W. John, A. Hinz and R. Felix, *Int. J. Hyperthermia*, 1993, **9**, 51–68.
59. M. Johannsen, A. Jordan, R. Scholz, M. Koch, M. Lein, S. Deger, J. Roigas, K. Jung and S. Loening, *J. Endourol.*, 2004, **18**, 495–500.
60. A. Jordan, R. Scholz, K. Maier-Hauff, F. K. H. van Landeghem, N. Waldoefner, U. Teichgraeber, J. Pinkernelle, H. Bruhn, F. Neumann, B. Thiesen, A. von Deimling and R. Felix, *J. Neuro-Oncol.*, 2006, **78**, 7–14.
61. K. Hayashi, M. Nakamura, H. Miki, S. Ozaki, M. Abe, T. Matsumoto, W. Sakamoto, T. Yogo and K. Ishimura, *Theranostics*, 2014, **4**, 834–844.
62. Z. Zhang and S. Song, *Biomaterials*, 2017, **132**, 16–27.
63. M. Moros, A. Ambrosone, G. Stepien, F. Fabozzi, V. Marchesano, A. Castaldi, A. Tino, J. M. de la Fuente and C. Tortiglione, *Nanomedicine*, 2015, **10**, 2167–2183.
64. B. Kozissnik, A. C. Bohorquez, J. Dobson and C. Rinaldi, *Int. J. Hyperthermia*, 2013, **29**, 706–714.
65. A. Chiu-Lam and C. Rinaldi, *Adv. Funct. Mater.*, 2016, **26**, 3933–3941.
66. H. Huang, S. Delikanli, H. Zeng, D. M. Ferkey and A. Pralle, *Nat. Nanotechnol.*, 2010, **5**, 602–606.
67. L. Asín, G. F. Goya, A. Tres and M. R. Ibarra, *Cell Death Dis.*, 2013, **4**, e596.
68. J. O. E. W. Grisham, D. S. Greenberg, D. G. Kaufman and G. J. Smith, *Proc. Natl. Acad. Sci. USA*, 1980, **77**, 4813–4817.
69. P. J. McCormick and J. S. Bertram, *Proc. Natl. Acad. Sci. U. S. A.*, 1982, **79**, 4342–4346.

70. C. R. Hunt, R. K. Pandita, A. Laszlo, R. Higashikubo, M. Agarwal, T. Kitamura, A. Gupta, N. Rief, N. Horikoshi, R. Baskaran, J.-H. Lee, M. Lobrich, T. T. Paull, J. L. Roti Roti and T. K. Pandita, *Cancer Res.*, 2007, **67**, 3010–3017.
71. V. Milani, E. Noessner, S. Ghose, M. Kuppner, B. Ahrens, A. Scharner, R. Gastpar and R. D. Issels, *Int. J. Hyperthermia*, 2002, **18**, 563–575.
72. M. Creixell, A. C. Bohórquez, M. Torres-Lugo and C. Rinaldi, *ACS Nano*, 2011, **5**, 7124–7129.
73. V. Connord, P. Clerc, N. Hallali, D. El Hajj Diab, D. Fourmy, V. Gigoux and J. Carrey, *Small*, 2015, **11**, 2437–2445.
74. M. R. Calatayud, M. P. Asin, L. Tres, A. Goya and G. F. Ibarra, *Curr. Nanosci.*, 2016, **12**, 372–377.
75. R. T. Gordon, J. R. Hines and D. Gordon, *Med. Hypotheses*, 1979, **5**, 83–102.

CHAPTER 7

Light-responsive Drug Delivery Systems

X. TAN, E. L. BURCHFIELD AND K. ZHANG*

Northeastern University, Department of Chemistry and Chemical Biology, 360 Huntington Ave., Boston, MA 02115, USA
*E-mail: k.zhang@northeastern.edu

7.1 Introduction

In recent years, medical applications have promoted the development of various kinds of nanomaterials.[1,2] Efforts have been spent on the design and engineering of nanoparticles (NPs) capable of serving as efficient diagnostic or imaging agents.[3,4] In particular, many NPs are developed as drug delivery systems (DDSs) for cancer treatment because of the enhanced permeability and retention (EPR) effect, which amplifies drug concentration near the leaky vasculature of tumor tissues, thereby reducing systemic side effects.[5,6] Using NPs loaded with drugs for passive, diffusion-driven release, however, may not be ideal. Certain situations require the therapeutic agents to be released from the DDS promptly upon local injection or cellular internalization.[7] Such a DDS should thus release the drug when a specific stimulus in the environment is present or an external one is supplied, in order to minimize premature release. Towards this idea, different stimuli, such as temperature, pH, redox agents, enzymes, light, magnetic field and ultrasound, *etc.*, have been studied extensively in the last few decades.[8–11]

Biomaterials Science Series No. 1
Stimuli-responsive Drug Delivery Systems
Edited by Amit Singh and Mansoor M. Amiji

Published by the Royal Society of Chemistry, www.rsc.org

Being a relatively noninvasive, externally applied stimulus, light is an outstanding trigger.[12] Light is not present *in vivo*, which means that an ideal light-induced drug release system can offload the drug selectively in irradiated areas with high contrast to non-irradiated areas, and drug dosage can be adjusted *via* light dosage. While *ex vivo* incident light, especially ultraviolet-visible (UV-vis) irradiation, has limited tissue penetration depth and may pose photo-damage to normal tissues,[12] it is still feasible to target deep tissues using specialized fiber-optic light-guiding equipment and/or near-infrared (NIR) light, of which the penetration depth reaches its maximum in the 650–1350 nm region.[12,13]

This chapter will first discuss major light-responsive systems of organic or inorganic compositions for drug release, and then address the designs and applications of organic-inorganic hybrid materials. NIR light-activated materials are given priority, as these systems are better-suited for translation. The materials discussed are summarized in Figure 7.1. To allow for a deeper discussion, we have narrowed the scope of this review to light-activated DDSs, excluding pure photothermal therapy (PTT) and photodynamic therapy (PDT) from discussion, as reviews of these broad topics are already available.

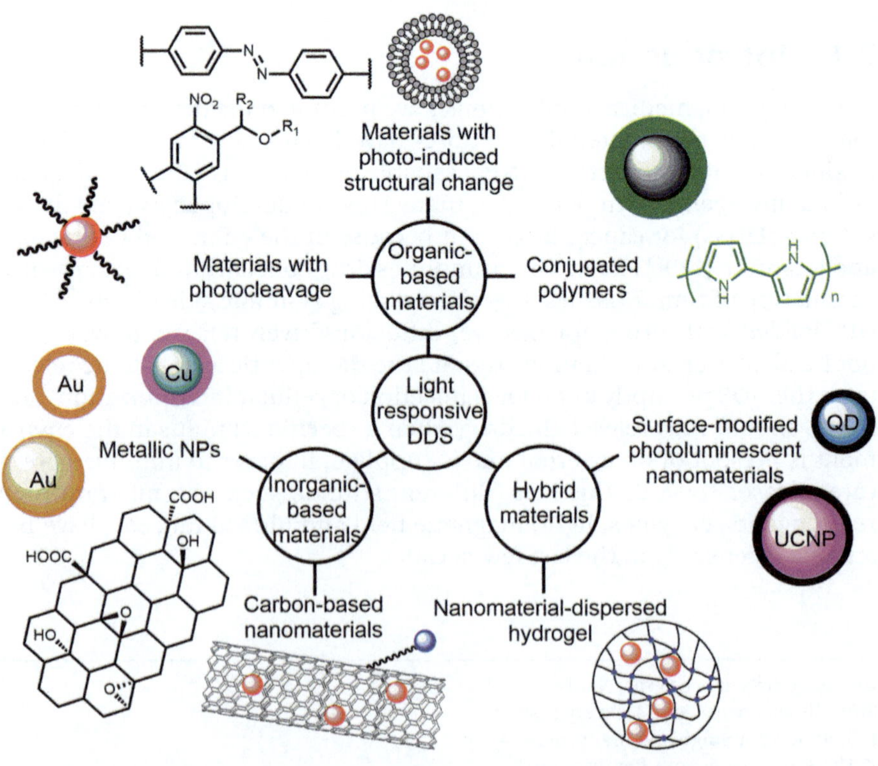

Figure 7.1 Class of materials explored for light-activated drug delivery.

7.2 Light-responsive Organic Nanomaterials

The use of organic nanomaterials in DDSs has advanced significantly in the last decade. While the design and composition of these DDSs vary greatly, light responsiveness is granted typically by adding certain functional groups (FGs) to the system. The structures of these photo-sensitive FGs and their reactions are summarized in Scheme 7.1. UV-vis light is generally used to trigger bond-cleavage reactions and subsequent drug release, as covalent bond breakage using high-energy irradiation is more straightforward in pure organic systems than using lower-energy photons.

7.2.1 Drug Release by Photo-induced Bond Cleavage

One way a DDS may release the drug is by breaking covalent bonds connecting the DDS with the drug using incident light. Among all the photocleavable FGs, *O*-nitrobenzyl, coumarin and pyrene are the three most studied moieties. With the help of substituent effects, they can absorb incident light with varying absorption maxima (mostly within the UV-vis region) and exhibit unique light-responsiveness.[14] In addition, all three moieties are hydrophobic, which provides additional possibilities in the design of the delivery system. In this section, we categorize recent literature related to drug release by photo-induced bond cleavage using three major release mechanisms.

Photolysis of a prodrug is the first and the most straightforward method. When a photolabile linkage is used between the drug and the DDS, light irradiation is able to cleave the prodrug and induce on-demand release.[15–18] For example, a UV-triggered biocompatible DDS using coumarin-functionalized block copolymers has been reported.[15] In this study, hydrophilic bromide-terminated polyethylene glycol (PEG) is used as an atom transfer radical polymerization (ATRP) macroinitiator to randomly copolymerize *n*-butyl methacrylate and coumarin methacrylate. The anticancer drug, 5-fluorouracil (5-FU), is then covalently conjugated to the coumarin moiety *via* [2+2] cycloaddition under UV irradiation (wavelength > 310 nm). Controlled release of free 5-FU from the PEG-*block*-drug micelles is achieved using 254 nm UV irradiation, resulting in excellent drug efficacy in cancer cells (see Figure 7.2(a)). Other than reversible cycloaddition, coumarin can participate in different photo-induced cleavage reactions. A single component, photo-responsive DDS consisting of a fused-ring structure of spiropyran and coumarin can release chlorambucil, which is linked to the hydroxymethyl group *via* an ester bond, under visible light irradiation (≥410 nm).[16] The coumarin ring is initially locked by the spiropyran and its fluorescence is diminished. When the conjugate is in the proximity of cancer tissues, the reduced pH causes ring opening of the acidochromic spiropyran, generating its merocyanine isomer form and restoring the fluorescence of coumarin. The appearance of blue fluorescence makes tumor imaging possible, and allows the prodrug to

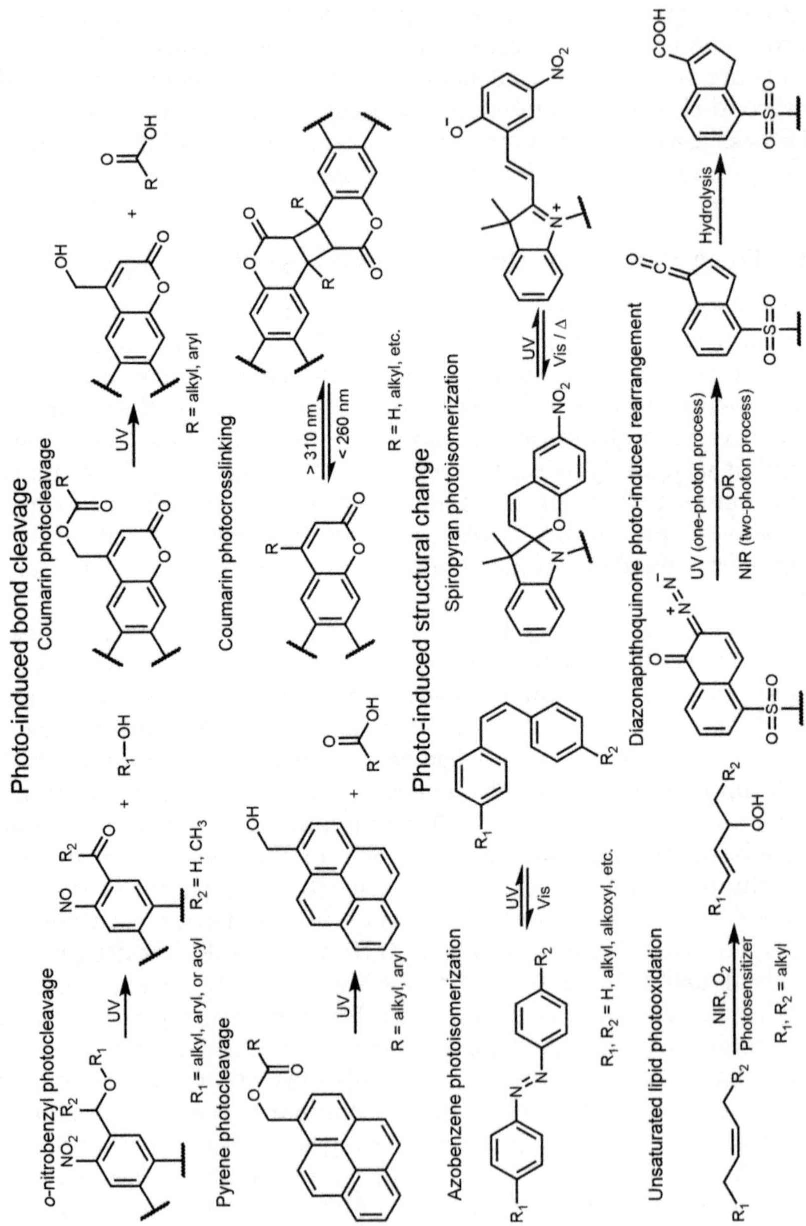

Scheme 7.1 Structures and light-induced reactions of representative FGs.

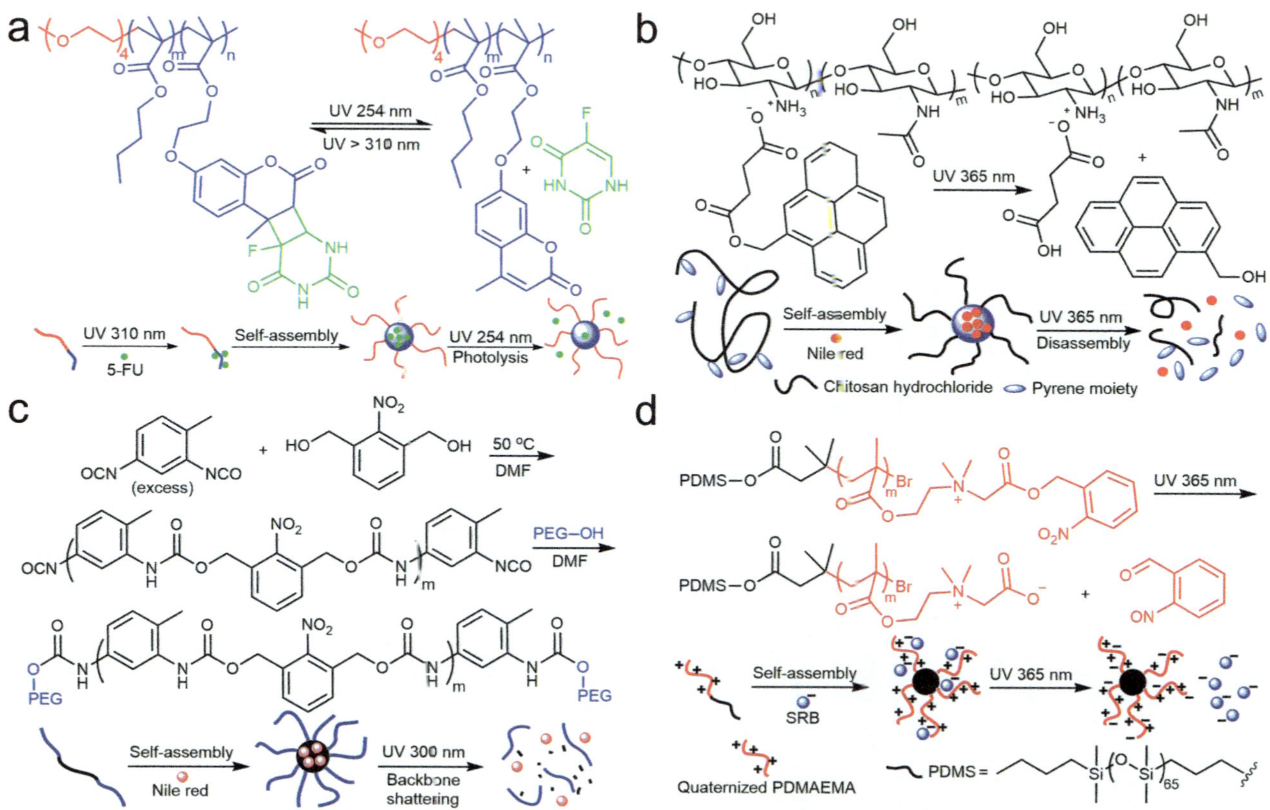

Figure 7.2 Examples of different drug release strategies for organic nanomaterials using photo-induced bond cleavage, which include: (a) photolysis of a coumarin-linked prodrug (ref. 15), (b) hydrophilicity change of a pyrene–chitosan complex (ref. 19), (c) backbone shattering of the *O*-nitrobenzyl-incorporated polymeric micelles (ref. 23), and (d) charge removal from the PDMS-*b*-PDMAEMA diblock copolymers (ref. 31).

be cleaved in a dose-controlled manner using an external light source. Adopting the *O*-nitrobenzyl FG, Johnson *et al.* demonstrated orthogonally triggered release of camptothecin (CPT), doxorubicin (DOX), and cisplatin (CDDP) from a single organic NP synthesized by ring-opening metathesis polymerization (ROMP),[18] wherein DOX is released by photolysis using UV light (405 nm), while the other two drugs are released by ester hydrolysis or a redox reaction, respectively.

A second mechanism by which light may be used to trigger drug release is through the disassembly of the DDS into individual small molecules or macromolecule building blocks. This disassembly is driven by an abrupt change in the physicochemical properties of the building blocks after light exposure, resulting in enhanced leakage or even burst release of the loaded guest molecules. Two general strategies are often used. The first is to disturb the assembly by changing the hydrophobicity-to-hydrophilicity ratio of the amphiphiles that constitute the DDS, which can be done by making the hydrophobic block more hydrophilic after light irradiation. Photocleavable FGs including *O*-nitrobenzyl, coumarin, and pyrene groups are proper candidates for this purpose because they can easily act as hydrophobic pendant groups.[19–21] Lu *et al.* have adopted this method by synthesizing an amphiphilic pyrene–chitosan complex.[19] The starting material, 1-pyrenemethanol, is modified with succinic anhydride to possess an anionic carboxylate group. The anionic pyrene compound is then complexed to the cationic chitosan hydrochloride *via* Coulombic interaction, forcing the chitosan to collapse onto pyrene to form NPs, and a model drug, Nile red (NR), is then encapsulated in the NPs. The nanocarrier becomes hydrophilic once the ester bond adjoining the pyrene group is photolyzed by UV light, and enhanced release of NR is observed as a consequence of the disassembly event (see Figure 7.2(b)).

While changing hydrophobicity can be incorporated into micelle-based DDSs, light-induced backbone-shattering is a more universal method since micelles may or may not be involved.[22–25] The Zhang group has reported the design and synthesis of a carrier-free UV-responsive system to co-deliver DNA and CPT intracellularly.[22] In this study, three CPT molecules are connected to a self-immolative core, which is tethered to an antisense DNA sequence *via* a single *O*-nitrobenzyl group. Self-assembly of the conjugate creates structures analogous to the spherical nucleic acid (SNA), which exhibit enhanced cellular uptake. In response to 365 nm UV irradiation, free DNA is liberated from the NP and the three CPT molecules are released *via* a self-immolative process, leading to nearly complete restoration of the cytotoxicity of CPT. Instead of using a single photolabile junction, Zhao *et al.* prepared an amphiphilic ABA triblock copolymer, of which *O*-nitrobenzyl moieties are positioned repeatedly along the hydrophobic segment, block B.[23] Block B is synthesized by condensation polymerization of 2-nitro-1,3-benzenedimethanol with toluene 2,4-diisocyanate in slight excess. The amphiphilic triblock copolymers are then made by reacting the two terminal isocyanates with hydroxyl-terminated PEG. Upon 300 nm UV irradiation, burst release of the encapsulated NR is observed, which is caused by the shattering of the polymer backbone and fast disintegration of the micelle (see Figure 7.2(c)). Similarly, Cheng *et al.* have designed chain-shattering polymeric therapeutics (CSPTs) and demonstrated light-induced anticancer activity *in vitro*.[25] This design involves copolymerizing light-active monomers with drug monomers. The light-active monomer consists of 2,6-bis(hydroxymethyl) aniline with the amine capped by an *O*-nitrobenzyl group. This diol is used as a monomer to copolymerize with either bifunctional 10-hydroxycamptothecin (HCPT) or 9-aminocamptothecin (ACPT), using phosgene to form carbonate or carbamate linkages, respectively. These CSPTs are self-immolative, meaning that the backbone can depolymerize simultaneously if the polymer chain is scissored at the chain end or in the middle. The CSPTs are loaded into PEG-*b*-poly(L-lactide) (PLLA) micelles in water by co-precipitation. Upon cleavage of the *O*-nitrobenzyl capping group by external UV light, the free aromatic amino groups on the backbone induces chain shattering, leading to either 1,4-elimination or 1,8-elimination, releasing free drug molecules. The authors show that both HCPT- and ACPT-incorporated micelles exhibit reduced cytotoxicity, and their cell-killing power is restored upon UV treatment.

Instead of using backbone fragmentation to release drugs, de-crosslinking can also lead to destabilization of the carrier and augmentation of drug release. This method is often applied in light-responsive micelles, hydrogels, and nanogels.[26–28] Coumarin is a common linker in this case due to the reversibility of the photo-initiated dimerization. Crosslinking or de-crosslinking of the drug carrier can be readily accomplished by shifting the wavelength of the incident light. Zhao *et al.* have demonstrated loading and controlled release of NR in crosslinked micelles by optically tuning the degree of crosslinking and de-crosslinking.[27] Diblock copolymers, PEG-*b*-poly[2-(2-methoxyethoxy) ethyl methacrylate-*co*-4-methyl-[7-(methacryloyl)oxyethyloxy]coumarin] (PEG-*b*-P(MEOMA-*co*-CMA)), are used to form micelles when the temperature is higher than the lower critical solution temperature (LCST) of P(MEOMA-*co*-CMA). Crosslinking is then performed by dimerizing inter-chain coumarin groups under UV irradiation ($\lambda > 310$ nm). Once the cyclobutane bridges are established, the swelling of the nanogel can be controlled by either decreased temperature or incident UV light ($\lambda < 260$ nm), and the reversible volume transition can be used to take up or expel drug molecules.

A third common mechanism for drug release is *via* charge removal or reversal. By disturbing the Coulombic complexation between the guest molecules and the DDS, controlled drug release can be attained. Charge removal of the DDS using the *O*-nitrobenzyl linker is a relatively common approach because of its easily derivatized structure.[29–31] Plasmid DNA, for instance, can be electrostatically loaded into a photolabile gene carrier, which consists of *O*-nitrobenzyl moieties that link the cationic material polyethyleneimine (PEI) with a polymethacrylamide backbone.[29] In the absence of UV irradiation, the nanocarrier stays intact and shields the DNA from degradation.

When UV cleavage using 365 nm light is induced, the PEI and the plasmid DNA are irreversibly released from the polymeric carrier, which results in a three-fold increase in transgene expression. While this example involves disintegration of the nanocarrier, charge removal may exert the desirable effect with only subtle structural changes. Meier *et al.* exemplified this point by preparing a cationic-to-zwitterionic switchable diblock copolymer.[31] In this study, a hydrophobic bromide-terminated poly(dimethylsiloxane) (PDMS) is used as a macroinitiator to polymerize 2-(dimethylamino)ethyl methacrylate (DMAEMA) *via* ATRP. Quaternization of the tertiary amine groups of PDMS-*b*-PDMAEMA using 2-nitrobenzyl-2-chloroacetate produces cationic, amphiphilic diblock copolymers, which assemble into micelles in solution. The *O*-nitrobenzyl groups mask the zwitterionic nature of the hydrophilic segment by capping the carboxylates adjacent to the quaternary amines, allowing an anionic dye, sulforhodamine B (SRB), to be loaded into the system electrostatically. With 365 nm UV light irradiation, the micelles lose the *O*-nitrobenzyl groups and become zwitterionic, which results in weakened Coulombic interaction between the SRB and the micelle and subsequent leakage of SRB (see Figure 7.2(d)). In this example, light exposure only changes the electrostatic interactions, but not the particle structure itself, which may be useful for a more tempered release profile and better spatiotemporal control.

7.2.2 Drug Release by Photo-induced Chemical Structure Change

Light-responsiveness of DDSs may be achieved by changing the chemical structure in ways other than bond cleavage. In these cases, the payload molecules are generally noncovalently associated with the delivery system, and NIR light is often involved, since lower energy photons are energetic enough to induce those non-covalent structural changes. In this section, two general release mechanisms, *trans*-to-*cis* transformation and hydrophilicity change, are summarized by analyzing recently published literature. These studies cover three types of light-triggered chemical reactions, namely photoisomerization, photooxidation, and photo-induced rearrangement.

Trans-to-*cis* transformation is a very versatile reaction that can interfere with drug release in many different ways. Owing to the zigzag structure, the trans isomers of unsaturated long alkyl chains can pack themselves more efficiently than the *cis* isomers in the three-dimensional space. By switching the configuration of the double bond of the DDS *via* light, the crystalline domains of the DDS can be made more amorphous, resulting in enhanced leakage of guest molecules. Azobenzene, one of the most studied photo-sensitive FGs, can photoisomerize and the two isomers can switch back and forth using light of different wavelengths.[32] The *trans*-to-*cis* transformation of azobenzene has been utilized to destabilize liposome-like catanionic vesicles.[33,34] One study involves the cationic surfactant,

4-cholesterocarbonyl-4′-(*N,N,N*-triethylamine butyloxyl bromide) azobenzene (CAB), and the anionic surfactant, sodium dodecyl sulfate (SDS), which co-assemble to form either unilamellar or multilamellar vesicles.[33] As a result of the Coulombic and the hydrophobic interactions, the bilayer consists of an even mixture of CAB and SDS. Upon UV light irradiation, isomerization occurs, increasing the inter-lamellar spacing of the *cis*-vesicles by 0.2–0.3 nm, which leads to enhanced release of rhodamine B and DOX. As a result, increased cytotoxicity of DOX-loaded CAB/SDS vesicles to NIH 3T3 and Bel 7402 cells is observed. Another way of utilizing azobenzene is through the terminal benzene ring, which can be inserted into the cavity of cyclodextrins (CDs) to form the azobenzene-CD inclusion complex.[35] The strong hydrophobic interaction weakens when the azobenzene is in its *cis* form due to the increased steric hindrance, allowing the two moieties to be separated. This photo-switchable host-guest chemistry has been coupled with a de-crosslinking mechanism to release drugs.[36,37] In another example, a red-light-responsive hydrogel has been prepared by the Wu group to illustrate the controlled release of proteins.[37] In this work, tetra-*ortho*-methoxy-substituted azobenzene (mAzo) and β-CD with reactive amino groups are first grafted onto a poly(acrylic acid) (PAA) backbone to yield PAA-mAzo and PAA-β-CD, respectively, which form a hydrogel upon mixing. A model protein, bovine serum albumin (BSA), is loaded into the PAA-mAzo/PAA-β-CD hydrogel. Light induces isomerization and subsequent de-crosslinking, and the weakened hydrogel yields faster BSA release. Due to the methoxy substitution, the isomerization wavelength of the azo group is shifted into the red light regions, which is beneficial for *in vivo* applications. Azobenzene is also frequently explored in inorganic materials, in particular, mesoporous silica NPs (MSNs). Compounds of interest can be loaded into the interior mesopores of the MSN, with azobenzene-containing molecules serving as gatekeepers near the surface of the pores to prevent non-specific leakage. Upon light irradiation, the pores can be closed or opened by switching the configuration of the azobenzene moiety, which leads to controlled release of guest molecules.[38–40] For instance, azobenzene-modified DNA (Azo-DNA)/MSN nanostructures have been exploited as drug carriers by the Tan group.[38] Short, single-stranded DNA (Arm-DNA) is first immobilized at the pore mouth of the MSN, and a longer, Azo-DNA hybridizes across two adjacent Arm-DNA strands, capping pores as a result. Due to the presence of azobenzene moieties in the Azo-DNA strand, UV light can be used to switch hybridization off while visible light irradiation can turn on hybridization. This is because when the azobenzene is in its extended trans form, hybridization is unhindered but the *cis* isomer is associated with steric hindrance which destabilizes DNA duplexes. The authors demonstrated controlled loading and release of model payload molecules Rhodamine 6G and DOX (see Figure 7.3).

Hydrophilicity change is another mechanism by which drug release can be achieved without photolysis. One example involves the photooxidation of an unsaturated phospholipid, which is commonly used in liposome-based DDSs. The singlet oxygen-mediated oxidation causes the formation of allylic

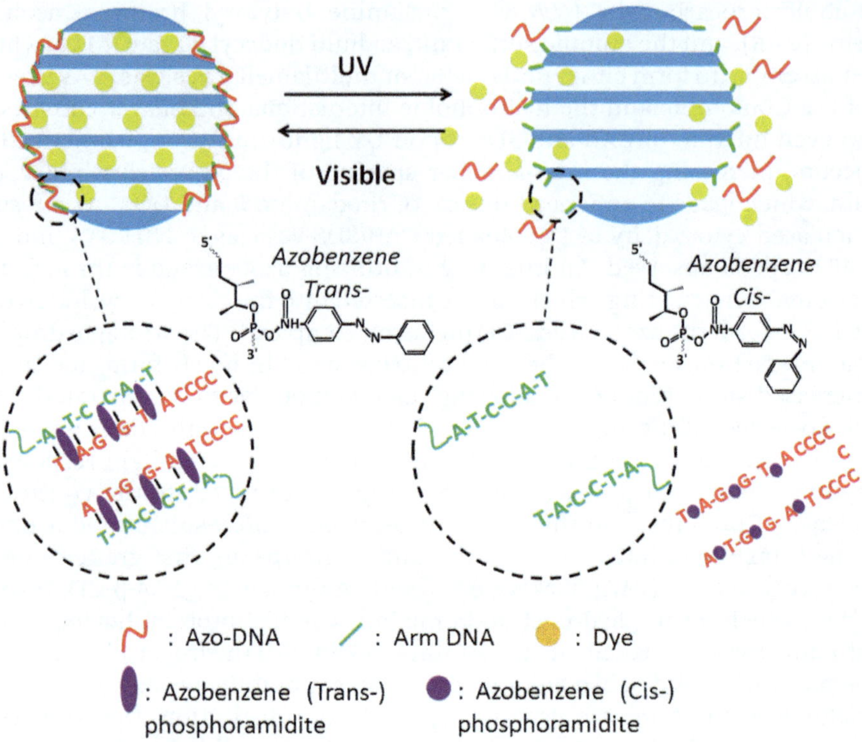

Figure 7.3 Schematics for light-induced drug release from mesoporous silica particles with dsDNA capping. Trans azobenzene isomerizes to the *cis* form by 365 nm UV irradiation, leading to DNA dehybridization, pore opening, and release of guest molecules. Reprinted with permission from ref. 38. Q. Yuan, Y. Zhang, T. Chen, D. Lu, Z. Zhao, X. Zhang, Z. Li, C. H. Yan and W. Tan, *ACS Nano*, 2012, **6**(7), 6337. Copyright 2012 American Chemical Society.

hydroperoxides, which decreases the hydrophobicity of the unsaturated phospholipids, leading to destabilization and permeabilization of the liposome.[41,42] To generate reactive singlet oxygen *in situ*, the co-delivery of photosensitizers is required, which can be implemented by either encapsulating them inside the hydrophilic interior or anchoring them into the hydrophobic bilayer region of the liposome.[42,43] Lovell *et al.* have demonstrated the use of NIR light to release DOX in a controlled manner, using the unsaturated phospholipid dioleoylphosphatidylcholine (DOPC) and porphyrin-phospholipid (PoP) in the saturated 1,2-distearoyl-*sn*-glycero-3-phosphocholine (DSPC)/cholesterol bilayers.[43] When 665 nm light irradiation is applied to the system, the photosensitizer porphyrin generates singlet oxygen, and accelerated release of DOX is observed (see Figure 7.4(a)). The involvement of singlet oxygen in the mechanism is corroborated by the control experiment, showing that an oxygen scavenger or an antioxidant can suppress light-triggered drug release. When the materials are tested *in vivo*, a significant growth

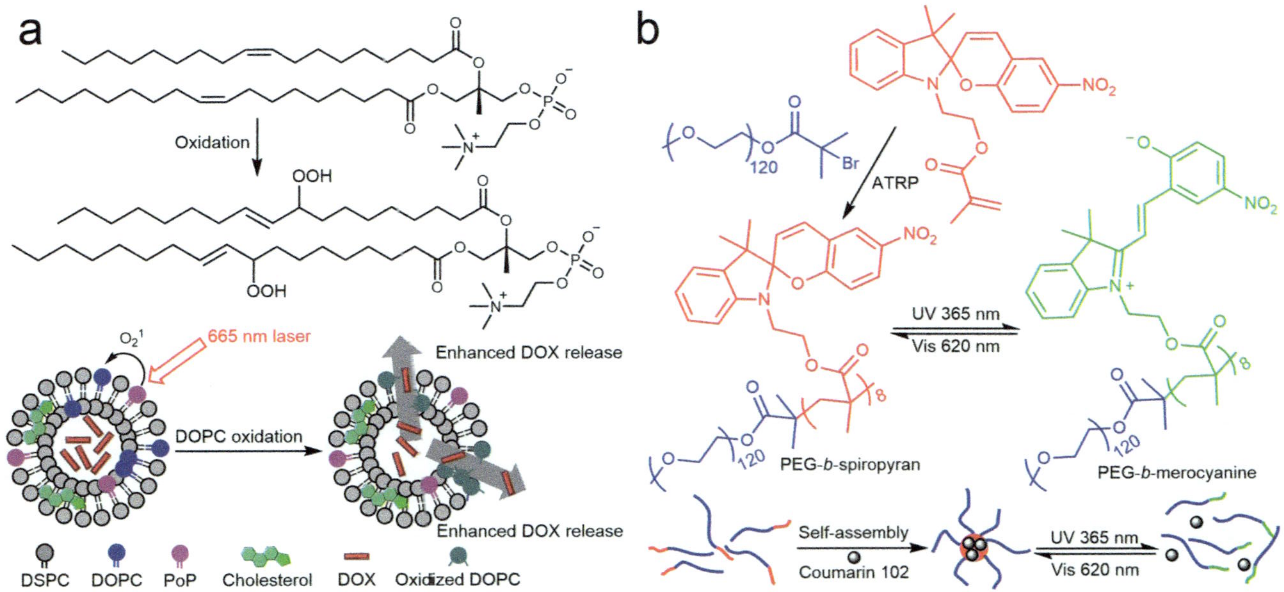

Figure 7.4 Examples of non-photolytic drug release from organic nanomaterials. (a) Light produces singlet oxygen, which leads to the oxidation of DOPC (an unsaturated lipid), destabilization of the liposome, and enhanced DOX release from the DDS (ref. 43). (b) Light isomerizes spiropyran from the hydrophobic form to a hydrophilic merocyanine form, which leads to the disassembly of micellar drug carriers and rapid release of the loaded molecule Coumarin 102 (ref. 46).

suppression of the human pancreatic xenograft in mice with a single chemo-phototherapy treatment is observed. Other frequently used chemical species to induce structural change include spiropyran and diazonaphthoquinone (DNQ). Upon UV irradiation, the photochromatic spiropyran isomerizes from a non-charged, hydrophobic form to a zwitterionic, hydrophilic form, merocyanine. The reverse reaction can occur with exposure to heat or visible light. DNQ undergoes an irreversible UV- (one-photon process) or NIR-sensitive (two-photon process) Wolff rearrangement to form a ketene derivative, and following hydrolysis generates a hydrophilic 3-indenecarboxylic acid as the final product.[44] The incorporation of spiropyran and DNQ as the photosensitive FGs has been adopted by several groups.[45–48] Matyjaszewski *et al.* demonstrated reversible drug loading and release using spiropyran-containing polymeric micelles.[46] These structures are constructed by polymerizing spiropyran-containing methacrylate monomers using ATRP from a bromide-terminated PEG macroinitiator. The amphiphilic diblock copolymer, PEG-*b*-spiropyran, then self-assembles in the presence of coumarin 102, forming a micelle encapsulating the model payload molecule. With 365 nm UV irradiation, the pendant spiropyran photoisomerizes to merocyanine, turning the amphiphilic polymer into a hydrophilic one, and releasing coumarin 102 as the polymer solubilizes. Reformation of the micelle and partial re-encapsulation can occur with 620 nm visible light, which restores the amphiphilic structure (see Figure 7.4(b)). Similarly, Zhu *et al.* have designed Janus-type dendritic poly(amido amine) (PAMAM)-based amphiphiles, with the hydrophobic segment incorporating DNQ and the hydrophilic segment containing lactose.[47] The micellar structure of the assembled amphiphile is responsive to both NIR (808 nm laser) and UV (365 nm) triggers, which lead to Wolff rearrangement, micelle disassembly, and the release of encapsulated DOX.

7.2.3 Conjugated Polymers

Conjugated polymers represent a unique class of organic materials for DDS. Unlike light-responsive small molecules, conjugated polymer-containing DDSs typically involve photothermal mechanisms for drug release. Photothermal conversion typically results from the non-irradiated transition of excited electrons to the ground state, apart from the plasmonic effect.[49,50] Many conjugated polymers with negligible fluorescent emission have been exploited for photothermal conversion in the first NIR optical window (NIR-I) covering 750–1000 nm, and more recently the second NIR optical window (NIR-II, 1000–1700 nm).[51–53]

Among the many polymers, copolymers, or polymer blends for photo-induced drug release, polyaniline (PANI) and polypyrrole (PPy) stand out as the two versatile and commonly used materials for photothermal drug delivery.[54–57] Zhu *et al.* have synthesized a lecithin/PEG-poly(caprolactone) (PCL) liposome NP to release the anticancer drug CDDP.[55] To enable light activation, PANI is loaded into the lecithin hydrophobic tail region during the synthesis of the NPs. The outer carboxyl-terminated PEG chain is conjugated with the antibody Trastuzumab to target Her-2 overexpressing tumor cells. After incubating with cells, 808 nm irradiation is applied to excite PANI, and the local heating compromises the liposome structure, accelerating the release of the encapsulated CDDP. PPy can be prepared facilely by oxidation of pyrrole, and is often incorporated into complex structures by *in situ* polymerization. A core–shell poly(lactic-*co*-glycolic acid) (PLGA)-PPy NP has been prepared using this method by Xie *et al.*[56] In this study, a chemotherapeutic drug, curcumin (Cur), is first encapsulated into the PLGA particles, which serves as nanoscopic initiators to template the polymerization of pyrrole by H_2O_2, generating the drug-loaded core–shell PLGA-PPy NPs. Slow Cur leakage is observed under physiological conditions due to diffusion and the hydrolysis of the PLGA ester backbone, but pronounced, pulsed drug release can be achieved by intermittent 808 nm NIR irradiation, which is attributed to local hyperthermia. A similar strategy to incorporate PPy in DDSs has also been reported by the Liu group.[57] In this work, magnetic Fe_3O_4 nanoclusters are first dispersed and stabilized into the solution using sodium dodecylbenzenesulfonate (SDS) and poly(vinyl alcohol) (PVA) as emulsifiers and stabilizers, respectively. PPy coating is achieved by mixing with the monomer pyrrole and subsequent *in situ* oxidative polymerization using Fe^{3+} as the oxidant. Finally, an amphiphilic polymer, PEG-grafted poly(maleic anhydride-*alt*-1-octadecene) (C18PMH-PEG), is added to noncovalently decorate the Fe_3O_4/PPy NPs *via* hydrophobic interactions in order to enhance physiological stability. When DOX is loaded into the hydrophobic shell of PPy, controlled release can be achieved with 808 nm NIR light. Interestingly, enhanced cell uptake of the composite particles is observed with external magnetic fields and NIR irradiation, due to mild magnetothermal and photothermal heating, respectively (see Figure 7.5).

7.3 Inorganic Nanomaterials for Photothermal Drug Delivery

Inorganic materials have enormous potential as drug delivery vectors.[58] These materials may access unique shapes, sizes, good homogeneity, high surface area, plasmonic and magnetic phenomena, and various biological interactions. Many methods have been developed to functionalize the surface of inorganic materials with surface ligands, which can impart increased colloidal stability and affinity towards target molecules or cells. Meanwhile, inorganic particles may consist of inert atoms and well-defined crystal lattices, which render them highly stable even under diluted physiological conditions. Notably, many inorganic particles are able to interact with light irradiation themselves without organic ligands, and can

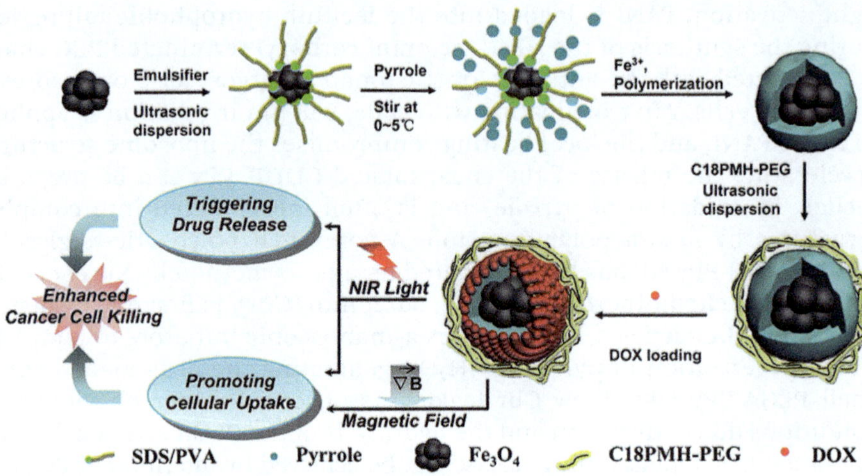

Figure 7.5 Schematic illustration of the synthesis of Fe_3O_4/PPy NPs, subsequent DOX loading, and the controlled intracellular release of DOX induced by an external magnetic field and NIR light. Reprinted with permission from ref. 57. C. Wang, H. Xu, C. Liang, Y. Liu, Z. Li, G. Yang, L. Cheng, Y. Li and Z. Liu, *ACS Nano*, 2013, 7(8), 6782. Copyright 2013 American Chemical Society.

disperse the photo-energy in different pathways, making these particles good systems to explore for light-induced drug delivery.[59] In this section, we address inorganic nanomaterial-based DDSs that are capable of drug release with vis-NIR light.

7.3.1 Metallic NPs

Metallic NPs consist of a metallic core and typically a surface coating that can be controlled by different synthetic procedures.[60] Due to the plasmonic resonance effect, metallic NPs can exhibit strong absorption of incident NIR light, followed by a series of non-radiative processes that efficiently convert the absorbed light into heat.[61] Various metallic NPs such as gold, silver, palladium, and copper have been reported to exploit photothermal properties, and are used for a range of purposes spanning PTT, imaging, and drug delivery.[61,62]

Several novel metallic light-triggered DDSs have been developed to mediate controlled drug release,[63] with drug loading occurring at different locations of the material.[64–66] For example, Li *et al.* reported a DOX-encapsulated hollow gold nanosphere that is capable of both photothermal ablation of cancer cells and NIR light-induced drug release.[64] Due to the porous and negatively-charged nature of the hollow gold nanospheres, high DOX loading is attained under physiological conditions. Once NIR light is applied, the electrostatic complexation between the carrier and DOX is disrupted, leading to enhanced drug release. If the metallic NP is non-porous, the exterior surface of the NPs is often used for drug attachment. Lin *et al.* have reported NIR-responsive $Cu_{1.8}S/mSiO_2$ nanocomposites, which consist of mesoporous silica-coated $Cu_{1.8}S$ NPs that are surface-modified with DNA to cap $mSiO_2$-loaded Cur, while the DNA double helices themselves are loaded with DOX by intercalation.[65] The dsDNA undergoes thermal dehybridization with 980 nm laser irradiation, which leads to the simultaneous release of the DNA-loaded DOX and the mesopore-loaded drug Cur. Another interesting location to load the drug is the inter-particle cavity created when the particles form an aggregation. Ijiro *et al.* have designed such a DDS using crosslinked gold nanoparticle vesicles (AuNVs), which are synthesized from individual AuNPs modified with surface semi-fluorinated ligands.[66] These particles pre-assemble to form AuNVs in tetrahydrofuran, which are subsequently crosslinked and stabilized using thiol-bifunctionalized PEG. Guest molecules are loaded into the inter-particle nanogaps at an elevated temperature. Upon cooling, the gaps contract, trapping the guest molecules inside. As the DDS is exposed to short-term laser irradiation (5 min, 532 nm), the gaps again expand due to the photothermal effect, allowing the entrapped guest molecules to be released (see Figure 7.6).

AuNPs have recently been extensively studied as vehicles for light-triggered gene delivery, due to the large surface area, efficient photothermal conversion, and the relatively labile Au–S bonds. AuNPs may collapse to form a fused gold cluster in response to photothermal heating, and subsequently release surface-attached moieties. For example, gold nanorods (AuNRs) complexed with the plasmid DNA encoding for enhanced green fluorescence protein (EGFP) have been used to enable light-activation of the plasmid.[67] When irradiated with femtosecond NIR light, the DNA-AuNR complex undergoes morphology transformation to form spherical AuNPs, which weakens the strong interactions between the DNA and the NPs, resulting in the release of free EGFP plasmids. The study shows that GFP gene expression can be induced in cells that are locally exposed to NIR irradiation. This technique has implications in many biotechnological applications as well as drug delivery. Another form of AuNP, the hollow gold nanosphere (a thin gold wall with a hollow interior), has been used for light-induced delivery of thiolated siRNA, which is anchored to the NP by the Au–S bond.[68] With NIR exposure, enhanced siRNA activity is observed. One possible mechanism for the enhanced gene knockdown is the Au–S bond cleavage and release of the siRNA as a result of local heating, which improves the accessibility of the siRNA. Apart from inducing particle collapse or bond cleavage, the photothermal effect can also be utilized to cause duplex oligonucleotide to dehybridize, which releases a therapeutic single-stranded oligonucleotide. For example, a dsDNA-AuNR conjugate with an aspect ratio of 3.5 has been utilized to release non-thiolated ssDNA.[69] Each particle has a large surface area to accommodate approximately 250 molecules of dsDNA, which releases ssDNA upon 753 nm light activation. A similar study conducted by the Halas group shows that 4′,6-diamidino-2-phenylindole (DAPI)-dsDNA conjugates can be anchored to AuNP surfaces, and be released in a controlled fashion intracellularly with

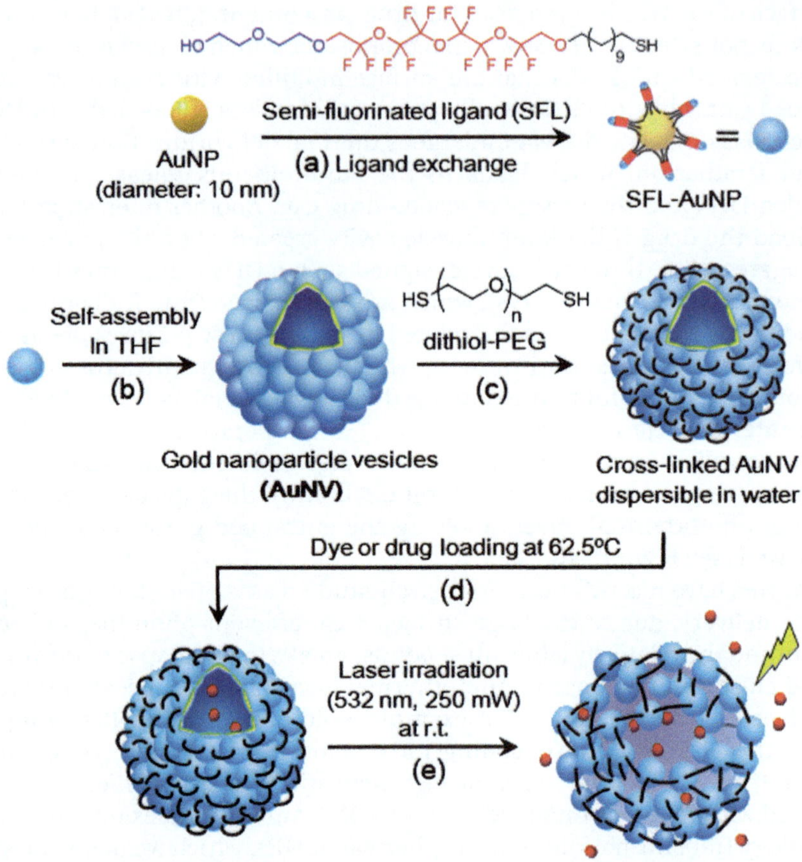

Figure 7.6 Synthesis of AuNVs, loading of guest molecules, and visible light-triggered photothermal release. Reprinted with permission from ref. 66. K. Niikura, N. Iyo, Y. Matsuo, H. Mitomo and K. Ijiro, *ACS Appl. Mater. Interfaces*, 2013, 5(9), 3900. Copyright 2013 American Chemical Society.

NIR light irradiation.[70] DAPI in this case enables the visualization and monitoring of the release event.

7.3.2 Carbon-based Nanomaterials

Carbon-based materials have been of interest for incorporation into light-triggered DDSs. Facile surface modification, structure rigidity, hydrophobicity, electron-richness, and good thermal conductivity are some of the advantageous properties useful for designing DDSs.[71] Carbon nanotubes (CNTs) and graphene are the two best-studied materials, which represent the canonical three-dimensional and two-dimensional carbon-based nanostructures, respectively. While both are used to design light-responsive DDSs, applications differ based on their distinctive structures.

CNTs are cylindrical molecules consisting of a hexagonal arrangement of sp^2 hybridized carbon atoms. They can also be considered as single or multiple layers of graphene sheets that are rolled up, which produces single-walled carbon nanotubes (SWCNTs) and multi-walled carbon nanotubes (MWCNTs), respectively. Both the exterior and the interior of CNTs can be exploited as drug loading sites. For instance, SWCNTs have been used by Chen *et al.* for intracellular protein delivery (see Figure 7.7).[72] SWCNTs have a distinct needle-like shape, which enables them to enter cells with minimal damage. Alkyl chain-modified streptavidin (SA) is anchored onto the surface of SWCNTs *via* hydrophobic interactions, and proteins of interest can be bound *via* a SA-desthiobiotin (SA-DTB) linkage. The DTB binds less tightly to SA than biotin (Kd = 10^{-11} M *versus* Kd = 10^{-15} M, respectively) but nonetheless provides a high level of specificity.[73] The weaker linkage allows NIR light to photothermally cleave it, releasing the payload protein from the SWCNTs. Similar long alkyl chain anchors have also been reported to attach various drug molecules to CNTs.[74,75] Another method to link payload molecules to CNTs is *via* π–π interactions. For example, SWCNT has been used as the nanocarrier to load Ru(II) polypyridyl complexes.[76] In this case, PTT is triggered upon 808 nm laser irradiation, accompanied with the photothermal release of the loaded Ru(II) complexes. The released Ru(II) polypyridyl complexes can then produce singlet oxygen by absorbing two 808 nm photons, initiating two-photon photodynamic therapy (TPPDT). As a result, the bimodal

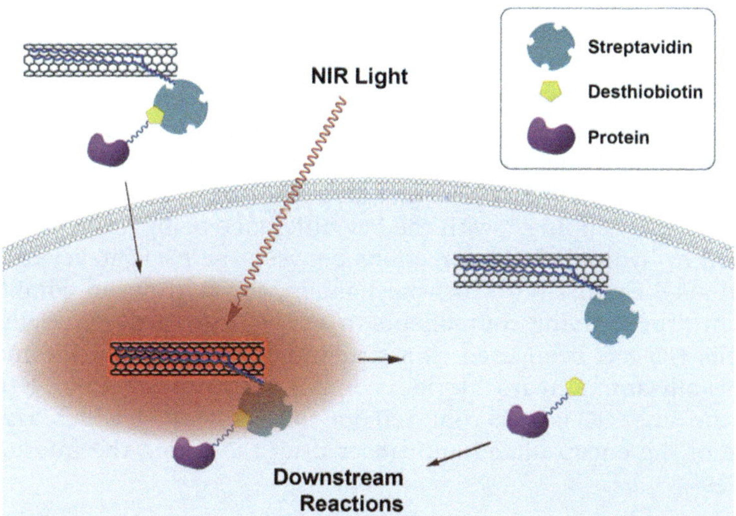

Figure 7.7 Design for the NIR-triggered intracellular release of protein molecules from protein-SWCNT conjugates. Upon cell internalization, NIR irradiation on SWCNTs generates local heating, which disrupts the SA-DTB linkage and releases the payload protein. Reprinted with permission from ref. 72. H. Li, X. Fan and X. Chen, *ACS Appl. Mater. Interfaces*, 2016, **8**(7), 4500. Copyright 2016 American Chemical Society.

PTT-TPPDT works synergistically to enhance the cell-killing efficacy. The π–π stacking interactions have been used to directly load drug molecules containing π systems such as DOX and nucleic acid to CNTs.[77–79] Apart from the surface attachment, encapsulation is also possible. For example, simulation has shown that the drug ciprofloxacin can be inserted into the cavity of CNTs, and NIR-induced local hyperthermia of CNTs causes increased diffusion that can assist the release of encapsulated drugs, making it possible to intracellularly deliver this highly polar drug.[80]

Unlike CNTs, graphene has a two-dimensional honeycomb structure of a monolayer of carbon atoms. Unmodified graphene is rarely used in drug delivery, but analogues such as graphene oxide (GO) and reduced graphene oxide (rGO) are often investigated as delivery vehicles. The oxygen imperfections introduce epoxy bridges, hydroxyl, and carboxyl groups, which are very useful to covalently conjugate molecules. However, although many types of drugs have been reported to load onto GO or rGO nanosheets *via* various mechanisms (covalent conjugation, electrostatic complexation, π–π stacking, and hydrophobic interactions *etc.*), direct light-triggered release from the nanosheet is not common, due to the low photothermal conversion efficiency of graphene materials. Nonetheless, with unique, unusual designs, the photothermal effect can still be utilized. For example, instead of loading drug molecules onto its surface, the flexible nanosheet can warp to seal or release drug molecules entrapped in cavities, resembling the function of a gatekeeper. Chen *et al.* utilized this method to create GO-wrapped MSN.[81] The surface of the MSN is modified with amino groups, and the negatively charged GO nanosheets adhere to the particle surface *via* Coulombic interactions, locking the loaded DOX molecules within the mesopores of the MSN. The design also involves a targeting aptamer sequence, which is tethered to the GO surface *via* π–π stacking. Controlled drug release is achieved by photothermally peeling the GO gatekeeper off the particle surface using NIR irradiation. A similar strategy is also reported using alkyl chain-grafted MSN with rGO capping,[82] with the key difference being that the capping is achieved by hydrophobic interactions between the rGO gatekeeper and the grafted alkyl chains. With this mechanism, GO or rGO can combine with different drug-carrying components to achieve light-activated delivery. For example, GO has been used as a structural component in the multilayer GO-poly(allylamine hydrochloride) composite capsules.[83] Upon NIR irradiation, the outer GO layer is ruptured due to local heating, which triggers the release of the encapsulated anticancer drug DOX from the interior of the capsules.

7.4 NIR Light-responsive Organic-inorganic Hybrid Materials

Under certain scenarios, neither organic nor inorganic materials have all the desired properties of a light-responsive DDS. Organic materials generally do not respond to NIR irradiation well, while inorganic materials face short- and/or long-term clearance and toxicity difficulties. However, their non-overlapping drawbacks also imply possible solutions. In this section, we discuss light-responsive organic–inorganic hybrid materials with several examples to demonstrate how their combination circumvents the disadvantages associated with each.

7.4.1 Surface-modified Photoluminescent Inorganic Nanomaterials

One key disadvantage of using organic-based FGs for light-activated drug release is the use of high-energy UV-vis irradiation, which can induce phototoxicity and has poor penetration of human tissue. NIR light, while carrying little phototoxicity and being more tissue-permeable, is of relatively low energy. Therefore, it is desirable to use NIR light to carry out activities that require UV-vis light. One solution is to use photoluminescent nanomaterials, which dissipate absorbed NIR energy in a radiative pathway, emitting photons of higher energy compared with the incident photons.[84] As such, they have been adopted as the light-responsive component in DDSs in the last decade. Due to the small size and rigid metal-doped nanocrystal core structure, surface modification is the most commonly used method to attach or load drug components when using photoluminescent inorganic nanomaterials.

In photon upconversion, two or multiple photons of lower energy are absorbed, and one photon with higher energy is emitted as an anti-Stokes emission. Upconverting NPs (UCNPs), which typically involve lanthanide-doped (usually Ln^{3+}) nanocrystals, can absorb NIR irradiation and act as a second light source to emit UV-vis light and initiate light-triggered events. All the FGs-related drug release mechanisms such as direct photolysis of prodrugs,[85,86] structure disassembly,[87–90] charge removal,[91] and *trans*-to-*cis* transformation[92] have been combined with UCNPs to attain NIR-triggered controlled release. For instance, $LiYF_4:Tm^{3+}/Yb^{3+}$–UCNPs have been reported to release DOX upon NIR irradiation by the Hartman team.[86] The photocleavable *O*-nitrobenzyl moiety is used to tether DOX with the bis-carboxylated ligand glutamic acid, which anchors to the UCNP surface by forming a strong coordinative complex with the exposed Ln^{3+} ions. With 980 nm laser light, the upconverted UV light photolyzes the *O*-nitrobenzyl linkage and free DOX is observed immediately. In addition to photolysis, the release of noncovalently encapsulated drug is also often explored using UCNPs as the photon absorber. Lu *et al.* have reported a silica-coated, Yb^{3+}/Tm^{3+}-doped UCNP system,[87] which is surrounded by a shell of spiropyran- and PEG-containing organic materials, with the drug molecules encapsulated in the hydrophobic region of the shell. Upon intracellular delivery and NIR activation, the hydrophobic spiropyran isomerizes to the hydrophilic merocyanine form by the secondary high-energy UV light, which causes disruption of the self-assembled coating and the release of the encapsulated drugs. Similarly, light-induced de-crosslinking can be coupled with UCNPs as well. In a study conducted by the

Xing group,[89] Yb^{3+}/Tm^{3+}-doped $NaYF_4$ UCNPs are coated with mesoporous silica, which has further modified folic acids as a targeting moiety and *O*-nitrobenzyl-linked terminal alkenes, which are crosslinked by metathesis using Grubbs' catalyst. The synthetic procedure forms a polymer coating, which caps the mesopores of the silica. These NPs can target folate receptor-overexpressing cancer cells, and the model drug DOX, which is loaded inside the silica layer, can be released upon photo-decrosslinking of the polymer layer triggered by secondary UV irradiation (see Figure 7.8).

Two-photon absorption is another nonlinear optical process that has been explored for light-induced drug delivery. Semiconducting quantum dots (QDs) are a class of strong two-photon absorbing species. Ranging in particle size from 2–10 nm, QDs are extremely tunable with respect to their optical properties, which is achieved by changing both reaction conditions and choice of precursors and surfactants.[93,94] After excitation, various non-radiative relaxation processes such as Förster resonance energy transfer (FRET) and photo-induced electron transfer (PET) can occur to dissipate the absorbed light energy. While QDs have been intensively studied as a photothermal and an energy donor material for PTT and PDT,[59] respectively, a notable use of QDs for light-activated DDSs is through their photoluminescent properties. For instance, Gui *et al.* have developed a glutathione-capped Mn^{2+}-doped ZnSe QDs that can release nitric oxide (NO) upon 1130 nm laser irradiation.[95] A NO precursor, Roussin's black salt (RBS, formula $NaFe_4S_3(NO)_7$) is loaded onto the surface of the QDs *via* electrostatic binding. The anion $Fe_4S_3(NO)_7^-$ releases NO after absorption of visible light, which is produced photoluminescently from the QD core. Utilizing a very similar strategy, Wang *et al.* have reported a nanocomposite containing silica-coated, amine-modified Mn^{2+}-doped ZnS QDs and carboxylate-modified MoS_2 QDs, which are bound electrostatically.[96] The photo-sensitive RBS is also loaded onto the amine-containing particle surface. Under excitation by 808 nm irradiation, the nanocomposite shows dual-emissive, two-photon photoluminescence from both QDs that induces the photolysis of the RBS to release NO *in situ*.

7.4.2 Nanomaterial-dispersed Hydrogel

Consisting of crosslinked hydrophilic polymeric networks, hydrogels represent a collection of materials capable of absorbing a large amount of water or biological fluids. When properly designed, hydrogels can simulate living systems with excellent biocompatibility owing to their high water content, porosity, and mechanical properties matching that of human tissues.[97] By physically or chemically encapsulating water, living cells, growth factors, and drug molecules in the three-dimensional space, hydrogels open up a wide range of opportunities for drug delivery.[97] Sustained release is one distinctive property of hydrogels since guest materials can slowly diffuse out of the crosslinked network. However, this sustained release profile may not always be desirable. In certain cases, burst or pulsatile drug release can have clear

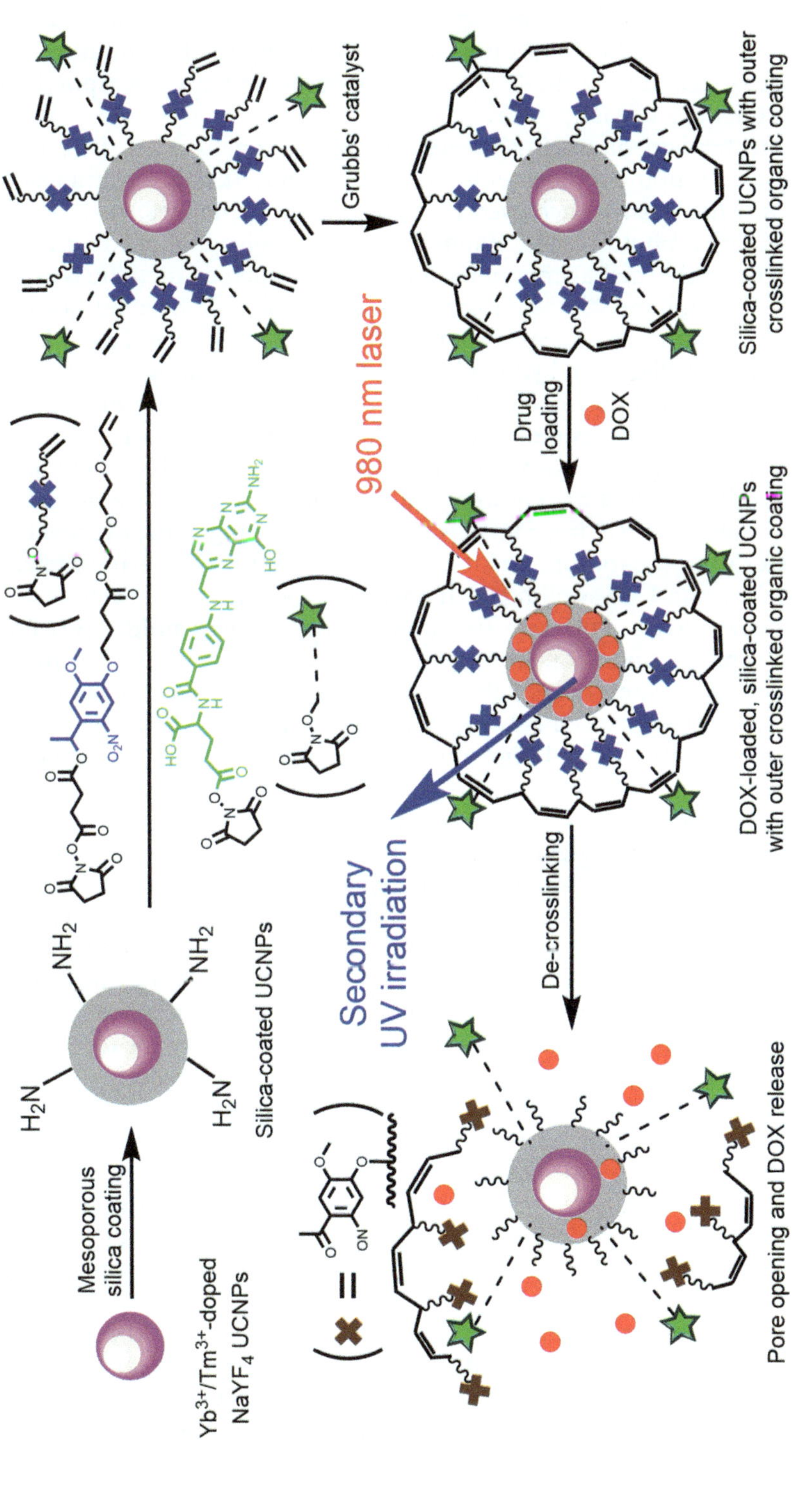

Figure 7.8 NIR-activated intracellular release of DOX using light-responsive, crosslinked UCNPs. Secondary UV irradiation cleaves *O*-nitrobenzyl linkages to peel off the outer organic coating, leading to pore deblocking and drug release (ref. 89).

benefits over conventional dosage forms.[98] By dispersing nanomaterials inside the hydrogel, the nanocomposite DDS can acquire the stimuli-responsiveness associated with the NPs, and gain significant spatiotemporal control over the drug release profile.

One mechanism of forcing the release of hydrogel-loaded drug is the reversible entrapping and expelling of water inside the three-dimensional networks, which can be achieved by environmental changes, such as temperature. Thus, photothermal nanomaterials can play a vital role in these designs. Photothermal agents mentioned earlier, such as metallic NPs,[99,100] CNTs,[101–103] graphene,[104,105] and conjugated polymers,[106] have all been utilized to facilitate cargo release from hydrogels. For instance, Lv *et al.* have designed a NIR-triggered, hydrogel-based DDS by combining a thermosensitive PCL-PEG-PCL polymer with MWCNTs.[103] The aromatic model drug, DOX, is noncovalently loaded onto chitosan-modified MWCNTs, which are then entrapped by the hydrogel. The composite gel exhibits slow, diffusion-controlled release at physiological temperature. With NIR activation, the photothermal effect from MWCNTs induces swelling of the hydrogel and accelerates drug separation, making it the dominant mechanism of release. On the other hand, de-swelling of the hydrogel can also be a mechanism to trigger drug release. Polymers like poly(*N*-isopropylacrylamide) (PNIPAM) undergo a phase transition from a swollen, hydrated state to a shrunken, dehydrated state when heated above their LCST, losing most of the volume and expelling the enclosed substances. West *et al.* have taken advantage of this phase transition and designed a hydrogel system containing PNIPAM-*co*-polyacrylamide and silica-gold nanoshell as a platform for pulsed delivery of cancer therapeutics.[100] The composite hydrogel has a LCST of ~40 °C, slightly above physiological temperature. As such, NIR light can be used to generate local heating and the mild increase in temperature can enhance the release of the payload molecules, in this case DOX and DNA. The two payloads, due to having vastly different physical properties and molecular weight, exhibit distinctive release profiles. Also adopting the PNIPAM hydrogel, Wang *et al.* have designed a photothermal release system using a narrow-bandgap conjugated polymer, poly(diketopyrrolopyrrole-*alt*-3,4-ethylenedioxythiophene) (PDPPEDOT).[106] PDPPEDOT can be readily fabricated as thin films, or be made as colloidal NPs by nanoprecipitation with the commercial product Pluronic F-127. PDPPEDOT exhibits a strong photothermal effect upon absorbing NIR light. By integrating either PDPPEDOT films or NPs with PNIPAM hydrogel, reversible swelling/de-swelling of the composite hydrogels is observed with repeated cycles of off/on 808 nm NIR light irradiation, and is accompanied by the uptake or release of DOX from the hydrogel (see Figure 7.9(a)).

As the polymeric chains of hydrogels are crosslinked by either noncovalent interactions or covalent bonds, a popular strategy to release the cargo from hydrogels is to disrupt the crosslinked network. Besides the aforementioned noncovalent interactions (electrostatic complexation, π–π stacking, hydrophobic interactions), hydrogen bonding is commonly used to stabilize

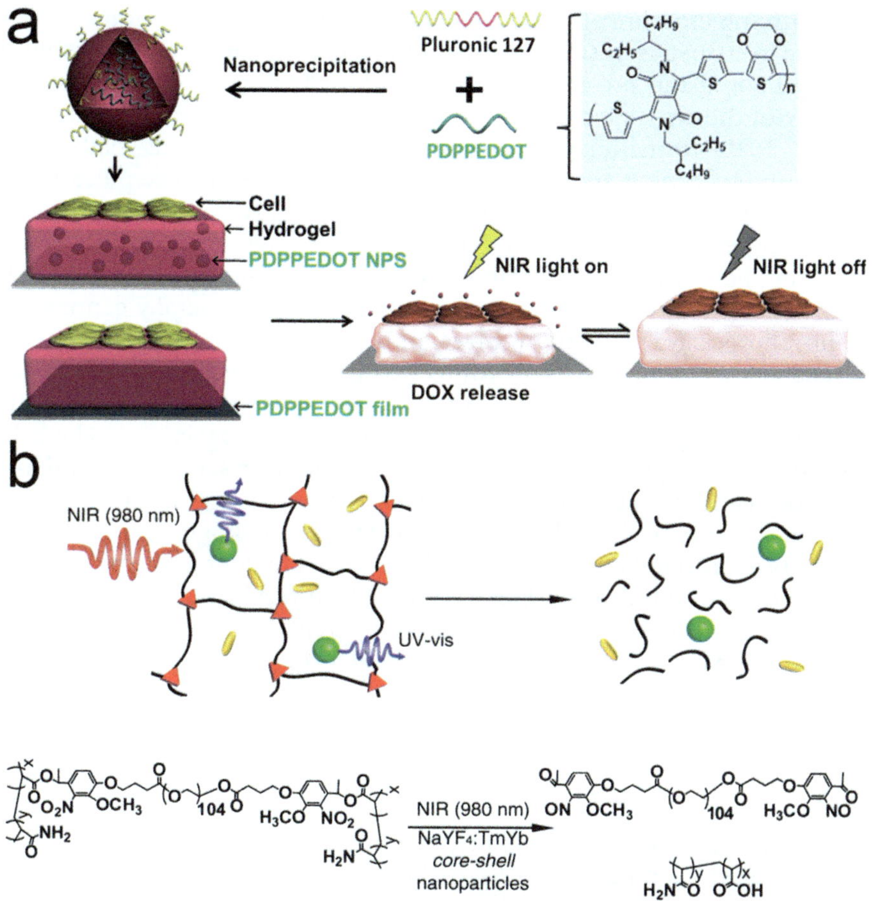

Figure 7.9 Examples of NIR-responsive hydrogels for controlled cargo release. (a) NIR-activated release of DOX from the PDPPEDOT/PNIPAM composite hydrogel *via* a reversible de-swelling/swelling mechanism. Reprinted with permission from ref. 106. Y. Wu, K. Wang, S. Huang, C. Yang and M. Wang, *ACS Appl. Mater. Interfaces*, 2017, **9**(15), 13602. Copyright 2017 American Chemical Society. (b) NIR-triggered de-crosslinking of a photolabile, covalently-linked composite hydrogel by secondary UV irradiation from Yb^{3+}/Tm^{3+}-doped $NaYF_4$ UCNPs. Reprinted with permission from ref. 110. B. Yan, J. C. Boyer, D. Habault, N. R. Branda and Y. Zhao, *J. Am. Chem. Soc.*, 2012, **134**(40), 16558. Copyright 2012 American Chemical Society.

hydrogels. Biomolecules with well-defined hydrogen bonding characters such as peptide and DNA have been used to create photolabile bridging in hydrogels.[107,108] For instance, Wang *et al.* reported a GO-peptide hybrid hydrogel containing a repeating glycine–alanine (GA) peptide sequence terminated with a pyrene motif and a tyrosine residue.[107] The peptides noncovalently link to the surface of GO sheets by the π–π stacking between pyrene and

GO, forming crosslinked junctions. Subsequently, the formation of antiparallel β-sheets between GA repeating units facilitate inter-sheet crosslinking, strengthening the pre-assembled networks. Finally, covalent photo-crosslinking of the tyrosine motif is initiated by a ruthenium-catalyzed reaction, leading to the formation of strong, hybrid hydrogels. The drug-loaded hydrogel exhibits very slow spontaneous drug release, but enhanced pulsatile drug release can be observed once exposed to a 808 nm NIR laser, which elevates local temperature and partially unfolds the β-sheet and disrupts the π–π stacking, both serving to destabilize the crosslinking. For covalently linked hydrogels, low-energy irradiation such as NIR light generally cannot cause bond photocleavage. In these cases, photoluminescent inorganic nanomaterials can be added to the hydrogel to generate secondary, high-energy photons. Recently published work indicates great potential using UCNPs as a tool to trigger drug release from composite hydrogels.[109,110] The hybrid UCNP-hydrogel system developed by Zhao *et al.* represents the first demonstration using Yb^{3+}/Tm^{3+}-doped $NaYF_4$ UCNPs to trigger structural changes and biomacromolecule release in photo-sensitive hydrogels.[110] The hydrogel consists of crosslinked polyacrylamide-PEG backbone that is covalently held by *O*-nitrobenzyl linkages. Upon 980 nm NIR irradiation, de-crosslinking of the structure is observed, which leads to the release of the entrapped model protein, BSA, whose bioactivity is recovered during the gel–sol transition (see Figure 7.9(b)).

7.5 Conclusion

In summary, we have reviewed recent literature on a variety of light-responsive materials as potential DDSs, including organic small-molecule linkers, conjugated organic polymeric nanostructures, inorganic photothermal materials for drug loading and release, and organic-inorganic hybrid materials. Each class of materials has their distinct mechanisms for light activation and drug release, as well as advantages and disadvantages associated. Organic materials show excellent flexibility and mechanistic diversity for light-triggered drug release, but activation typically requires high-energy photons for bond cleavage or altering, which is a critical shortcoming. Inorganic structures possess useful intrinsic properties such as NIR absorption, high photothermal conversion, and non-linear optical responses. Nonetheless, these particles may require additional modification in order to carry drug molecules, and their *in vivo* safety may require further investigation. Organic-inorganic hybrid nanomaterials to some extent circumvent those disadvantages and enable new possibilities. Collectively, these materials allow for increased spatiotemporal control over drug release, which can be combined with other drug delivery techniques such as utilization of the EPR effect and cell surface receptor targeting, to represent a unique branch of drug delivery research.

References

1. S. Marchesan and M. Prato, *ACS Med. Chem. Lett.*, 2013, **4**(2), 147.
2. S. Nazir, T. Hussain, A. Ayub, U. Rashid and A. J. MacRobert, *Nanomedicine*, 2014, **10**(1), 19.
3. O. N. Oliveira Jr, R. M. Iost, J. R. Siqueira Jr, F. N. Crespilho and L. Caseli, *ACS Appl. Mater. Interfaces*, 2014, **6**(17), 14745.
4. J. A. Barreto, W. O'Malley, M. Kubeil, B. Graham, H. Stephan and L. Spiccia, *Adv. Mater.*, 2011, **23**(12), H18.
5. R. A. Petros and J. M. DeSimone, *Nat. Rev. Drug Discovery*, 2010, **9**(8), 615.
6. J. Shi, P. W. Kantoff, R. Wooster and O. C. Farokhzad, *Nat. Rev. Cancer*, 2017, **17**(1), 20.
7. S. Ganta, H. Devalapally, A. Shahiwala and M. Amiji, *J. Controlled Release*, 2008, **126**(3), 187.
8. S. Mura, J. Nicolas and P. Couvreur, *Nat. Mater.*, 2013, **12**(11), 991.
9. J. R. Lopes, G. Santos, P. Barata, R. Oliveira and C. M. Lopes, *Curr. Pharm. Des.*, 2013, **19**(41), 7169.
10. D. Wang, T. Zhao, X. Zhu, D. Yan and W. Wang, *Chem. Soc. Rev.*, 2015, **44**(12), 4023.
11. D. Wang, Y. Jin, X. Zhu and D. Yan, *Prog. Polym. Sci.*, 2017, **64**, 114.
12. X. Ai, J. Mu and B. Xing, *Theranostics*, 2016, **6**(13), 2439.
13. A. M. Smith, M. C. Mancini and S. Nie, *Nat. Nanotechnol.*, 2009, **4**(11), 710.
14. M. J. Hansen, W. A. Velema, M. M. Lerch, W. Szymanski and B. L. Feringa, *Chem. Soc. Rev.*, 2015, **44**(11), 3358.
15. Q. Jin, F. Mitschang and S. Agarwal, *Biomacromolecules*, 2011, **12**(10), 3684.
16. S. Barman, J. Das, S. Biswas, T. K. Maiti and N. D. P. Singh, *J. Mater. Chem. B*, 2017, **5**(21), 3940.
17. S. Barman, S. K. Mukhopadhyay, K. K. Behara, S. Dey and N. D. P. Singh, *ACS Appl. Mater. Interfaces*, 2014, **6**(10), 7045.
18. L. Liao, J. Liu, E. C. Dreaden, S. W. Morton, K. E. Shopsowitz, P. T. Hammond and J. A. Johnson, *J. Am. Chem. Soc.*, 2014, **136**(16), 5896.
19. W. Cui, X. Lu, K. Cui, J. Wu, Y. Wei and Q. Lu, *Nanotechnology*, 2011, **22**(6), 065702.
20. J. Babin, M. Pelletier, M. Lepage, J. F. Allard, D. Morris and Y. Zhao, *Angew. Chem., Int. Ed.*, 2009, **48**(18), 3329.
21. Y. Li, Y. Qian, T. Liu, G. Zhang and S. Liu, *Biomacromolecules*, 2012, **13**(11), 3877.
22. X. Tan, B. B. Li, X. Lu, F. Jia, C. Santori, P. Menon, H. Li, B. Zhang, J. J. Zhao and K. Zhang, *J. Am. Chem. Soc.*, 2015, **137**(19), 6112.
23. D. Han, X. Tong and Y. Zhao, *Macromolecules*, 2011, **44**(3), 437.
24. A. Soleimani, A. Borecki and E. R. Gillies, *Polym. Chem.*, 2014, **5**(24), 7062.

25. Y. Zhang, Q. Yin, L. Yin, L. Ma, L. Tang and J. Cheng, *Angew. Chem., Int. Ed.*, 2013, **52**(25), 6435.
26. C. de Gracia Lux, J. Lux, G. Collet, S. He, M. Chan, J. Olejniczak, A. Foucault-Collet and A. Almutairi, *Biomacromolecules*, 2015, **16**(10), 3286.
27. J. He, X. Tong and Y. Zhao, *Macromolecules*, 2009, **42**(13), 4845.
28. P. M. Kharkar, K. L. Kiick and A. M. Kloxin, *Polym. Chem.*, 2015, **6**(31), 5565.
29. M. S. Kim, J. Gruneich, H. Jing and S. L. Diamond, *J. Mater. Chem.*, 2010, **20**(17), 3396.
30. H. J. Li, H. X. Wang, C. Y. Sun, J. Z. Du and J. Wang, *RSC Adv.*, 2014, **4**(4), 1961.
31. I. A. Dinu, J. T. Duskey, A. Car, C. G. Palivan and W. Meier, *Polym. Chem.*, 2016, 7(20), 3451.
32. H. M. D. Bandara and S. C. Burdette, *Chem. Soc. Rev.*, 2012, **41**(5), 1809.
33. S. Geng, Y. Wang, L. Wang, T. Kouyama, T. Gotoh, S. Wada and J. Y. Wang, *Sci. Rep.*, 2017, 7, 39202.
34. Y. C. Liu, A. L. M. Le Ny, J. Schmidt, Y. Talmon, B. F. Chmelka and C. T. Lee Jr, *Langmuir*, 2009, **25**(10), 5713.
35. P. Bortolus and S. Monti, *J. Phys. Chem.*, 1987, **91**(19), 5046.
36. Q. Jiang, Y. Zhang, R. Zhuo and X. Jiang, *Colloids Surf., B*, 2016, **147**, 25.
37. D. Wang, M. Wagner, H. J. Butt and S. Wu, *Soft Matter*, 2015, **11**(38), 7656.
38. Q. Yuan, Y. Zhang, T. Chen, D. Lu, Z. Zhao, X. Zhang, Z. Li, C. H. Yan and W. Tan, *ACS Nano*, 2012, **6**(7), 6337.
39. J. Lu, E. Choi, F. Tamanoi and J. I. Zink, *Small*, 2008, **4**(4), 421.
40. X. Mei, S. Yang, D. Chen, N. Li, H. Li, Q. Xu, J. Ge and J. Lu, *Chem. Commun.*, 2012, **48**(80), 10010.
41. A. Y. Rwei, J. J. Lee, C. Zhan, Q. Liu, M. T. Ok, S. A. Shankarappa, R. Langer and D. S. Kohane, *Proc. Natl. Acad. Sci. U. S. A.*, 2015, **112**(51), 15719.
42. A. Pashkovskaya, E. Kotova, Y. Zorlu, F. Dumoulin, V. Ahsen, I. Agapov and Y. Antonenko, *Langmuir*, 2010, **26**(8), 5726.
43. D. Luo, N. Li, K. A. Carter, C. Lin, J. Geng, S. Shao, W. C. Huang, Y. Qin, G. E. Atilla-Gokcumen and J. F. Lovell, *Small*, 2016, **12**(22), 3039.
44. A. P. Goodwin, J. L. Mynar, Y. Ma, G. R. Fleming and J. M. J. Fréchet, *J. Am. Chem. Soc.*, 2005, **127**(28), 9952.
45. X. Wang, J. Hu, G. Liu, J. Tian, H. Wang, M. Gong and S. Liu, *J. Am. Chem. Soc.*, 2015, **137**(48), 15262.
46. H. I. Lee, W. Wu, J. K. Oh, L. Mueller, G. Sherwood, L. Peteanu, T. Kowalewski and K. Matyjaszewski, *Angew. Chem., Int. Ed.*, 2007, **46**(14), 2453.
47. L. Sun, X. Ma, C. M. Dong, B. Zhu and X. Zhu, *Biomacromolecules*, 2012, **13**(11), 3581.
48. G. Y. Liu, C. J. Chen, D. D. Li, S. S. Wang and J. Ji, *J. Mater. Chem.*, 2012, **22**(33), 16865.

49. J. Geng, C. Sun, J. Liu, L. D. Liao, Y. Yuan, N. Thakor, J. Wang and B. Liu, *Small*, 2015, **11**(13), 1603.
50. R. Huschka, J. Zuloaga, M. W. Knight, L. V. Brown, P. Nordlander and N. J. Halas, *J. Am. Chem. Soc.*, 2011, **133**(31), 12247.
51. X. D. Zhang, H. Wang, A. L. Antaris, L. Li, S. Diao, R. Ma, A. Nguyen, G. Hong, Z. Ma, J. Wang, S. Zhu, J. M. Castellano, T. Wyss-Coray, Y. Liang, J. Luo and H. Dai, *Adv. Mater.*, 2016, **28**(32), 6872.
52. S. Diao, J. L. Blackburn, G. Hong, A. L. Antaris, J. Chang, J. Z. Wu, B. Zhang, K. Cheng, C. J. Kuo and H. Dai, *Angew. Chem., Int. Ed.*, 2015, **54**(49), 14758.
53. Y. Cao, J. H. Dou, N. J. Zhao, S. Zhang, Y. Q. Zheng, J. P. Zhang, J. Y. Wang, J. Pei and Y. Wang, *Chem. Mater.*, 2017, **29**(2), 718.
54. B. Xia, B. Wang, J. Shi, Y. Zhang, Q. Zhang, Z. Chen and J. Li, *Acta Biomater.*, 2017, **51**, 197.
55. C. You, H. Wu, M. Wang, S. Wang, T. Shi, Y. Luo, B. Sun, X. Zhang and J. Zhu, *Nanotechnology*, 2017, **28**(16), 165102.
56. M. Liu, N. Xu, W. Liu and Z. Xie, *RSC Adv.*, 2016, **6**(87), 84269.
57. C. Wang, H. Xu, C. Liang, Y. Liu, Z. Li, G. Yang, L. Cheng, Y. Li and Z. Liu, *ACS Nano*, 2013, **7**(8), 6782.
58. I. Ojea-Jiménez, J. Comenge, L. García-Fernández, Z. A. Megson, E. Casals and V. F. Puntes, *Curr. Drug Metab.*, 2013, **14**(5), 518.
59. U. Chitgupi, Y. Qin and J. F. Lovell, *Nanotheranostics*, 2017, **1**(1), 38.
60. H. M. Chen and R. S. Liu, *J. Phys. Chem. C*, 2011, **115**(9), 3513.
61. H. Liao, C. L. Nehl and J. H. Hafner, *Nanomedicine*, 2006, **1**(2), 201.
62. X. Huang and M. A. El-Sayed, *J. Adv. Res.*, 2010, **1**(1), 13.
63. M. P. Melancon, M. Zhou and C. Li, *Acc. Chem. Res.*, 2011, **44**(10), 947.
64. J. You, G. Zhang and C. Li, *ACS Nano*, 2010, **4**(2), 1033.
65. Y. Zhang, Z. Hou, Y. Ge, K. Deng and J. Lin, *ACS Appl. Mater. Interfaces*, 2015, **7**(37), 20696.
66. K. Niikura, N. Iyo, Y. Matsuo, H. Mitomo and K. Ijiro, *ACS Appl. Mater. Interfaces*, 2013, **5**(9), 3900.
67. C. Chen, Y. Lin, C. Wang, H. Tzeng, C. Wu, Y. Chen, C. Chen, L. Chen and Y. Wu, *J. Am. Chem. Soc.*, 2006, **128**(11), 3709.
68. W. Lu, G. Zhang, R. Zhang, L. G. Flores II, Q. Huang, J. G. Gelovani and C. Li, *Cancer Res.*, 2010, **70**(8), 3177.
69. S. E. Lee, G. L. Liu, F. Kim and L. P. Lee, *Nano Lett.*, 2009, **9**(2), 562.
70. R. Huschka, O. Neumann, A. Barhoumi and N. J. Halas, *Nano Lett.*, 2010, **10**(10), 4117.
71. C. Cha, S. R. Shin, N. Annabi, M. R. Dokmeci and A. Khademhosseini, *ACS Nano*, 2013, **7**(4), 2891.
72. H. Li, X. Fan and X. Chen, *ACS Appl. Mater. Interfaces*, 2016, **8**(7), 4500.
73. J. D. Hirsch, L. Eslamizar, B. J. Filanoski, N. Malekzadeh, R. P. Haugland, J. M. Beechem and R. P. Haugland, *Anal. Biochem.*, 2002, **308**(2), 343.
74. L. Wang, M. Zhang, N. Zhang, J. Shi, H. Zhang, M. Li, C. Lu and Z. Zhang, *Int. J. Nanomed.*, 2011, **6**, 2641.

75. Z. Liu, K. Chen, C. Davis, S. Sherlock, Q. Cao, X. Chen and H. Dai, *Cancer Res.*, 2008, **68**(16), 6652.
76. P. Zhang, H. Huang, J. Huang, H. Chen, J. Wang, K. Qiu, D. Zhao, L. Ji and H. Chao, *ACS Appl. Mater. Interfaces*, 2015, **7**(41), 23278.
77. Z. Liu, A. C. Fan, K. Rakhra, S. Sherlock, A. Goodwin, X. Chen, Q. Yang, D. W. Felsher and H. Dai, *Angew. Chem., Int. Ed.*, 2009, **48**(41), 7668.
78. S. Liu, A. C. T. Ko, W. Li, W. Zhong and M. Xing, *J. Mater. Chem. B*, 2014, **2**(9), 1125.
79. M. Zheng, A. Jagota, E. D. Semke, B. A. Diner, R. S. Mclean, S. R. Lustig, R. E. Richardson and N. G. Tassi, *Nat. Mater.*, 2003, **2**(5), 338.
80. V. V. Chaban, T. I. Savchenko, S. M. Kovalenko and O. V. Prezhdo, *J. Phys. Chem. B*, 2010, **114**(42), 13481.
81. Y. Tang, H. Hu, M. G. Zhang, J. Song, L. Nie, S. Wang, G. Niu, P. Huang, G. Lu and X. Chen, *Nanoscale*, 2015, **7**(14), 6304.
82. D. He, X. Li, X. He, K. Wang, J. Tang, X. Yang, X. He, X. Yang and Z. Zou, *J. Mater. Chem. B*, 2015, **3**(27), 5588.
83. R. Kurapati and A. M. Raichur, *Chem. Commun.*, 2013, **49**(7), 734.
84. A. M. Ibarra-Ruiz, D. C. R. Burbano and J. A. Capobianco, *Adv. Phys.*, 2016, **1**(2), 194.
85. L. Zhao, J. Peng, Q. Huang, C. Li, M. Chen, Y. Sun, Q. Lin, L. Zhu and F. Li, *Adv. Funct. Mater.*, 2014, **24**(3), 363.
86. M. M. Dcona, Q. Yu, J. A. Capobianco and M. C. T. Hartman, *Chem. Commun.*, 2015, **51**(40), 8477.
87. Q. Xing, N. Li, Y. Jiao, D. Chen, J. Xu, Q. Xu and J. Lu, *RSC Adv.*, 2015, **5**(7), 5269.
88. B. Yan, J. C. Boyer, N. R. Branda and Y. Zhao, *J. Am. Chem. Soc.*, 2011, **133**(49), 19714.
89. Y. Yang, B. Velmurugan, X. Liu and B. Xing, *Small*, 2013, **9**(17), 2937.
90. S. Chen, Y. Gao, Z. Cao, B. Wu, L. Wang, H. Wang, Z. Dang and G. Wang, *Macromolecules*, 2016, **49**(19), 7490.
91. Y. Yang, F. Liu, X. Liu and B. Xing, *Nanoscale*, 2013, **5**(1), 231.
92. J. Liu, W. Bu, L. Pan and J. Shi, *Angew. Chem., Int. Ed.*, 2013, **52**(16), 4375.
93. Y. Jin and X. Gao, *Nat. Nanotechnol.*, 2009, **4**(9), 571.
94. C. Chen, J. Peng, S. R. Sun, C. W. Peng, Y. Li and D. W. Pang, *Nanomedicine*, 2012, 7(3), 411.
95. Z. Xu, Z. Wu, J. Sun and R. Gui, *Mater. Chem. Phys.*, 2015, **162**, 286.
96. H. Jin, R. Gui, J. Sun and Y. Wang, *Anal. Chim. Acta*, 2016, **922**, 48.
97. E. Caló and V. V. Khutoryanskiy, *Eur. Polym. J.*, 2015, **65**, 252.
98. X. Huang and C. S. Brazel, *J. Controlled Release*, 2001, **73**(2–3), 121.
99. A. Shiotani, T. Mori, T. Niidome, Y. Niidome and Y. Katayama, *Langmuir*, 2007, **23**(7), 4012.
100. L. E. Strong, S. N. Dahotre and J. L. West, *J. Controlled Release*, 2014, **178**, 63.
101. E. Miyako, H. Nagata, K. Hirano and T. Hirotsu, *Small*, 2008, **4**(10), 1711.
102. Z. Cheng, R. Chai, P. Ma, Y. Dai, X. Kang, H. Lian, Z. Hou, C. Li and J. Lin, *Langmuir*, 2013, **29**(30), 9573.

103. X. Dong, C. Wei, J. Liang, T. Liu, D. Kong and F. Lv, *Colloids Surf., B*, 2017, **154**, 253.
104. W. Li, J. Wang, J. Ren and X. Qu, *Adv. Mater.*, 2013, **25**(46), 6737.
105. F. Teodorescu, Y. Oz, G. Quéniat, A. Abderrahmani, C. Foulon, M. Lecoeur, R. Sanyal, A. Sanyal, R. Boukherroub and S. Szunerits, *J. Controlled Release*, 2017, **246**, 164.
106. Y. Wu, K. Wang, S. Huang, C. Yang and M. Wang, *ACS Appl. Mater. Interfaces*, 2017, **9**(15), 13602.
107. J. Wu, A. Chen, M. Qin, R. Huang, G. Zhang, B. Xue, J. Wei, Y. Li, Y. Cao and W. Wang, *Nanoscale*, 2015, **7**(5), 1655.
108. H. Kang, A. C. Trondoli, G. Zhu, Y. Chen, Y. Chang, H. Liu, Y. Huang, X. Zhang and W. Tan, *ACS Nano*, 2011, **5**(6), 5094.
109. G. Jalani, R. Naccache, D. H. Rosenzweig, L. Haglund, F. Vetrone and M. Cerruti, *J. Am. Chem. Soc.*, 2016, **138**(3), 1078.
110. B. Yan, J. C. Boyer, D. Habault, N. R. Branda and Y. Zhao, *J. Am. Chem. Soc.*, 2012, **134**(40), 16558.

CHAPTER 8

Integrated Polymer Composites for Electro-responsive Drug Delivery

P. PRADEEP, P. KUMAR, Y. E. CHOONARA AND V. PILLAY*

Wits Advanced Drug Delivery Platform Research Unit, Department of Pharmacy and Pharmacology, School of Therapeutic Sciences, Faculty of Health Sciences, University of the Witwatersrand, Johannesburg, South Africa
*E-mail: viness.pillay@wits.ac.za

8.1 Introduction

Owing to the tremendous advances in pharmaceutical sciences, a broad range of delivery systems with specialised architectures and properties have been developed. These delivery systems not only effectively target the diseased area, but do so in an 'on-demand' fashion.[1] These 'on-demand' delivery systems release the therapeutic agent with precise control through a stimuli response that identifies the changes in the immediate microenvironment and reacts towards it (Figure 8.1).[2,3] This approach requires biocompatible materials that could control the drug bio-distribution in response to a specific physical stimulation, which could be anything from variations in temperature, magnetic field, ultrasound intensity, light, electric environment, changes in pH, enzyme concentration or redox gradients.[4] In the present

Biomaterials Science Series No. 1
Stimuli-responsive Drug Delivery Systems
Edited by Amit Singh and Mansoor M. Amiji

Published by the Royal Society of Chemistry, www.rsc.org

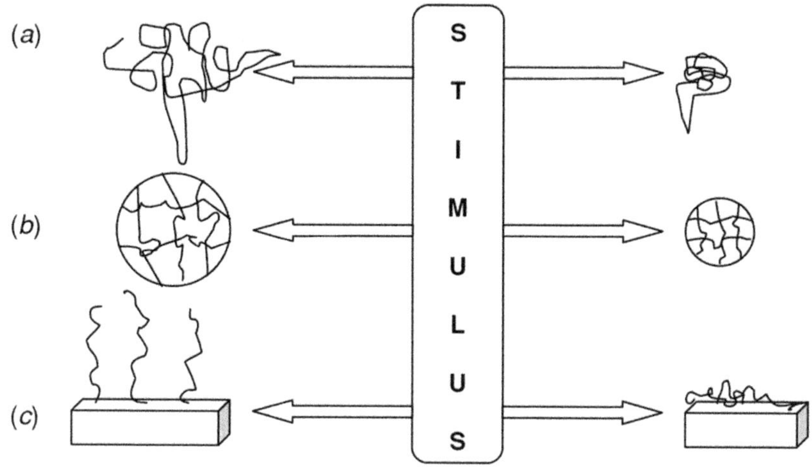

Figure 8.1 Classification of the polymers by their physical form: (a) linear free chains in solution where polymer undergoes a reversible collapse after an external stimulus is applied; (b) covalently cross-linked reversible gels where swelling or shrinking of the gels can be triggered by environmental change; and (c) chain adsorbed or surface-grafted form, where the polymer reversibly swells or collapses on surface, once an external parameter is changed. (Reprinted from Progress in *Polymer Science*, **32**, A. Kumar, A. Srivastava, I. Y. Galaev and B. Mattiasson, Smart polymers: Physical forms and bioengineering applications, 1205–1237, Copyright 2007, with permission from Elsevier).[2]

book chapter, we are going to discuss recent advances made in the development of electro-responsive drug delivery (ERDD) systems.

Hydrogels are three-dimensional high-molecular weight networks which comprise of a polymer backbone, water and a crosslinking agent. Hydrogels do not dissolve in water at physiological conditions and swell considerably in an aqueous medium due to their high-water imbibing property into their polymeric network, which provides them with a certain degree of structural integrity and elasticity.[5] Their hydrophilicity could be attributed to the presence of functional groups such as –OH, –COOH, $-CONH_2$, and $-SO_3H$ which are capable of absorbing water without undergoing dissolution. The hydrogels usually exhibit a phase transition in response to the change in external conditions such as pH, electric current, temperature, ionic concentration and are known as "stimuli-responsive" or "smart" gels. Being insoluble, the polyelectrolyte hydrogels, owing to their resemblance to biological tissues, have been widely used as electro-responsive drug depots where the delivery of drug from the gel matrix is governed by an electric stimulation (Figure 8.2).[6] Due to alteration of the hydrogel charge, the drug release is promoted by the built-in electrostatic interactions and the osmotic pressure. Due to their high water/biological content (up to 99%), these natural or synthetic polyelectrolyte complexes are highly sensitive to electrical currents as they

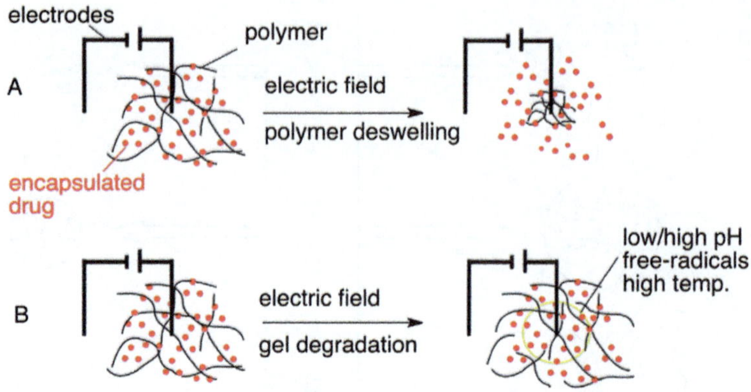

Figure 8.2 Electric field-stimulated drug delivery. (A) Polyelectrolyte gels de-swell towards oppositely charged nodes in the presence of electric fields. During de-swelling, water and therapeutic compounds are expelled. The de-swelling is a passive, diffusion-controlled process. (B) pH and chemical changes accompanying an electric filed can degrade or de-gel the depot, leading to release of therapeutic compounds (Reprinted from *Journal of Controlled Release*, **219**, Y. Brudno and D. J. Mooney, On-demand drug delivery from local depots, 8–17, Copyright 2015, with permission from Elsevier).[6]

align themselves parallel to the electrodes when stimulated electrically and allow for efficient transport of electrical ions. Under the influence of the electric field water is osmotically removed from the system, which leads to the de-swelling of the polyelectrolyte and thus forcing it to eject the drug.[7] The amount of drug release in response to the electrical fiend is closely correlated to the loss of gel-mass. Precise control of the magnitude of electric field pulses, duration of pulse and the interval between the pulsed is controlled using external medical equipment. A range of poly-anionic polymers are employed for the synthesis of these polyelectrolyte gels *such as* hyaluronic acid, hydrolyzed polyacrylamide, alginate, chondroitin sulphate agarose and xanthan. On the other hand, electro-responsive polyelectrolyte hydrogels made from poly-cationic or neutral polymers are relatively rare.[6,8]

Synthetic, as well as naturally occurring polymers either alone or in combination, have also been used. Hydrogels made from natural polymers offer numerous advantageous properties such as inherent biocompatibility, biodegradability and biologically recognizable moieties that support cellular activities. However, they may not always provide sufficient mechanical strength and may evoke immune/inflammatory responses in the presence of higher pathogens. Synthetic hydrogels, on the other hand, do not possess any inherent bioactive properties, which is compensated for by their well-defined structures that could be modified to yield tailored degradability and functionality.[4] Additionally, hydrogels could be fabricated in a multitude of shapes, such as, rods, disks, films, microparticles, liable on the intended application and the site of administration.[9]

8.2 An Ideal Electroresponsive Hydrogel Device

Precise and optimal control of drug concentration in the plasma could be achieved if the hydrogel drug delivery system responds in a reproducible and predictable fashion towards the external electrical stimulus, where the release of the drug is directly related to the magnitude of the stimuli.[10] The polyelectrolyte hydrogel should be biocompatible, non-irritant, with ease of administration and biodegradable (it avoids the need to remove the hydrogel once the entire drug-load has been depleted). Generally, the electro-responsive hydrogel systems are used for transdermal delivery systems where they could be easily applied as a film. If administered internally a multitude of factors needs to be considered. Thermo-intelligent hydrogels, which also have electro-conductive properties could be proposed as ideal delivery systems where the implant is needed inside the body. The polymer could be injected into the required site and it would gel *in situ* once in the desired place. This would remove the need of surgical intervention to implant the hydrogel, thus improving patient compliance and cost of treatment.[11]

In 1982, Tanaka and co-workers for the first time explained experimentally how a minuscule change in electric potential across a polyelectrolyte gel can produce a very pronounced, reversible volume change. Due to which, the volume of the collapsed gel could be up to several hundred times smaller (up to 500 folds less) compared to that of the swollen gel. The authors demonstrated that the phase transition could be induced by the application of an electric field across the gel and the electric forces on the charged sites of the network produce a stress gradient along the electric field lines in the gel. There exists a critical stress below which the gel is swollen and above which the gel collapses. The volume change at the transition is either discrete or continuous, depending on the degree of ionization of the gel and on the solvent composition.[12] As early as in 1993, Banga and co-workers established the relationship of electric induced (iontotherapeutic) drug release from a hydrogel matrix. The authors demonstrated the release of three peptides, these were, insulin, calcitonin and vasopressin from a polyacrylamide hydrogel-based iontotherapeutic matrix system. The release of the peptides from the hydrogel matrix was observed to follow a Q *vs* $t^{1/2}$ equation under passive diffusion conditions, which shifted to a Q *vs* t equation under the influence of electric potential. Additionally, the release flux ($\mathrm{d}Q/\mathrm{d}t$) of the peptides was seen to decline when the electric stimuli was withdrawn and the release was resumed when the current was *turned on.*[13] —Further to this, Jensen and co-workers, in 2002, highlighted one major concern of electro-responsive hydrogels, which is the back-flow of the drug into the gel.[14,15] Occasionally, a back-flow of drug from the gel environment into the gel has been observed once the electric field is switched off, due to lower drug concentration in the hydrogel. This back-flow of the drug into the gel occurs by the phenomenon of convection, contrary diffusion, which rather seems that the mechanism of the drug concentration inside the hydrogel was lower than the drug concentration in the surrounding medium. The authors demonstrated the back-flow of the released drug into chondroitin sulphate gel once the electric field was switched off for the very first time. The size of the drug molecule was also found to play a crucial part in this phenomenon smaller molecules tend to be more prone to back-flow compared to bigger molecules, as the molecules with a larger structure become restricted by the pore size of the de-swelled hydrogels.[14,16]

8.3 Recent Advances in Electroresponsive Drug Delivery: Representative and Leading Examples

Electro-responsive hydrogels have been found to be very effective in transdermal and/or subcutaneous implants. Incorporation of various additives such as polyethylenedioxythiophene, polypyrrole, polyaniline, redox-based conductors and carbon nanotubes (both single walled and multiwalled) to form nanocomposite hydrogels further improve the mechanical and electrical properties of these hydrogels (Table 8.1).[17–20] A multitude of polymeric materials have been fabricated in a way to impart electro-conductive properties to the resulting hydrogel, however often it has been observed that the conductive properties are not high enough to obtain the desired pulsating drug release. This leads to the need to incorporate an additional conductive material into the polymeric matrix of the hydrogel.[21]

8.3.1 Polymer-carbon Nanotube Composites

The composites of polymers and carbon nanotubes (CNT) have been pursued vigorously in the formation of EDDA pertaining to the noteworthy properties of CNT (strength, flexibility, electrical conductivity, and high aspect ratio). However, due to the remarkably small size of CNTs it is challenging to control their spatial organization once in the composite.[22] Once the surrounding environment of CNTs is studied it becomes easier to control the interactions between CNTs and their surrounding environment.[23] In this regard, CNTs, either single-walled (SWNTs) or multi-walled (MWNTs), have been widely used and have been shown to enhance the conductive properties and strength of the hydrogel. Carbon nanotubes are pure carbon polymers, where each carbon atom in the lattice is bound covalently to only three surrounding carbon atoms because of which, a symmetric beam-and-truss structure is formed which not only provides strength but also maintains flexibility. Although, the remaining valence electron is shared and mobile, which provides the conductive property to the nanotubes. CNT-polymer hybrid materials have been shown to have a rather peculiar property which none of the composing materials show otherwise when alone. There have been reports where highly electro-responsive delivery systems have been fabricated (even employing polymers that are not polyelectrolyte) with regulated drug release profiles. Larger hydrophobic surfaces of CNT

Table 8.1 Electrochemically controlled drug delivery based on intrinsically conducting polymers (Reproduced with permission from ref. 20 ©2013 WILEY PERIODICALS, INC.).

Electroactive polymer(s) used	Bioactive	Salient finding(s)
Polypyrrole coating	Neurotrophin-3; dexamethasone	The amount of bioactive incorporated into and released from the polymer can be controlled by varying the polymerization time, with longer growth periods incorporating more drug and releasing a lower percentage of the incorporated molecule
Poly(3,4-ethylene-dioxythiophene) poly(styrenesulfonate) fibers	Ciprofloxacin hydrochloride	The drug retained its antibacterial properties during fibre fabrication and electrochemical treatment for controlled release of the drug
Polypyrrole	Neurotrophin-3 and brain-derived neurotrophic factor	The conducting polymer films provided a biocompatible substrate for storage and release of neurotrophins and encouraged neurite outgrowth toward the electrodes
Poly(*p*-phenylene vinylene)/poly-acrylamide (PPV/PAAM)	Salicylic acid (SA)	Drug was released *via* combined mechanisms: The expansion of PPV chains inside the hydrogel; the reduction reaction under a negative potential driving the anionic SA through the PAAM matrix; and the expansion of the matrix pore
Spherical gelatin/CNTs hybrid microgels	Diclofenac sodium salt	Covalently inserted different amounts of nanotubes (up to 35% by weight) may confer the highest electric sensitivity to the composite microspheres
Polypyrrole scaffolds	Risperidone	The high surface area polypyrrole inverse opal scaffolds exhibited enhanced drug loading and releasing capabilities compared to conventional nonporous PPy films
Polypyrrole/poly(acrylic acid) hydrogel	Sulfosalicylic acid	The drug diffusion coefficient decreases with increasing drug size/mesh size ratio, irrespective of the presence of the conductive polymer as the drug carrier
Cross-linked poly(vinyl alcohol) hydrogel	Sulfosalicylic acid	The diffusion coefficients of drug-loaded PVA hydrogels decrease with increasing cross-link ratio and depend critically on the electric field strength, as well as on the electrode polarity

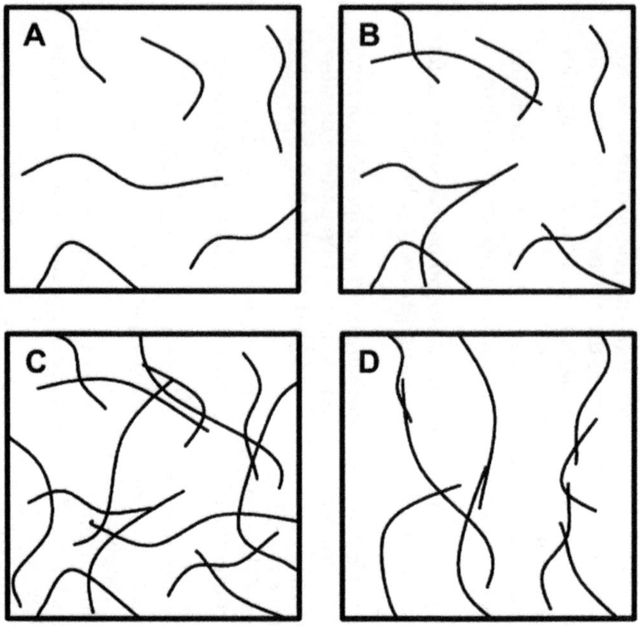

Figure 8.3 Schematic diagram of electrical percolation in fibre-reinforced composite materials. (A–C) The increased fibre loading level leads to longer conducting pathways through the sample. (D) How long conductive pathways and electrical percolation can be achieved at lower loading levels by aligning the fibre phase (Reprinted from, *Acta Biomaterialia*, **4**, R. A. MacDonald, C. M. Voge, M. Kariolis and J. P. Stegemann, Carbon nanotubes increase the electrical conductivity of fibroblastseeded collagen hydrogels, 1583–1592, Copyright 2008, with permission from Elsevier).[24]

could be used as a functional element for drug loading *via* π–π stacking and when an external electric field is applied such interactions are disturbed, leading to either attraction or repulsion depending on the charge of the drug and ultimately the release of the drug.[24] MacDonald and co-workers (2008) described a unique phenomenon applicable to almost every composite structure doped with a conducting nanomaterial. The authors concluded that for an efficient electrical percolation, longitudinal and continuous conductive pathways should be created (Figure 8.3). Such pathways can be achieved either by high loading of the reinforcing material or by aligning the reinforcing phase, given the high aspect ratio of the CNTs. Additionally, good and uniform distribution of the reinforcing phase is crucial to the electroactuability of the composite.[24]

Voge and co-workers (2013) extensively studied CNTs (both single walled and multi walled) as reinforcements for electrical conductivity in hydrogels. SWNT and human dermal fibroblast cells (collagen) composite were fabricated by the authors. The resulting constructs were then evaluated for their electrical properties in the frequency range 5–500 kHz and were found to be uniformly increased with increasing wt% of SWNT. MWCNTs were also employed as reinforcements in the composites of extracellular matrix proteins, which have demonstrated conductive properties. MWCNTs were functionalization with a carboxyl group and polymer wrapped with either Pluronic F127 or gelatin in the aqueous media. Polymer wrapping resulted in increased surface adsorbed oxygen (Pluronic F127) and nitrogen (gelatin), respectively, along with augmentation of the stability of MWCNTs. Functionalization and wrapping were also associated with maintenance of fibroblast function within protein–MWCNTs.[25]

MWNTs as electro-conducting elements were hybridised with gelatin, employing a modified grafting approach to form a composite hydrogel as demonstrated by Spizzirri and co-workers. MWNTs significantly enhanced the electro-conductive properties of the hydrogel appropriate to modulating the release of diclofenac sodium salt when under the influence of an external direct current voltage. Functionalization of MWNTs was carried out by carboxylation and later coated with Pluronic VR or gelatin. Eventually, the modified MWCNTs were used to make 3D collagen–fibrin–MWCNT composite materials. The delivery system proved to show higher conductivity and improved cytocompatibility.[26]

- Curcio and co-workers (2015) fabricated electro-responsive hybrid hydrogel devices for pulsating delivery of diclofenac sodium salt (DS) and ciprofloxacin (CP), as anionic and cationic model drugs, respectively. Polymeric hydrogel was synthesised from UV induced polymerization of acrylamide (monomer) and *N*,*N*-ethylene bisacrylamide (as a crosslinking agent). To impart the electro-responsive properties and to control the release of the ionized therapeutics (by the modification of the hybrid charge), MWNTs were covalently inserted into the polymeric matrix. Hydrogel properties were found to depend strongly on the MWNT content. Hybrid materials with higher amounts of MWNT showed very interesting electro-responsivity.[27]
- In their study, Im and co-workers (2010) showed the synthesis of an electrically conductive and electro-sensitive polyethylene oxide/pentaerythritol triacrylate transdermal film which was reinforced using MWCNTs for even better electrical conductivity and biocompatibility for the transdermal delivery of ketoprofen. MWCNTs were electrospun with the polymeric fibres as a conductive additive. Drug release was found to be efficiently increased with applied electrical voltage, pertaining to the outstanding conductivity of the MWCNTs.[28]
- Yun and co-workers also managed to develop an electroresponsive transdermal delivery system by electrospinning of poly(vinyl alcohol)/poly(acrylic acid)/MWCNTs nanocomposites. Furthermore, the MWCNTs were surface treated by oxyfluorination to introduce functional groups on the rather hydrophobic MWCNTs with an aim of improving the distribution of hydrophobic MWCNTs in hydrophilic hydrogel matrices. Electrospun hydrogel nanofibers containing PVA/PAA/MWCNT

were then observed to have increased drug release due to increased surface area. The conductivity of nanofibers also improved, pertaining to an increase in the content of MWCNTs and performing oxyfluorination with a higher oxygen content. Uniform distribution of the oxyfluorinated MWCNTs in the nanofibers was crucial to the electro-responsive swelling, and thus drug releasing performance, of the nanofibers.[29]

Several chronic inflammatory diseases lead to destructive erosion of bone (osteolysis) as a serious complication and bone implants have been implied in various instances. Occasionally, these orthopaedic implants fail, which could be due to the generation of metallic or polymeric wear and these loosened particles then trigger an immune response. To this end, Sirivisoot and co-workers (2011) suggested the on-demand delivery of penicillin, streptomycin and dexamethasone *in situ* from the PPy surface on the bone implants. MWNT–Ti and were used as templates to electrodeposit PPy, which was bound with either penicillin, streptomycin or dexamethasone. It was observed that these anionic drugs, immobilized electrostatically inside the PPy thin film, were released when a negative voltage was applied. Furthermore, PPy can also be coated on electrodes or integrated within implantable chips to introduce an electrical signal into the biological environment. It has been observed that PPy could be overoxidized and subsequently lose its electroactivity at higher potentials, or when voltages are applied for prolonged periods, but the incorporation of MWCNTs helps in sustaining the electroactivity of PPy because of their outstanding conductivity properties.[30]

8.3.2 Polypyrrole Based Scaffolds

Apart from CNTs, which are extensively used in the fabrication of electro-conductive hydrogels, the use of electrically conductive polymers has also been reported for the fabrication of conductive delivery systems. PPy is an organic conductive polymer and can be easily synthesized either chemically or electrochemically by oxidative polymerization of pyrrole and has been shown to display inherent good electrical conductivity, long-term stability, and biocompatibility. PPy has been seen as a useful tool for a magnitude of applications such as biosensor and drug delivery systems. Yang and co-workers (2016) synthesized a novel conductive hydrogel composed of pyrrole incorporated hyaluronic acid (HA) and PPy. Which were soft, porous, structurally stable and electrically conductive. Furthermore, electrical conductivity was found to be directly proportional to the concentration of PPy.[31] For the treatment of non-infectious posterior uveitis, corticosteroids such as dexamethasone are frequently prescribed, which are either administered *via* intravitreal injection or with ocular implants. Intravitreal injections are associated with poor treatment adherence and complications such as endophthalmitis, whereas ocular implants provide sustained corticosteroid release at predetermined rates but lack the ability for dose individualisation. To this end, Seyfoddin and co-workers demonstrated the use of PPy, loaded with dexamethasone sodium phosphate (DexP), an anti-inflammatory corticosteroid, for the fabrication of a biocompatible and electrically conductive vitreal drug delivery system. Colloidal crystal films possessing 370 nm polymethylmethacrylate colloids were first deposited on ITO coated glass substrates, and later used as sacrificial templates for the synthesis of high surface area, 3-dimensionally ordered, macroporous PPy inverse opal (PPy IO) thin films. By utilising a sub-micron sized polymer colloids sacrificial template, PPy films having precise pores and well-ordered three-dimensional structure were fabricated. Ultimately, the sacrificial polymer template was removed leaving behind the microporous conductive PPy film. The microporous film provided significantly higher drug release as compared to conventional non-porous films. The drug release from the polymer film followed the typical diffusion driven sustained release profile which can further be actuated by electrical stimulation to provide intermittent burst release for on demand delivery of dexamethasone.[32]

8.3.3 Semi conductive, Polymer-Peptide-Hydrogel (PPH) Nanocomposite

Biocompatibility is a pertinent issue when it comes to nanocomposites. Strategies such as incorporation of biocompatible polymer coating or modification with peptides can provide much needed immuno-conformity to the drug delivery system.[33] Pourjavadi and Doroudian (2015) described a complex nano-hydrogel composite wherein hydrolysed collagen (the peptide component) was first modified with polycaprolactone (the polymer component) followed by *in situ* incorporation into a polyacrylic acid matrix (the hydrogel component). To render the PPH system conductive, aniline was polymerized *in situ* within the swollen hydrogel to form "conductive nanofiber pathways". The surface resistivity of the developed films before and after applying potential (3 V for 10 min) was within the biomedical range of 1.5×10^{-4} to 3×10^{-5} S cm^2. Additionally, the swelling ratio of the hydrogel films, and hence drug release, increased with increasing potential due to nanofibrous polyaniline induced pore size expansion.[34]

8.3.4 Smart Montmorillonite-polypyrrole Scaffolds

Smart clay-polymer composites are capable of providing a large surface area for higher incorporation of a dopant drug.[35] Wang and co-workers (2016) developed a unique clay-polymer scaffold wherein PPy was synthesized in the presence of montmorrilonite (Mt) *via* electropolymerization. Innovatively, an anionic drug (aspirin) was included in the polymerization process as the model drug, as well as a dopant to neutralize the positively charged PPy chains. The drug was thus also intercalated within the layered Mt architecture and interacted with the cations in Mt. With aspirin being an anion, the electrical stimulation was carried out at a negative potential of −0.6 V.

Interestingly, the cumulative aspirin release (at 230 min) for PPy-Mt hybrid was 70.4 μg mL^{-1} as compared to 57.3 μg mL^{-1} in the case of PPy-alone scaffolds, confirming the role of Mt in significantly improving the electro-responsive drug release from the PPy matrix.[36]

8.3.5 Electrically Responsive Microreservoirs (ERMR) for Bone Tissue Engineering

Glucocorticoids such as dexamethasone (Dex) are required for the suppression of immune responses, as well as for the stimulation of osteogenic differentiation post-implantation. However, uncontrolled or irregular release of glucocorticoids may lead to complications such as glucocorticoid-induced osteoporosis.[37–39] Paun and co-workers (2017) reported electro-responsive microreservoirs consisting of vertical IP-L780 microtubes produced by a two photon polymerization–laser direct writing (2PP_LDW) technique. The conductive element (PPy)/drug (Dex) was incorporated into the microreservoirs *via* a simple immersion technique and the ERMR produced were sealed with a PLGA layer employing Matrix Assisted Pulsed Laser Evaporation (MAPLE) (Figure 8.4). When stimulated electrically at −1 V and +1 V cycles, the ERMR

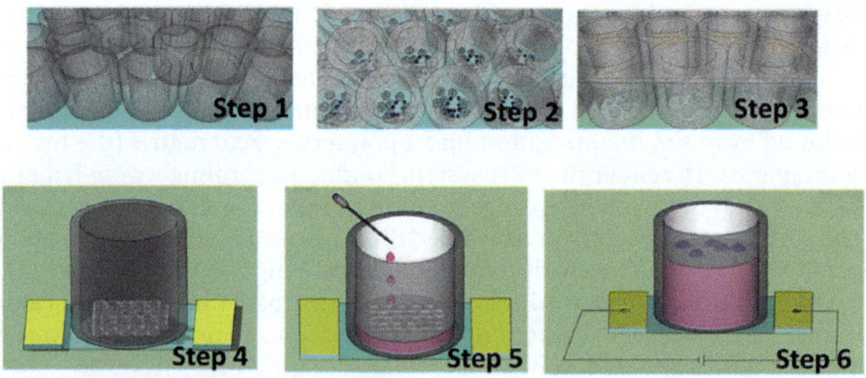

Figure 8.4 Schematic representation of ERRs preparation and electrically controlled Dex release: Step 1: Arrays of vertical microtubes produced by 2 PP LDW. Step 2: Microtubes loaded with PPy/Dex-Gly (PPy: gray spheres; Dex: black spheres). Step 3: Electrically responsive microreservoires (ERRs): PPy/Dex-Gly loaded microtubes covered with a thin PLGA layer deposited by MAPLE. Step 4: Polypropylene rings are glued around the ERRs, to create wells for cell seeding. Step 5: MG-63 cells seeded in the sample wells. Step 6: Electrical stimulation of the ERRs controllers the kinetics of Dex release (Reprinted from *Applied Surface Science*, **392**, I. A. Paun, M. Zamfirescu, C. R. Luculescu, A. M. Acasandrei, C. C. Mustaciosu, M. Mihailescu and M. Dinescu, Electrically responsive microreservoires for controllable delivery of dexamethasone in bone tissue engineering, 321–331, Copyright 2017, with permission from Elsevier).[40]

released 450–460 nM Dex following three voltage cycles within a time scale of 150–275 h. The dimensions of microreservoirs played a significant role in the amount of drug released over a given period. The osteocompatibility studies confirmed that the osteoblast-like MG-63 cell cultures retained their viability and adhesion properties and the polygonal morphology was preserved.[40]

8.3.6 Peptide-conjugated Hydrogel Nanoparticles

The blood brain barrier is a major impediment to CNS drug delivery and even when this barrier is overcome, the delivery and release of bioactives within the neurospace cannot be ascertained. Neurocompatible functionalization and electrostimulation in conjugation may provide the solution to the above status quo.[41] In a study published in Angewandte Chem International Edition, Ying and co-workers (2014) developed angiopep-2 (ANG; a brain targeting peptide) modified electro-responsive hydrogel nanoparticles (ERHNPs) for neuroefficient delivery of phenytoin across the blood brain barrier. The multipolymeric hydrogel nanoparticles were synthesized *via* co-polymerization of 2-dimethylamino ethyl methacrylate (DMAEMA), sodium 4-vinylbenzene sulfonate (NaSS), styrene (ST), and acrylate-poly(ethylene glycol)-*N*-hydroxysuccinimidylester (ACLT-PEG-NHS). The polyelectrolyte, poly(sodium 4-vinylbenzene sulfonate) (PSS), formed within the hydrogel demonstrated electrostimulation properties. On application of an electric field of 100 and 200 μA, 60.8% and 87.3% of accumulated phenytoin was released, respectively, as compared to 34.6% with no electrical stimulation. This finding was very significant, as an extracellular change in the electrical potential at the onset of epileptic seizures may lead to the triggered release of phenytoin, thus providing an "*in situ* on-demand" delivery of an anti-epileptic intervention.[42]

8.3.7 Reduced Graphene Oxide Composite Hydrogel

Carbon nanostructures such as CNTs and graphene can be incorporated into hydrogel structures forming nanocomposites with a large surface area, electrical conductivity and adjustable physicomechanical properties. Specialized structures such as reduced graphene oxide (rGO) can further provide redox functionalization *via* active oxygen-containing functional groups.[43] Kenna and co-workers (2015) incorporated rGO within a jeffamine polyetheramine and polyethylene glycol diglycidyl ether (PEGDGE) hydrogel system. The researchers included a negatively charged model compound, methyl orange, to elucidate the effect of electrical stimulation on small molecule release from the rGO-hydrogel. Given the anionic nature of the molecule, it was expected that the MO and rGO molecules would show electrostatic repulsion at a negative applied potential and hence an accelerated release from the matrix. The opposite was true for an applied positive potential. The incorporation of rGO not only provided the much needed mechanical and electrochemical properties but also rendered "on-demand controlled release" properties to the hydrogel, wherein the concentration of rGO within the matrix (up to 20%) and polarity and magnitude of the electrode (−6 V to +6 V) delivered slow to rapid and low to high concentration of molecular release.[44]

8.4 Molecular mechanism(s) of Electro-actuable Drug Release

The authors have published several research and review reports describing the applicability of electroresponsive polymers for drug delivery applications. In one of the initial studies, Tsai and co-workers (2011) developed a unique electroactuable drug-eluting system composed of diethylacetamidomalonate (DAA)-crosslinked polyvinyl alcohol (PVA) as the hydrogel component and polyaniline (PANi) as the electroconductive component. Interestingly, the crosslinker (DAA) provided the required structural stability to the hydrogel architecture while maintaining the swellability of the PVA matrix. The electroconductive hydrogels (ECH) so developed were capable of "ON–OFF" switchable drug release at an applied potential of as low as 0.3 V proving the electroresponsiveness of the developed drug delivery system. Additionally, the drug release under the influence of an electrical field was significantly affected by formulation factors such as the degree of crosslinking, PVA/PANi ratio, and the drug loading. Molecular mechanistic simulations of the PVA-PANi hydrated polymer matrix under the influence of electrical field corroborated the experimental observations and concluded that the electro-actuable release from the PVA/PANi matrix was due to the structural and network instability caused by the external electrical field. This increased instability with increasing electrical stimulation was attributed to the destabilization of bond torsional and bond angle energy components.[45]

The diethylacetamidomalonate (DAA)-crosslinked polyvinyl alcohol (PVA) electroresposnive system was further developed into a complex co-polymeric implantable gel *via* the incorporation of polyethylene glycol (PEG) and sodium polystyrene sulphonate (NaPss) and was evaluated for potential electro-liberation of the model drug (diclofenac sodium). The electro-active gel exhibited superior structural stability across various electroactuation cycles and the inherent homogeneous morphology of the matrix was maintained. An on demand "ON–OFF" pulsatile drug release of 37% (t_{30min}) to 94% (t_{180min}) was obtained on electrostimulation, confirming the applicability of the PEG/NaPss/PVA ERP-Gel as an injectable drug delivery implant. Under molecular mechanics simulations, the following mechanism of electroactuable drug release was proposed; when an electric potential is applied, the distribution of charge within the polymer network causes the hydrogel matrix to concentrate towards the outer shell creating the critical concentration paradigm required for polymer erosion, this was corroborated by the fact that some matrix loss was observed after application of the electric field. The torsional and stretching energy contributions discussed above by Tsai and co-workers were also observed within PEG/NaPss/PVA ERP-gels in addition to ion–ion

Table 8.2 Electro-influenced Geometrical Organization-ReOrganization Theory for bipolymeric interfacially plasticized electroresponsive hydrogel (Reprinted from ref. 47, *International Journal of Pharmaceutics*, **462**, S. Indermun, Y. E. Choonara, P. Kumar, L. C. du Toit, G. Modi, R. Luttge, V. Pillay, An interfacially plasticized electro-responsive hydrogel for transdermal electro-activated and modulated (TEAM) drug delivery, 52–65, Copyright 2014, with permission from Elsevier).

The O = organization	The re-organization
Polymeric chains organize with respect to the direction and strength of electric field: Electric field application → polymer chains organization → increase in static energy due to electron transfer reaction → molecular alignment → planar structural conformation → reduced networking → electroresponsive drug release	Polymeric chains in assumptions 1 reorganize with respect to surrounding polymer molecules/plasticizer/solvent molecules *via* "local oriental correlations (LOCs)": Intrinsic interactions → local oriental correlations → change in reaction co-ordinates → solvent relaxation → polymer chains reorganization → decrease in static energy values → increased networking → drug retention

dipole translational forces due to electrical interactions. Additionally, such tortuous differentiations within the matrix may lead to localised expansions within the crosslinked and rigid intermolecular complex and hence network swelling and loosening. The directional energetic destabilization of the matrix architecture, enhanced matric swelling, and resulting matrix erosion can be associated with enhanced drug release *via* electrical stimulation.[45,46]

Continuing the mechanistic elucidation of electro-conductive hydrogels (ECHs), Indermun and coworkers provided a pioneering theory to explain the electro-responsive drug release behaviour of semi-interpenetrating networks (semi-IPNs) ECHs synthesized from polyacrylic acid (PAA) and poly(vinyl alcohol) PVA) and doped with poly(ethyleneimine) (PEI) and 1-vinylimidazole (VI) for electroconductivity. The bipolymeric interfacially plasticized electro-responsive hydrogel (BiPErG) underwent an electro-influenced geometrical organization and reorganization to release the small drug molecule from the matrix architecture. The assumptions and observations of the theory are depicted in Table 8.2.[47–49]

8.5 Conclusions and Future Directions

It is evident from the above discussion that integrating materials composites with electroactive hydrogels provides an edge to ERDDs in terms of enhanced electro-actuability, higher drug loading, and better control of burst release per actuation. However, there are several limitations within this section of drug delivery such as: (i) availability of only a few biocompatible candidates that qualify for electro-responsiveness; (ii) the disintegration of the network architecture on application of electric potential, limiting the implantability of the delivery system; and (iii) the practical applicability of the ERDDs in a clinical setting.[50] Most of the ERDD based research reports did not provide pre-clinical/clinical data to confirm the *in vivo* effectiveness of such delivery vehicles. In order to advance the electro-responsive capabilities and to revolutionize this "on-demand" delivery technology, more research efforts are required to focus on the design of novel electroactive polymers that are biocompatible and even 3D printable. If the 3D printability of these polymers can be realised in the near future, the possibility of designing an electromodulated device incorporating electro-responsive hydrogels capable of being actuated portably by the patient him/herself can be achieved.

Acknowledgements

This study was funded by the National Research Foundation (NRF) of South Africa.

References

1. J. Kost and R. Langer, *Adv. Drug Delivery Rev.*, 2001, **46**, 125.
2. A. Kumar, A. Srivastava, I. Y. Galaev and B. Mattiasson, *Prog. Polym. Sci.*, 2007, **32**, 1205.
3. I. Y. Galaev and B. Mattiasson, *Trends Biotechnol.*, 1999, **17**, 335.
4. A. K. Bajpai, S. K. Shukla, S. Bhanu and S. Kankane, *Prog. Polym. Sci.*, 2008, **33**(11), 1088.
5. F. Song, X. Li, Q. Wang, L. Liao and C. Zhang, *J. Biomed. Nanotechnol.*, 2015, **11**(1), 40.
6. Y. Brudno and D. J. Mooney, *J. Controlled Release*, 2015, **219**, 8.
7. A. Kikuchi and T. Okano, *Adv. Drug Delivery Rev.*, 2002, **54**, 53.
8. G. Filipcsei, J. Feher and M. Zrinyi, *J. Mol. Struct.*, 2000, **554**, 109.
9. S. Murdan, *J. Controlled Release*, 2003, **92**(1–2), 1.
10. N. K. Guimard, N. Gomez and C. E. Schmidt, *Prog. Polym. Sci.*, 2007, **32**, 876.
11. R. Ravichandran, S. Sundarrajan, J. R. Venugopal, S. Mukherjee and S. Ramakrishna, *J. R. Soc., Interface*, 2010, 7, S559.
12. T. Tanaka, I. Nishio, S. T. Sun and S. Ueno-Nishio, *Science*, 1982, **218**(4571), 467.
13. A. K. Banga and Y. W. Chien, *Pharm. Res.*, 1993, **10**(5), 697.
14. M. Jensen, P. Birch Hansen, S. Murdan, S. Frokjaer and A. T. Florence, *Eur. J. Pharm. Sci.*, 2002, **15**(2), 139.
15. K. Sawahata, M. Hara, H. Yasunaga and Y. Osada, *J. Controlled Release*, 1990, **14**, 253.
16. R. Tomer, D. Dimitrijevic and A. T. Florence, *J. Controlled Release*, 1995, **33**, 405.
17. P. A. McCarthy, J. Juang, S. Yang and H. Wang, *Langmuir*, 2002, **18**, 259.
18. P. R. Bidez, S. Li, A. G. Macdiarmid, E. C. Venancio, Y. Wei and P. I. Lelkes, *J. Biomater. Sci., Polym. Ed.*, 2006, **17**, 199.

19. J. Pelto, S. Haimi, E. Puukilainen, P. G. Whitten, G. M. Spinks, M. Bahrami-Samani, M. Ritala and T. Vuorinen, *J. Biomed. Mater. Res., Part A*, 2010, **93A**, 1056.
20. V. Pillay, T. S. Tsai, Y. E. Choonara, L. C. du Toit, P. Kumar, G. Modi, D. Naidoo, L. K. Tomar, C. Tyagi and V. M. K. Ndesendo, *J. Biomed. Mater. Res., Part A*, 2014, **102A**, 2039.
21. I. Singh, A. K. Rehni, P. Kumar, M. Kumar and H. Y. Aboul-Enein, *Fullerenes, Nanotubes, Carbon Nanostruct.*, 2009, **17**(4), 361.
22. H. Zhang, H. X. Li and H. M. Cheng, *J. Phys. Chem. B*, 2006, **110**, 9095.
23. O. C. Farokhzad and R. Langer, *ACS Nano*, 2009, **3**, 16.
24. R. A. MacDonald, C. M. Voge, M. Kariolis and J. P. Stegemann, *Acta Biomater.*, 2008, **4**(6), 1583.
25. C. M. Voge, J. Johns, M. Raghavan, M. D. Morris and J. P. Stegemann, *J. Biomed. Mater. Res., Part A*, 2013, **101**(1), 231.
26. U. G. Spizzirri, S. Hampel, G. Cirillo, F. P. Nicoletta, A. Hassan, O. Vittorio, N. Picci and F. Iemma, *Int. J. Pharm.*, 2013, **448**(1), 115.
27. M. Curcio, U. G. Spizzirri, G. Cirillo, O. Vittorio, N. Picci, F. P. Nicoletta, F. Iemma and S. Hampel, *RSC Adv.*, 2015, **5**, 44902.
28. J. S. Im, B. C. Bai and Y. S. Lee, *Biomaterials*, 2010, **31**(6), 1414.
29. J. Yun, J. S. Im, Y. S. Lee and H. I. Kim, *Eur. Polym. J.*, 2011, **47**(10), 1893.
30. S. Sirivisoot, R. Pareta and T. J. Webster, *Nanotechnology*, 2011, **22**(8), 085101.
31. J. Yang, G. Choe, S. Yang, H. Jo and J. Y. Lee, *Biomater. Res.*, 2016, **20**, 31.
32. A. Seyfoddin, A. Chan, W. T. Chen, I. D. Rupenthal, G. I. Waterhouse and D. Svirskis, *Eur. J. Pharm. Biopharm.*, 2015, **94**, 419.
33. D. Denning, M. T. Abu-Rub, D. I. Zeugolis, S. Habelitz, A. Pandit, A. Fertala and B. J. Rodriguez, *Acta Biomater.*, 2012, **8**, 3073.
34. A. Pourjavadi and M. Doroudian, *Polymer*, 2015, **76**, 287.
35. Y. Kong, H. L. Ge, J. X. Xiong, S. X. Zuo, Y. Wei, C. Yao and L. H. Deng, *Appl. Clay Sci.*, 2014, **99**, 119.
36. R. Wang, Y. Peng, M. Zhou and D. Shoud, *Appl. Clay Sci.*, 2016, **134**, 50.
37. B. K. Wershil, G. T. Furuta, J. A. Lavigne, A. R. Choudhury, Z. S. Wang and S. J. Galli, *J. Immunol.*, 1995, **154**, 1391.
38. J. J. Yoon, J. H. Kim and T. G. Park, *Biomaterials*, 2003, **24**, 2323.
39. S. Walsh, G. R. Jordan, C. Jefferiss, K. Stewart and J. N. Beresford, *Rheumatology*, 2001, **40**, 74.
40. I. A. Paun, M. Zamfirescu, C. R. Luculescu, A. M. Acasandrei, C. C. Mustaciosu, M. Mihailescu and M. Dinescu, *Appl. Surf. Sci.*, 2017, **392**, 321.
41. H. Xin, X. Sha, X. Jiang, L. Chen, K. Law, J. Gu, Y. Chen, X. Wang and X. Fang, *Biomaterials*, 2012, **33**, 1673.
42. X. Ying, Y. Wang, J. Liang, J. Yue, C. Xu, L. Lu, Z. Xu, J. Gao, Y. Du and Z. Chen, *Angew. Chem., Int. Ed.*, 2014, **53**, 12436.
43. H.-W. Liu, S.-H. Hu, Y.-W. Chen and S.-Y. Chen, *J. Mater. Chem.*, 2012, **22**, 17311.
44. N. M. Kenna, P. Calvert, A. Morrin, G. G. Wallace and S. E. Moulton, *J. Mater. Chem. B*, 2015, **3**, 2530.

45. T. S. Tsai, V. Pillay, Y. E. Choonara, L. C. du Toit, G. Modi, D. Naidoo and P. Kumar, *Polymers*, 2011, **3**, 150.
46. D. Bijukumar, Y. E. Choonara, P. Kumar, L. C. du Toit and V. Pillay, *Pharm. Dev. Technol.*, 2016, **21**(7), 875.
47. S. Indermun, Y. E. Choonara, P. Kumar, L. C. du Toit, G. Modi, R. Luttge and V. Pillay, *Int. J. Pharm.*, 2014, **462**, 52.
48. S. Indermun, Y. E. Choonara, P. Kumar, L. C. du Toit, G. Modi, S. van Vuuren, R. Luttge and V. Pillay, *Int. J. Pharm.*, 2015, **496**, 351.
49. S. Indermun, Y. E. Choonara, P. Kumar, L. C. du Toit, G. Modi, S. van Vuuren, R. Luttge and V. Pillay, *J. Pharm. Sci.*, 2017, **106**, 1111.
50. L. C. du Toit, P. Kumar, Y. E. Choonara and V. Pillay, in *Industrial Applications for Intelligent Polymers and Coatings*, ed. M. Hosseini and A. S. H. Makhlouf, Springer International Publishing, Switzerland, 2016, p. 51.

CHAPTER 9

Enzyme-responsive Drug Delivery Systems

EMILIE SECRET AND JENNIFER S. ANDREW*

Dept. of Materials Science and Engineering, University of Florida, Gainesville, FL, USA
*E-mail: jandrew@mse.ufl.edu

9.1 Introduction

Over the past 15 years interest has been rising in the application of enzyme-responsive materials for drug delivery. These enzyme-responsive drug delivery systems present a modified chemical structure so that the action of an enzyme can trigger, target or enhance the effect of the drug release. Enzymes play a major role in cell regulation, and are often overexpressed in disease states making them an attractive target for therapeutic and diagnostic applications. Enzymes are proteins that have the ability to catalyze chemical reactions inside cells. In the absence of enzymes, most biochemical reactions are too slow to occur in physiological conditions. Enzymes accelerate the kinetics of these reactions by over a million-fold, so they can occur in fractions of a seconds if catalyzed by the correct enzyme. A variety of types of enzymes exist, and they can be classified into 6 categories based on their mechanism: oxidoreductases, transferases, hydrolases, lyases, isomerases and ligases. For enzyme-responsive drug delivery systems, the enzyme must be able to catalyze the delivery vector such that it degrades, changes its shape or its structure, or breaks some bond that allows the release of the drug. This is

Biomaterials Science Series No. 1
Stimuli-responsive Drug Delivery Systems
Edited by Amit Singh and Mansoor M. Amiji

Published by the Royal Society of Chemistry, www.rsc.org

Table 9.1 Types of enzymes used in enzyme sensitive drug delivery systems.

Oxidoreductases	Horseradish peroxidase Glucose reductase Azo-reductase Nitrate reductase
Hydrolases	Proteases Esterase Glycosidase

why among these families of enzymes, only two are really interesting and studied for drug delivery systems, because of their ability to break bonds: oxidoreductases and hydrolases, see Table 9.1.

Oxidoreductases are enzymes which catalyze oxidoreduction reactions. Examples of oxidoreductase enzymes are horseradish peroxidase, glucose reductase, or azo-reductase. Those three enzymes are particularly interesting for drug delivery systems due to their role in disease pathology and cell regulatory processes.

Hydrolases are enzymes that catalyze the cleavage of C–O, C–N or C–C bonds. As hydrolases lead to the cleavage of molecules, they are the most studied enzymes in the field of enzyme-responsive drug delivery systems. The hydrolase family is composed of different types of enzymes including proteases, which are able to cleave proteins, esterases, which are able to cleave ester groups, glycosidases, which are able to hydrolyze glycosidic bonds in sugars, or ureases, which are able to catalyze the degradation of urea into CO_2 and NH_3.

Figure 9.1 provides an overview of the typical vectors for enzyme responsive drug delivery, which include hydrogels, liposomes or capped mesoporous silica nanoparticles. In each of these platforms the drug is retained within a material, until an enzymatic reaction occurs to either open up the structure or degrade the material in the case of enzyme responsive hydrogels (Figure 9.1a), or to destabilize the structure of a liposome or vesicle (Figure 9.1b), or to open the pore of a mesoporous silica based particle (Figure 9.1c).

9.2 Oxidoreductase-sensitive Drug Delivery Systems

Oxidoreductase enzymes represent a class of enzymes that catalyze reactions in which one molecule is oxidized and the other is reduced. These enzymes can be further specified as being oxidases, peroxidases, reductases, or dehydrogenases, where the name specifies the reaction that is catalyzed by each particular enzyme type. Sensitivity to oxidoreductases has been leveraged for a variety of biological applications, including drug delivery and diagnostics. For example, horseradish peroxidase is one of the most prevalent labels for immunodetection used predominantly in enzyme-linked immunosorbent assays (ELISA). For drug delivery,

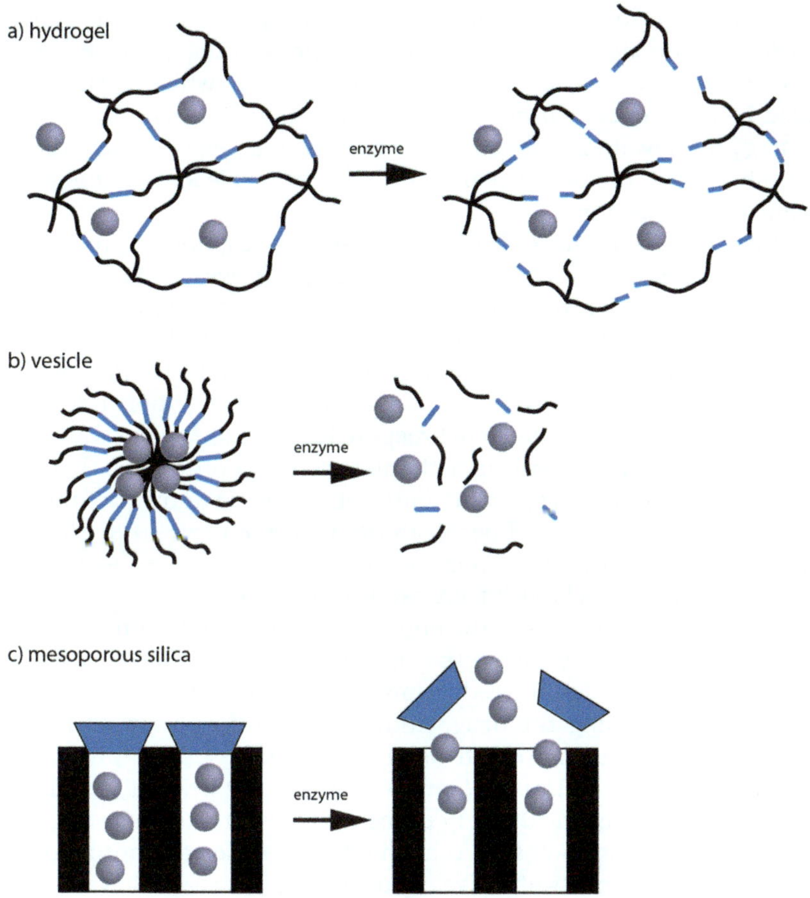

Figure 9.1 Schematic of different materials that are utilized in enzyme responsive drug delivery, including (a) hydrogels with an enzymatic degradable region, (b) vesicles, and (c) mesoporous silica where the pores are capped by an enzyme responsive material.

oxidoreductases are important targets in a variety of diseases such as cancer, neurodegenerative diseases, and other diseases where oxidative stress is involved.

Oxidoreductases targets have been utilized in drug delivery leveraging a variety of delivery vectors, including liposomes, micelles, polymers, nanoparticles, and mesoporous silica nanoparticles (MSNs). For example, Hubbell *et al.* fabricated vesicles formed by an ABA amphiphilic triblock copolymer, where the A-block was poly(ethylene glycol) (PEG) and the B-block was poly(propylene sulfide) (PPS). They further encapsulated glucose oxidase within the vesicle, such that the triggered release occurs when its substrate, glucose, is allowed to permeate through the membrane.[1] After glucose permeates through the membrane it is catalyzed by glucose oxidases, resulting in its conversion to gluconolactone and hydrogen peroxide. The generation of hydrogen peroxide then oxidizes the PPS block of the copolymer, converting it to more hydrophilic sulfoxides and sulfones, which destabilizes the vesicle, leading to its "self-destruction". This example by Hubbell represents a strategy for how liposomes are made enzyme-responsive, where the vesicle is destabilized by the presence of a specific enzyme, leading to the release of any encapsulated cargo.

Similarly, Dai *et al.* have developed a cascade-responsive prodrug micellar drug delivery platform for tumor therapy *in vivo*.[2] The cascade here describes a series of properties or functionalities imparted to the pro-drug and/or carrier in order to achieve enhanced delivery at the desired site. For example, the polymer-based pro-drug was conjugated with PEG as the outermost layer to provide a long circulation lifetime. The polymeric portion of the prodrug also included a pH-responsive layer as the middle layer. The micelle was also functionalized with folate acid receptors such that it would be taken up *via* folate-receptor mediated endocytosis upon accumulating in the tumor *via* the enhanced permeability and retention (EPR) effect. Once in the endosome, the pH-responsive layer of the micelle, poly(2-(diethylamino)ethyl methacrylate) (PDEA) would protonate leading to the escape of the prodrug micelles from the endosome or lysosome and into the cytoplasm. Once in the cytoplasm the high glutathione concentration of the cytoplasm would lead to the reduction of disulfide bonds, destabilizing the micelles and releasing the drug. Additionally, a porphyrin molecule was loaded and released that could be subsequently used for photodynamic therapy. This paper by Dai *et al.* outlines a strategy where the drug carrier is altered throughout the delivery process by the unique physiological conditions in each step, to ensure that the drug reaches the mitochondria of the cell where it is most effective, by a cascade of events that at each step creates a carrier that is optimized for delivery for that step, be it a PEG coating for long circulation, or pH-responsiveness for endosomal or lysosomal escape, followed by sensitivity to the reductive environment of the cell's cytoplasm. They demonstrated the effectiveness of this method using both *in vitro* and *in vivo* experiments. Xiang *et al.* also leveraged glutathione responsive micelles for the delivery of a chemotherapeutic alongside photodynamic therapy.[3]

Gold nanoparticles have also been utilized as carrier vehicles for glutathione responsive delivery. Park *et al.* functionalized the surface of gold nanoparticles with per-6-thio-β-cyclodextrin (SH-cyclodextrin) for the non-covalent loading of β-lapachone, an anti-cancer drug. They also added PEG moieties and anti-epidermal growth factor receptor (anti-EGFR) to the surface for enhanced circulation and targeting, respectively. When these functionalized gold nanoparticles were exposed to a glutathione rich environment, produced by a human lung cancer cell line (A549), ligand exchange occurs between SH-cyclodextrin and the glutathione in the intracellular matrix of the cancer cells leading to the release of the b-lapachone that was loaded within the SH-cyclodextrin shell.

Another delivery vector that has been used extensively in enzyme-responsive drug delivery systems are MSNs. MSN materials are particularly attractive for drug delivery due to their large surface area to volume ratio, as well as the pore volume that allows drugs to be loaded. These pores can then be "capped-off" or coated by an enzymatically responsive substrate. In this case, when the enzyme of interest is present, it can cleave the substrate that caps the pore opening, allowing drugs to be released. This is particularly attractive, as it allows the drugs to be protected during transport to the desired location, and only allows release in the presence of a particular enzyme that is characteristic of the disease state. The use of MSNs for enzymatic drug delivery is particularly useful for the treatment of diseases of the colon. This is because the MSN particles allow the drug to be protected within the pores as the delivery vector passes through the upper GI-tract. Additionally, utilizing a cap that is responsive to the particular enzymes found only in the colon will prevent the premature release or degradation of the drug.[4–6] Li *et al.* demonstrated just such an approach when they loaded MSNs with ibuprofen and capped the pores with azo benzene bridged silsequioxanes for the treatment of colon specific diseases.[4] When these carriers were exposed to simulated colonic fluid with azoreductase, the azo bonds of the azo-benzene bridges silsequioxanes were cleaved leading to the release of the ibuprofen. However, they also showed that the drug was not released in simulated gastric or intestinal fluid, showing that the carrier would hold the drug until it arrived further in the digestive tract, where bacteria native to the microflora of the colon generate azoreductase. This method of capping the pores of MSNs with a "gate-keeper" molecule, was also described by Mas *et al.*, where they loaded the pores of the MSN with either a dye, Rhodamine B, or a cytotoxic drug, camptothecin, and then capped the pores with an azo-containing molecule, here they used azopyridinium salts.[6] To release the loaded cargo the capped or gated MSNs were exposed to nitrate reductase, however no release was observed in the absence of enzyme, or upon exposure to esterase, an enzyme found in the stomach. MSNs have also been developed that have been capped with molecules that are responsive to glucose. In another example, Xiao *et al.* developed a capped MSN system that responded to an oxidation stimulus mechanism, as a result of an oxidoreductase catalyzed reaction or to an applied voltage.[7] Here, the MSN were first surface functionalized with ferrocene, which was covalently bound to the silica surface *via* carbodiimide coupling. After ferrocene functionalization the pores were loaded with Rhodamine B dye and the entire particle was capped with β-cyclodextrin. These MSNs were then placed in dialysis tubing and exposed to either glucose or hydrogen peroxide. Drug release then occurs upon addition of the enzymes, horseradish peroxidase and glucose oxidase, where the relative amounts of glucose oxidase and horseradish peroxidase were able to modulate the rate of drug release as these two enzymatic reactions occurred simultaneously, and affected the oxidation of ferrocene which when oxidized the β-cylodextrin dissociates from it, leading to the cap opening and drug being released.

9.3 Hydrolase-sensitive Drug Delivery Systems

Hydrolases are enzymes able to catalyze the cleavage of different bonds. This large family of enzymes includes, among others, esterases, which can hydrolyze ester bonds, as well as proteases or peptidases, which can hydrolyze peptide bonds, and glycosidases, which can hydrolyse oligo- or polysaccharides. These three sub-families of enzymes have been used extensively for the design of enzyme-sensitive drug delivery systems.

9.3.1 Proteases

Proteases, also called peptidases or proteinases, are in the hydrolase category of enzymes. They catalyze proteolysis reactions, hence the hydrolysis of peptide bonds. Proteases are among the most studied enzymes for enzyme-sensitive drug delivery systems. Proteases are particularly interesting for drug delivery because many of them are overexpressed in cases of illness. Some of them, such as the cancer-associated proteases, are overexpressed in cancers for example. Among them are many of the matrix metalloproteinases (MMPs), some of the cathepsins, or the urokinase plasminogen activator.[8,9] The general mechanism to make protease-sensitive materials is to add a specific peptide sequence to the material that can be cut by the targeted protease to release the drug encapsulated (Figure 9.1).

9.3.1.1 Matrix Metalloproteinases (MMPs)

Matrix metalloproteinases are some of the most used proteases for drug delivery. MMPs are a family of calcium-dependent enzymes containing zinc, which are able to degrade extracellular matrix proteins.[10] Typical substrates for MMPs are collagen, gelatin, laminin, fibronectin and aggrecan. Different categories of MMPs exist, depending on their substrate and amino acid sequences. MMP-1, MMP-8 and MMP-13 are the collagenases, MMP-2 and MMP-9 form the family of the gelatinases, MMP-3, MMP-10 and MMP-11 are the stromelysins; MMP-7 and MMP-26 are the matrilysins; MMP-14, MMP-15, MMP-16, MMP-17, MMP-24 and MMP-25 are the membrane-type MMPs. MMPs can also degrade non-matrix proteins such as cytokines, chemokines, cell–cell interaction proteins or latent growth factors, which makes them important for cell behavior, such as its proliferation, adhesion, migration, differentiation and apoptosis. MMPs are overexpressed in a number of different diseases, which makes them a target of choice for enzyme-sensitive drug delivery systems. For example, some types of MMPs are overexpressed in pulmonary diseases,[11] cancer,[12] arthritis,[13] in cardiovascular diseases,[14] and in ocular and gastrointestinal diseases.[15] The most studied MMPs in the field of enzyme-responsive drug delivery systems are the gelatinases MMP-2 and 9. Different types of material have been used in order to design MMP-sensitive drug delivery systems: polymers, silica, liposomes, and gelatin. This section describes the different systems studied, by type of material used.

Polymers are among the most studied materials for drug delivery in general, and also for MMP-responsive drug delivery. Indeed, in order to develop a MMP-sensitive material, the most widely used way consists of conjugating MMP substrate peptides to the material to trigger the release. The use of polymers is convenient, since a peptide can be added in the polymer sequence through straightforward chemistries. This explains why there are many examples of MMP-responsive drug delivery made from various polymers. The first to report on this type of MMP-sensitive polymer were West and Hubbell in 1999.[16] They synthesized PEG hydrogels with an Ala-Pro-Gly-Leu peptide in the backbone of the PEG chain to give MMP-sensitivity. They found that the peptide-containing hydrogel was completely degraded after 170 h in a collagenase solution at 0.2 mg mL^{-1} and after 100 h in a collagenase solution at 2 mg mL^{-1}. Since then, and because of the potential of MMPs as targets for drug delivery, numerous researchers have used the same strategy with a range of polymers, at different scales and with different peptides.[17]

An early example is the work of Garripelli *et al.*,[18] where they studied thermogelling polymers made of a pluronic-based polymer (PEG-PPG-PEG) functionalized with the peptide Gly-Pro-Val-Gly-Leu-Ile-Gly-Lys, for the release of paclitaxel, a chemotherapeutic agent. The solubility of this hydrophobic drug is enhanced by the polymer in solution and this aqueous solution can gel at body temperature. This polymer then forms a gel *in situ* that can be degraded by MMP-2 or 9 allowing the release of the drug. Another example of MMP-sensitive polymers for drug release is the work of Chen *et al.* they synthesized polymeric micelles made of Biotin-PEG-*b*-PLL-peptide-Dox.[19] Here, the micelles have the drug doxorubicin attached within their cores by a MMP2-sensitive peptide, while the biotin molecules are at the surface for targeting cancer cells.

In our group, we leveraged similar chemistries to those outlined above, to synthesize MMP responsive PEG microparticles for pulmonary disease therapy.[20,21] These microparticles were fabricated such that their hydrodynamic diameters were between 1 and 5 μm in order to be deposited in the upper airways after inhalation by the patient. The employed chemistry can be modified by changing the peptide sequence to allow sensitivity to different MMPs. The degradation kinetics of the particles was governed not only by the specific peptide, but also by the length of the PEG chains used to make the particles. While best suitable for the encapsulation and delivery of hydrophobic drugs, they could also encapsulate and deliver hydrophilic drugs and proteins, in a controlled manner at MMP-2 concentrations relevant to quantities measured in patients with pulmonary diseases.

In some diseases, such as chronic inflammation, tumor progression and cardiovascular disease, MMPs are overexpressed and contribute to disease progression through tissue remodeling. A strategy is then to inhibit MMPs by the use of specific inhibitors named TIMPs. The MMPs can then be used to trigger the degradation of a material to release its own inhibitor. In the work of Purcell *et al.*,[22] hyaluronic acid hydrogels, injectable and degradable by MMPs, encapsulated rTIMP-3, a tissue inhibitor of MMP for the treatment of the adverse left ventricle (LV) remodeling in patients following a myocardial infarction. The hydrogel is injected directly in the area of the heart to treat, and the overexpressed MMPs degrade it to release the inhibitor against MMPs, inducing a decrease of the adverse ventricle remodeling and an improvement of the heart functions after a myocardial infection.

Another widely studied strategy with MMP-sensitive polymers is their use in hybrid nanoparticles in order to enhance their penetration deeper in the tumor and/or their internalization into cells. Huang *et al.*[23] developed a cationic polymer non-viral vector functionalized with cell penetrating peptides (CPP) that is activated by the lower pH and the MMP-2s overexpressed in the tumor. A positively charged CPP, is attached to a negatively charged peptide through a third peptide that is degradable by MMP-2. By electrostatic interaction, the negatively charged peptide masks the CPP. When in the tumor, the lower pH eliminates the electrostatic interaction, and the overexpressed MMP-2 cleaves the peptide linker in order to release the CPP that stays on the surface of the nanoparticles. This leads to an enhanced internalization of the nanoparticles in the cancer cells, to co-deliver plasmid expressing interfering RNA targeting VEGF and doxorubicin, for antiangiogenesis and apoptosis combined. Zhu *et al.*[24] also used MMPs to obtain a better internalization of their polymer nanoparticles in order to co-deliver siRNA and the drug paclitaxel. In their work, they synthesized a PEG-peptide-PEI-PE that self-assembled into micelles that were able to encapsulate siRNA *via* static interactions with PEI and hydrophobic drugs in their cores. The PEG at the surface of the nanocarrier can make the particles stealthy and improve the circulation time, in order to reach the tumor, but it also interferes with the internalization process.[25] Once in the tumor environment, overexpressed MMP-2 can cleave the sensitive peptide, removing the PEG chain and subsequently allowing the particles to be readily internalized by exposure of the PEI chains that were previously hidden. The tumor targeting, cell internalization and synergistic anti-tumor effect of siRNA and the drug was reported *in vitro*. Li *et al.*[26] in their work used MMP-7 to develop a dual targeting of their polymeric nanoparticles to deliver siRNA. Their nanoparticles, are made of two kinds of amphiphilic polymers that form micelles together. The first amphiphilic polymer is terminated at the surface of the nanoparticles by a PEG chain through a MMP-sensitive peptide. The second one is terminated by folic acid. When in the tumor environment, MMP-7 removes the PEG chains by degrading the peptide, to decrease the size of the nanoparticles and expose the folic acid moieties that can react with the folate receptors at the surface of cells. This results in better targeting of cancer cells that overexpress both MMP-7 and folate receptors, and provides a better internalization of the particles by these cells. Recently Grünwald *et al.*[27] developed PLGA-PEG nanocarriers with maghemite nanoparticles encapsulated in their cores. These nanoparticles are also functionalized with PEG chains through a MMP-sensitive peptide. They tested the degradation of their chosen peptide with MMP-2, 3, 8 and 9 and showed that it was fully degraded by MMP-2 and 9 but - that MMP-3 and 8 were not able to degrade it. Once again, in the tumor environment, the overexpressed MMP-2 and 9 can remove the PEG from the particles resulting in smaller nanoparticles that can penetrate deeper within the tumor and are thus able to be internalized more easily by tumor cells. Finally, this strategy was also used by Ke *et al.*[28] with micellar nanoparticles made of PEG-peptide-poly(D,L-lactide) that can be de-PEGylated by MMP-2 to expose a RGD peptide at the surface at the nanoparticles. They showed that their nanocarrier was able to better accumulate at the tumor site, to be better taken up by tumor cells and they also showed the ability to induce a significant improvement of the activity of the paclitaxel drug encapsulated inside the micelles.

Mesoporous silica nanoparticles provide another platform for drug delivery that have also been well studied as vectors for MMP-sensitive drug delivery. For example, Singh *et al.*[29] developed core–shell nanoparticles formed with a MSN core and a shell made of PEG chains containing a peptide sequence. The release of doxorubicin was higher in the presence of collagenase *in vitro*, especially when the drug was loaded in the core of the nanoparticles, less when loaded in the shell. They also showed a higher doxorubicin-induced cell death *in vivo* with the PEG peptide coated MSNs than with the PEG coated MSNs, due to a higher drug release because of the MMPs. Liu *et al.*[30] also studied the use of MSNs as MMP-responsive drug delivery systems. They capped doxorubicin loaded MCM-41 nanoparticles with phenylboronic acid conjugated human serum albumin (PBA-HSA) *via* a polypeptide linker. This polypeptide was composed of two subunits: a polyarginine unit serving as a cell penetrating peptide and a MMP-2 substrate peptide. Here, the HSA acts as a capping agent, retaining the drug inside the pore of the MSNs and also as an anti-phagocytosis agent. In the micro-environment of the tumor, MMP-2 cleaves the polypeptide linker, detaching the PBA-HSA, and leaving only the CPP at the surface of the MNPs in order to give a better internalization of the particles. This system showed a good inhibition of tumor growth *in vivo*. Another example of the use of MSNs for MMP-sensitive drug delivery is the work of van Rijt *et al.*[31] In this study they developed cisplatin-loaded MSNs capped with avidin through a MMP-9 cleavable peptide for the treatment of lung cancer. The efficacy of their system was demonstrated in *ex vivo* mouse and human lungs, with a 10- to 25-fold more effective treatment on tumor tissue overexpressing MMP-9 than on healthy regions of the same tissue slices.

Liposomes represent another popular drug delivery vector that have gained disease specificity through protease sensitivity. For example, Ji *et al.*[32] developed MMP-2 responsive liposomes by modifying them with β-cyclodextrin through a MMP-2 cleavable peptide sequence. These nanosystems were loaded with an anti-fibrosis drug inside the cyclodextrin and gemcitabine was loaded within the liposome core, in order to inhibit fibrosis and improve the targeted delivery of drugs for pancreatic cancer therapy. Wan *et al.*[33] also developed MMP-2 responsive liposomes for the gene delivery. Adenovirus loaded anionic liposomes were functionalized with PEG-peptide chains with a MMP-2-cleavable peptide. When in the presence of MMP-2, the peptide is cleaved, liberating the liposome from the PEG, and hence allowing better cellular internalization. Higher *in vitro* gene expression was demonstrated while the toxicity was decreased compared to bare adenovirus. Another study published by Xu *et al.* used MMP-responsive liposomes, where they developed MMP9-sensitive liposomes made with soybean phosphatidylcholine, cholesterol and PEG_{1k}-peptide-vitamin E succinate, because of the peptide presence these liposomes are degradable by MMP-9.[34] Inside the liposome are encapsulated smaller polymeric micelles that hold the antitumor drug docetaxel which are sensitive to another enzyme: reductive glutathione (Section II). When the liposomes are in the presence of MMP-9, they release the smaller nanoparticles that can penetrate deeper into the tumor, where they subsequently release the docetaxel. These MMP-sensitive nanomaterials were able to simultaneously inhibit the growth and metastasis of 4T1 breast cancer, through to the responsiveness of two different enzymes.

Another strategy for MMP-responsive drug delivery systems is to directly use a carrier that is also an MMP substrate, for example gelatin. This is what Wong *et al.*[35] did by using gelatin nanoparticles. Gelatin is a natural substrate of gelatinases (MMP-2 and MMP-9). They developed multistage nanoparticles made of gelatin nanoparticles of 100 nm in diameter with their surface covered with quantum dots. MMP-2 can degrade the gelatin nanoparticles in the tumor environment, in order to release the quantum dots deeper in the tumor tissue. Using this same principle, Ruan *et al.*[36] developed gelatin nanoparticles covered with gold nanoparticles functionalized with doxorubicin and PEG chains. These 186 nm composite nanoparticles presented a longer circulation time in the blood and better tumor extravasation due to the EPR effect in mice. Once in the tumor environment, the overexpressed MMP-2 degraded the gelatin to deliver smaller functionalized gold nanoparticles, which could then accumulate deeper into the tumor delivering doxorubicin. The same group also studied the use of gelatin particles decorated with doxorubicin and RGD peptide conjugated dendritic poly-L–lysine.[37]

Table 9.2 summarizes the examples of MMP-sensitive materials presented in this section, emphasizing the different peptides used to provide MMP-sensitivity. Different research groups have employed different peptide sequences for the same enzyme. The chosen peptide sequence is important for the release kinetics of the drug from the delivery vector. Some work has been done to determine the affinity of various MMPs for different peptide substrates.[38,39] In our work on MMP-sensitive microparticles for the treatment of lung diseases we used the results presented by Nagase *et al.*[38] to find a better sequence for the degradation of the microparticles by MMP-2.

9.3.1.2 *Other Proteases*

By changing the peptide sequence, other types of proteases can be used to trigger drug delivery. In this part of the chapter we will present one example of a drug delivery system for each protease used. Along with MMPs, a few other proteases belong to the family of cancer-associated proteases. These proteases are usually absent or present at very low concentration in healthy tissues, but are overexpressed in cancerous tissues. Examples of these

Table 9.2 Examples in the literature of MMP-sensitive drug delivery systems.

Targeted enzyme	Type of material	Peptide sequence	References
MMP1	PEG microparticles	GLK	Secret, 2014[20]
MMP2	PEG microparticles	GPQGIFGQK	Secret, 2014,[20] et Secret, 2015[21]
MMP2	Pluronic triblock co-polymer	GPVGLIGK	Garripelli, 2011[18]
MMP2	Polymeric micelles	GPLGIAGQ	Zhu, 2014[24]
MMP2	Cationic polymer nonviral vector	PLGLAG	Huang, 2013[23]
MMP2	Polymeric micelles	CPLGLAGG	Chen, 2014[19]
MMP2	Amphiphilic block copolymer nanoparticles	GPLGVRGDG	Ke, 2017[28]
MMP7	Polymeric nanoparticles	VPLSLYSGCG	Li, 2015[26]
MMP9	PLGA-PEG nanoparticles	GERGPPGPQGARGFPGTPGL	Grünwald, 2016[27]
MMPs (MMP1, 8 and 13 *in vitro*)	MSNs	CGPQGIWGQGCR	Singh, 2011[29]
MMP2	MSNs	PVGLIG	Liu, 2015[30]
MMP9	MSNs	RSWMGLP	van Rijt, 2015[31]
MMP2	Liposomes	GPLGIAGQC	Wan, 2013[33]
MMP2	Liposomes	GPLGIAGQ	Ji, 2016[32]
MMP9	Liposomes	PLGLAG	Xu, 2015[34]
MMP2	Porous silicon microparticles functionalized with polymeric nanoparticles	AGFSGPLGMWSAGSFG	Mi, 2016[40]

proteases are some of the cathepsins or the urokinase plasminogen activator (uPA). This last enzyme has been targeted in the work of Basel *et al.*,[41] with polymer caged liposomes. In order to fabricate these they synthesized liposomes at high osmolarities so they can swell and release the drug they encapsulated at physiological osmolarity. Then a poly(acrylic acid) polymer functionalized with both a cholesterol anchor and a peptide sequence that is degraded by uPA is used to cage the liposome. After crosslinking of the polymer with ethylenediamine, they demonstrated a 53% increase in drug release compared to the same vectors without the peptide sequence. All of the drug was released from the uPA-sensitive polymer caged liposomes after a night, in the presence of 25 μg mL^{-1} of uPA. Cathepsin b, a lysosomal cysteine protease, is also overexpressed in various tumor cells. In the work of Li *et al.*,[42] this overexpression of cathepsin b in breast cancer cells was used to trigger the release of paclitaxel from nanoparticles made of the self-assembly of Janus PEGylated dendrimers. The dendrimers were functionalized with the drug *via* a Gly-Phe-Leu-Phe peptide linker, conferring the cathepsin-sensitivity to the nano-vector. These particles demonstrated a higher cytotoxicity against 4T1 cells and a lower cytotoxicity against normal cells than free paclitaxel, both *in vitro* and *in vivo*.

In the work of Habraken *et al.*,[43] two other proteases were targeted: elastase and thermolysin. They used vesicles and micelles made of copolymers of poly(*n*-butyl acrylate) and poly-L-glutamic acid embedded with various quantities of L-alanine. They demonstrated that both the quantity of L-alanine in the backbone, and the type of synthetic polymer used were important for the degradation by elastase and thermolysin. Indeed, the copolymers containing 50% of L-alanine showed a high degradation response compared to polymers containing less L-alanine, and the replacement of poly(*n*-butyl acrylate) by polystyrene did not show any enzymatic degradation. Elastase was also used by Aimetti *et al.*, where they synthesized a PEG-peptide hydrogel, made by thiol-ene photopolymerization, with two different kinds of peptides: one is degraded quickly by human neutrophil elastase, the other slowly.[44] By placing the slowly-degradable hydrogel in a sandwich between the quickly degradable one, they showed that they were able to obtain a better release profile of the proteins.

Thrombin is another enzyme that has been targeted for drug delivery.[45] Here, PEG hydrogels were made by click chemistry, and used to obtain three layers hydrogels. In the middle layer, a fluorescent protein is attached at the end of some of the PEG chains through a thrombin-sensitive peptide. The fluorescent protein can be released from the middle layer by the action of the thrombin enzyme, while the hydrogel retains its shape and structure because of the two external PEG layers. In their study, Candel *et al.*[46] demonstrated the use of amidase and pronase as enzymes for drug delivery with MSNs. In order to do that, after encapsulating their drug, safranine O or camptothecin, they capped the pores of the MSN with gluconamide derivatives. These gluconamide derivatives led to steric hindrance, blocking the release of the drug. In the presence of amidase or pronase enzymes, the amide bonds were hydrolyzed in the anchored gluconamide, triggering the release of the encapsulated drugs. They showed that no drug was released in the absence of enzymes. *In vitro*, in the presence of enzymes, a significant decrease of cell viability was demonstrated, but this cytotoxic effect could be improved. Penicillin G amidase was targeted by Harnoy *et al.*[47] to degrade micelles they fabricated using PEG-dendron hybrids. The dendron part of the hybrid is terminated by phenylacetamide groups that can be cleaved by penicillin G amidase. The cleavage of this hydrophobic group decreased the hydrophobicity of the dendron and destabilized the micelles, leading to their disassembly and the release of the drug.

Finally some groups used several enzymes to test the sensitive drug delivery systems. For example, Vicent *et al.*[48] used rat liver lysosomal enzymes to simultaneously release doxorubicin, a chemotherapy drug, and aminoglutethimide, an endocrine model. The drug delivery system is composed of *N*-(2hydroxypropyl)methacrylamide (HPMA) functionalized to the two drugs

Table 9.3 Examples in the literature of protease-sensitive (except MMP) drug delivery systems.

Targeted enzyme	Type of material	Sensitive sequence	References
Urokinase plasminogen activator	Polymer caged liposomes	GSGRSAGK	Basel, 2011 [41]
Cathepsin B and papain	PEGylated dendrimer-paclitaxel prodrug-based nanoparticles	GFLG	Li, 2017 [42]
Elastase and thermolysin	Copolymer micelles	Polypeptides of L-glutamic acid and L-alanine	Habraken, 2011 [43]
Human neutrophil elastase	PEG hydrogels	CGAAPVRGGGC and CGAAP(Nva) GGGGC	Aimetti, 2009 [44]
Thrombin	PEG hydrogels	LVPRGS	Guo, 2017 [45]
Amidase, pronase	MSNs	Gluconamide	Candel, 2012 [46]
Penicillin G amidase	PEG-dendron micelles	Phenylacetamide	Harnoy, 2014 [47]
Lysosomal enzymes	HPMA polymer	GFLG	Vicent, 2005 [48]
Proteases from *S. griseus*	MSNs	ε-poly-L-lysine	Mondragon, 2014 [49]

through Gly-Phe-Leu-Gly peptides that can be cleaved by lysosomal enzymes. They showed that they had almost no drug release from their polymers in the absence of the enzymes, but that the drug release was quick with the enzymes. Mondragón *et al.*[49] also used this strategy of using a mix of enzymes by using proteases from *S. griseus*. The MCM41 nanoparticles encapsulating camptothecine were coated with ε-poly-L-lysine to cap the pores. Almost no drug release occurred when the enzymes were absent, but a full release occurred after 8 to 24 h in the presence of 0.12 mg mL^{-1} of enzyme, due to the hydrolysis of the amide bonds in the polylysine.

Table 9.3 presents an outlook of the different enzymes used in the literature for protease-sensitive drug delivery system, emphasizing the sensitive part giving their sensitivity to the systems in the specific papers presented in this chapter.

9.3.2 Esterases

Esterases are another type of hydrolase that are commonly used in drug delivery systems. Esterases are enzymes that lead to the hydrolysis of ester bonds. Their high concentration in endosomes and lysosomes make them a target of choice when designing nanoparticles for drug delivery. Indeed, when entering a cell, nanoparticles are internalized by endocytosis. Once trapped inside endosomes or lysosomes, esterases can trigger the drug release from the nanoparticles. This approach has been studied with gated mesoporous silica nanoparticles. Zink's group use their α-cyclodextrin "snap-top" system, stopped by an ester-linked adamantyl.[50] When the nanoparticles are put in contact with esterases, the ester bond is hydrolyzed, releasing the adamantyl stopper, freeing the α-cyclodextrin, and releasing the drug from the pores. Fernando *et al.* also made esterase-sensitive mesoporous silica nanoparticles by gating the pores with poly(β-amino ester).[51] The release of camptothecin encapsulated inside the pores can be triggered either in acidic conditions or in the presence of esterases.

Other types of inorganic nanoparticles have also been used with this strategy. For example, CdSe/CdS nanorods conjugated with PEG chains and a galactose shell have been studied for the transport of dopamine into the brain.[52] The carbohydrate shell allowed the recognition of the GLUT-1 transporters located on the membrane of brain capillary endothelial cells, hence allowing a passage across the blood brain barrier, and dopamine was anchored to the nanorod through an ester bond. Once inside the brain, the nanoparticles will be internalized by endocytosis, and esterases will trigger the dopamine release for treatment of Parkinson's disease. Ding *et al.* synthesized gold nanoparticles functionalized with paclitaxel through a specific linker, allowing a synergistic release of the drug in the presence of two different enzymes: glutathione and esterases.[53] This linker is made of a bifunctional PEG with a glutathione-sensitive S–H bond on one side to link the gold surface and an ester bond on the other to link the drug. A PEG chain of 400–1000 Da is necessary to enhance the water solubility and to increase the amount of drug loaded by eliminating steric hindrance between the hydrophobic paclitaxel molecules. The two enzymes are necessary to completely release the drug. Promising *in vivo* results have been shown with these particles.

Polymers are a class of material often studied for esterase-sensitive drug delivery systems. Li *et al.* designed a zwitterionic polymer made of poly(carboxybetaine) (PCB) conjugated with camptothecin through an ester bond, sensitive to both pH and esterases.[54] With the addition of cationic lipids, these polymers formed liposomes that can encapsulate siRNA. In the presence of esterase inside lysosomes, both siRNA and camptothecin are released. The use of PCB is of interest here because of its capacity to be protonated in an acidic environment, which can destroy the endosomes to ensure a better endosomal or lysosomal escape of the drug. The use of esterases to escape endosomes or lysosomes has also been studied by Lee *et al.*[55] In this work they conjugated a polypeptide-based amphiphilic copolymer with benzyl ester bonds to a photosensitizer and to a black hole quencher 3. The conjugated polymers self-assembled and encapsulated paclitaxel. In the presence of esterase after internalization of the nanoparticles, the benzylester bonds are cleaved, which leads to the collapse of the nanocarrier, and the release of the drug and the restoration of the photoactivity of the photosensitizer as it drifts away from the quencher. A light excitation can then lead to the disruption of the endosome membrane to release the drug in the cytoplasm and ensure its improved efficacy. Recently, Awino *et al.*[56] designed micelles made of a trialkyl modified surfactant stabilized by cross-linking them with an esterified diazido molecule through click chemistry, and functionalized with a thiolated DNA. The micelles can encapsulate hydrophobic drugs such as camptothecin. The esterases can destabilize the micelles and release the drugs by hydrolyzing the cross-linker.

In the esterase family, cholinesterase is of particular interest for drug delivery, because of its overexpression in Alzheimer's disease. Guo *et al.*[57] designed a cholinesterase-responsive supramolecular vesicle, based on the concept of host-guest chemistry. Specifically, *p*-sulfonatocalix[4]arene was complexed to myristylcholine, decreasing the critical aggregation concentration of myristylcholine, this lead to the self-assembly of the complexes into binary vesicles. The cholinesterase enzyme can then hydrolyze the myristylcholine moiety into myristic acid and choline, resulting in the degradation of the vesicle and the release of the anti-Alzheimer drug tacrine, that was encapsulated inside the vesicles.

Another subfamily of esterase enzymes are the lipases. Lipases are enzymes that specifically catalyze the hydrolysis of ester functions of fat (lipid, triglyceride) into free fatty acids and monoglycerides. Due to their function, lipases are found in the digestive system (intestine, pancreas, stomach, liver), this is why they are a target of choice when designing an enzyme-sensitive drug delivery system for oral delivery. This approach is commonly used with lipid particles.[58–60] Mahjub *et al.* designed a sustained oral peptide delivery system made with self-nanoemulsifying peptides.[61] To that end, octreotide, a therapeutic peptide, was ion paired with anionic surfactants and incorporated in lipase-degradable micelles containing a triglyceride. This oral drug delivery system can be degraded by intestinal lipases to release the peptide. Another example of nanoparticles commonly used for drug delivery systems that have been rendered sensitive to lipases is mesoporous silica nanoparticles. Park *et al.* synthesized MSN with pores capped with cyclodextrin through a stalk part that is degraded in the presence of lipases, to free the cyclodextrin and release the encapsulated drug.[62] Lipases can also be secreted by bacteria, making them a target of choice for bacterial treatments. In their work, Xiong *et al.* synthesized polymeric triple layered nanogels.[63] The particles were made of a polyphosphoester core, coated with a poly(ε-caprolactone) fence and then with a poly(ethylene glycol) shell. The poly(ε-caprolactone) layer is sensitive to lipase, and its degradation triggers the release of an encapsulated drug, vancomycin, to kill *Staphylococcus aureus*.

Phospholipases are a specific class of lipases able to catalyze the hydrolysis of phospholipids. Due to their substrate specificity, phospholipases can easily degrade liposomes, which are commonly studied for drug delivery systems sensitive to these enzymes. Phospholipase A2 (PLA2) is the most used phospholipase for drug delivery. Andresen's group has been studying oxaliplatin-encapsulated liposomes and their reactivity towards secretory PLA2 outside cells.[64] Secretory PLA2 hydrolyze the phosphoglycerolipids that form the liposomes, releasing the drug. The hydrolysis products, the lyso-lipids and free fatty acids, can then act as membrane permeability enhancers, allowing a better drug transport across the cellular membrane. PLA2 is also a thrombus-relevant enzyme, making it a good target to treat blood clots. Pawlowski *et al.* synthesized platelet microparticle-inspired nanovesicles, made of DSPC (1,2-distearoyl-*sn*-glycero-3-phosphocholine) lipids, encapsulating thrombolytic drugs, and functionalized with a peptide that binds P-selectin, and a RGD peptide that binds stimulated integrins, to target the blood clot.[65] Once reached, the PLA2 enzyme degrades the DSPC lipids and the vesicle to release the thrombolytic drugs.

9.3.3 Glycosidases

Glycosidases represent another class of hydrolase, which catalyze the hydrolysis of glycosidic bonds that are found in sugars. These enzymes are particularly relevant for drug delivery due to their physiological roles in either normal cellular function or in manners that are disease specific. For example, it has been recently shown that the tumor microenvironment has significant changes in the expression of glycan related molecules, including both glycosyltransferases and glycosidases.[66] Recognition of glycan groups poses another approach for molecular recognition based therapies and diagnostics. One limitation to the use of glycan-binding proteins as targeting ligands is the relatively weak affinity in glycobiological interactions. However, this can be overcome through the use of multivalent functionalization, where the drug delivery vehicle is functionalized with more than one type of carbohydrate, better mimicking glycoclusters that are present on cell surfaces.[66] Additionally, many natural polysaccharides are used in drug delivery, including ones discussed earlier in this chapter such as alginates, chitosan, dextran, and hyaluronic acid.[66] Although, in most cases outlined previously, these materials have been used to impart biocompatibility and/or hydrophilicity they can also be leveraged for their recognition by glycosidases.

For example, Ferguson *et al.* developed a bio-responsive dextrin-phospholipase A_2 (PLA_2) conjugate that was able to reduce the systemic toxicity of PLA_2, a toxin found in snake venom, while still retaining its anti-tumor activity.[67] PLA_2 is a crotoxin and anti-tumor protein that is believed to act through interactions with EGFR. PLA_2 has shown activity in phase I clinical trials for breast cancer, but showed nonspecific neurotoxicity in 50% of the patients, which will likely limit its use. Therefore, there is a real need to develop a carrier for these types of therapeutics that limit off-target effects. To accomplish this, Ferguson *et al.* conjugated the PLA_2 with dextrin, which is rapidly degraded by α-amylase, which is overexpressed in the tumor. Additionally, the degradation rate of dextrin by α-amylase can be decreased through the introduction of pendant groups to the dextrin *via* succinoylation. Here, the dextrin conjugation was able to decrease the systemic toxicity of the PLA_2, where the activity was triggered by exposure to α-amylase. By conjugating the

PLA_2 with dextrin, Ferguson *et al.* were able to decrease systemic activity, but the activity of PLA_2 was restored upon α-amylase degradation of dextrin in the tumor environment.

Zhao *et al.* have developed polysaccharide-based biomaterials with on-demand release in response to nitric oxide, mediated by enzyme catalysis.[68] Nitric oxide is an essential signaling molecule in cardiovascular diseases that is produced by the activity of nitric oxide synthase (NOS). Nitric oxide deficiency has been implicated in both cardiovascular diseases and diabetes; thus, it has been proposed that compounds or molecules that can decompose in a controlled manner to release nitric oxide could have clinical applications. These molecules that break-down to release nitric oxide (NO) are typically referred to as NO-donor compounds. One way to control the release of the NO-donor compound is through protection with a polysaccharide molecule, such as galactose.[68,69] In this case the decomposition of the compound and release of NO only occurs in the presence of glycosidase. However, Zhao *et al.* have extended this type of platform to generate a comb-shaped polymer composed of chitosan functionalized with glycosylated NO pendant chains.[68] Here, the chitosan imparts stability, and provides the ability for the material to be processed into a membrane or as an injectable hydrogel for long-term release. Galactosidase- responsive conjugates have also shown promise for the treatment of acute myelogenous leukemia (AML), to reduce side effects and increase treatment efficacy.[70] Clarhaut *et al.* developed what they referred to as a galactosidase-responsive "Trojan Horse", where they prepared doxorubicin that was conjugated to a folate acid ligand and a galactoside "trigger".[70] The folate ligand is used to target AML cells, where doxorubicin is only released intracellularly in the presence of β-galactosidase.

Similarly, hyaluronic acid (HA) based nanocapsules have been prepared that are cleaved in the presence of hyaluronidase, which is overexpressed in the presence of bacteria, including *Staphylococcus aureus* and *Escheria coli.*[71] To this end, Baier *et al.* developed HA-based nanocapsules that contained an anti-microbial agent, polyhexanide, that would be released in the presence of hyaluronidase, and they were able to show the efficacy of their platform against both *S. aureus* and *E. coli.*[71] Additionally, they were able to show that the capsules were stable, showing minimal release in the absence of hyaluronidase.

Beyond polymer-based responsive materials, MSNs have also proven to be a promising platform for glycosidase-responsive carriers. Similar, to what has been seen above, MSNs can be prepared such that the pores are capped with an enzyme-responsive agent. Here, the pores are capped with polysaccharides that are glycosidase-responsive.[29,62,72] To give a few examples, Park *et al.* capped the pores with cyclodextrin, providing drug release when the carriers were exposed to α-amylase and lipase, which are overexpressed in a number of diseases including pancreatitis.[62] Bernardos *et al.*, grafted starch derivatives onto the surface of the MSNs which yielded release in the presence of β-galactosidase.[72]

9.4 Summary

Enzymes are overexpressed in a wide-range of diseases, and as a result provide a promising platform for drug delivery. By designing carriers that only release drugs in the presence of disease-specific enzymes, systemic toxicity and off-target effects can be minimized. The use of enzyme-responsive drug carriers enables the delivery of highly effective drugs that currently cannot be used due to cytotoxicity when delivered systemically or as a free-drug. Thus, enzyme-responsive platforms hold significant promise and have been leveraged using a wide range of drug delivery vehicles, including liposomes, pro-drug conjugates, nanoparticles, and mesoporous silica, among others.

References

1. A. Napoli, M. Boerakker, N. Tirelli, R. Nolte, N. Sommerdijk and J. Hubbell, Glucose-oxidase based self-destructing polymeric vesicles, *Langmuir*, 2004, **20**(9), 3487–3491.
2. L. Dai, R. Cai, M. Li, Z. Luo, Y. Yu, W. Chen, X. Shen, Y. Pei, X. Zhao and K. Cai, Dual-targeted cascade-responsive prodrug micelle system for tumor therapy *in vivo*, *Chem. Mater.*, 2017, **29**(16), 6976–6992.
3. H. Xiang, H. Chen, H. P. Tham, S. Z. F. Phua, J.-G. Liu and Y. Zhao, Cyclometalated iridium(III)-complex-based micelles for glutathione-responsive targeted chemotherapy and photodynamic therapy, *ACS Appl. Mater. Interfaces*, 2017, **9**(33), 27553–27562.
4. X. Li, T. Tang, Y. Zhou, Y. Zhang and Y. Sun, Applicability of enzyme-responsive mesoporous silica supports capped with bridged silsesquioxane for colon-specific drug delivery, *Microporous Mesoporous Mater.*, 2014, **184**, 83–89.
5. A. Popat, B. Ross, J. Liu, S. Jambhrunkar, F. Kleitz and S. Qiao, Enzyme-responsive controlled release of covalently bound prodrug from functional mesoporous silica nanospheres, *Angew. Chem., Int. Ed.*, 2012, **51**(50), 12486–12489.
6. N. Mas, A. Agostini, L. Mondragon, A. Bernardos, F. Sancenon, M. Marcos, R. Martinez-Manez, A. Costero, S. Gil, M. Merino-Sanjuan, P. Amoros, M. Orzaez and E. Perez-Paya, Enzyme-responsive silica mesoporous supports capped with azopyridinium salts for controlled delivery applications, *Chem.–Eur. J.*, 2013, **19**(4), 1346–1356.
7. Y. Xiao, T. Wang, Y. Cao, X. Wang, Y. Zhang, Y. Liu and Q. Huo, Enzyme and voltage stimuli-responsive controlled release system based on beta-cyclodextrin-capped mesoporous silica nanoparticles, *Dalton Trans.*, 2015, **44**(9), 4355–4361.
8. M. Duffy, Proteases as prognostic markers in cancer, *Clin. Cancer Res.*, 1996, **2**(4), 613–618.
9. T. Bogenrieder and M. Herlyn, Axis of evil: molecular mechanisms of cancer metastasis, *Oncogene*, 2003, **22**(42), 6524–6536.

10. K. Isaacson, M. M. Jensen, N. B. Subrahmanyam and H. Ghandehari, Matrix-metalloproteinases as targets for controlled delivery in cancer: An analysis of upregulation and expression, *J. Controlled Release*, 2017, **259**, 62–75.
11. A. Houghton, Matrix metalloproteinases in destructive lung disease, *Matrix Biol.*, 2015, **44–46**, 167–174.
12. C. Gialeli, A. Theocharis and N. Karamanos, Roles of matrix metalloproteinases in cancer progression and their pharmacological targeting, *FEBS J.*, 2011, **278**(1), 16–27.
13. P. Burrage, K. Mix and C. Brinckerhoff, Matrix metalloproteinases: Role in arthritis, *Front. Biosci.*, 2006, **11**, 529–543.
14. S. Agewall, Matrix metalloproteinases and cardiovascular disease, *Eur. Heart J.*, 2006, **27**(2), 121–122.
15. D. Vartak and R. Gemeinhart, Matrix metalloproteases: Underutilized targets for drug delivery, *J. Drug Targeting*, 2007, **15**(1), 1–20.
16. J. West and J. Hubbell, Polymeric biomaterials with degradation sites for proteases involved in cell migration, *Macromolecules*, 1999, **32**(1), 241–244.
17. R. Chandrawati, Enzyme-responsive polymer hydrogels for therapeutic delivery, *Exp. Biol. Med.*, 2016, **241**(9), 972–979.
18. V. Garripelli, J. Kim, S. Son, W. Kim, M. Repka and S. Jo, Matrix metalloproteinase-sensitive thermogelling polymer for bioresponsive local drug delivery, *Acta Biomater.*, 2011, **7**(5), 1984–1992.
19. W. Chen, G. Luo, Q. Lei, H. Jia, S. Hong, Q. Wang, R. Zhuo and X. Zhang, MMP-2 responsive polymeric micelles for cancer-targeted intracellular drug delivery, *Chem. Commun.*, 2015, **51**(3), 465–468.
20. E. Secret, S. J. Kelly, K. E. Crannell and J. S. Andrew, Enzyme-responsive hydrogel microparticles for pulmonary drug delivery, *ACS Appl. Mater. Interfaces*, 2014, 10313–10321.
21. E. Secret, K. Crannell, S. Kelly, M. Villancio-Wolter and J. Andrew, Matrix metalloproteinase-sensitive hydrogel microparticles for pulmonary drug delivery of small molecule drugs or proteins, *J. Mater. Chem. B*, 2015, **3**(27), 5629–5634.
22. B. Purcell, D. Lobb, M. Charati, S. Dorsey, R. Wade, K. Zellars, H. Doviak, S. Pettaway, C. Logdon, J. Shuman, P. Freels, J. Gorman, R. Gorman, F. Spinale and J. Burdick, Injectable and bioresponsive hydrogels for on-demand matrix metalloproteinase inhibition, *Nat. Mater.*, 2014, **13**(6), 653–661.
23. S. Huang, K. Shao, Y. Liu, Y. Kuang, J. Li, S. An, Y. Guo, H. Ma and C. Jiang, Tumor-targeting and microenvironment-responsive smart nanoparticles for combination therapy of antiangiogenesis and apoptosis, *ACS Nano*, 2013, **7**(3), 2860–2871.
24. L. Zhu, F. Perche, T. Wang and V. Torchilin, Matrix metalloproteinase 2-sensitive multifunctional polymeric micelles for tumor-specific co-delivery of siRNA and hydrophobic drugs, *Biomaterials*, 2014, **35**(13), 4213–4222.

25. Z. Amoozgar and Y. Yeo, Recent advances in stealth coating of nanoparticle drug delivery systems, *Wiley Interdiscip. Rev.: Nanomed. Nanobiotechnol.*, 2012, **4**(2), 219–233.
26. H. Li, M. Miteva, K. Kirkbride, M. Cheng, C. Nelson, E. Simpson, M. Gupta, C. Duvall and T. Giorgio, Dual MMP7-proximity-activated and folate receptor-targeted nanoparticles for siRNA delivery, *Biomacromolecules*, 2015, **16**(1), 192–201.
27. B. Grünwald, J. Vandooren, E. Locatelli, P. Fiten, G. Opdenakker, P. Proost, A. Kruger, J. Lellouche, L. Israel, L. Shenkman and M. Franchini, Matrix metalloproteinase-9 (MMP-9) as an activator of nanosystems for targeted drug delivery in pancreatic cancer, *J. Controlled Release*, 2016, **239**, 39–48.
28. W. Ke, Z. Zha, J. Mukerabigwi, W. Chen, Y. Wang, C. He and Z. Ge, Matrix metalloproteinase-responsive multifunctional peptide-linked amphiphilic block copolymers for intelligent systemic anticancer drug delivery, *Bioconjugate Chem.*, 2017, **28**(8), 2190–2198.
29. N. Singh, A. Karambelkar, L. Gu, K. Lin, J. Miller, C. Chen, M. Sailor and S. Bhatia, Bioresponsive mesoporous silica nanoparticles for triggered drug release, *J. Am. Chem. Soc.*, 2011, **133**(49), 19582–19585.
30. J. Liu, B. Zhang, Z. Luo, X. Ding, J. Li, L. Dai, J. Zhou, X. Zhao, J. Ye and K. Cai, Enzyme responsive mesoporous silica nanoparticles for targeted tumor therapy *in vitro* and *in vivo*, *Nanoscale*, 2015, **7**(8), 3614–3626.
31. S. van Rijt, D. Bolukbas, C. Argyo, S. Datz, M. Lindner, O. Eickelberg, M. Konigshoff, T. Bein and S. Meiners, Protease-mediated release of chemotherapeutics from mesoporous silica nanoparticles to *ex vivo* human and mouse lung tumors, *ACS Nano*, 2015, **9**(3), 2377–2389.
32. T. Ji, S. Li, Y. Zhang, J. Lang, Y. Ding, X. Zhao, R. Zhao, Y. Li, J. Shi, J. Hao, Y. Zhao and G. Nie, An MMP-2 responsive liposome integrating antifibrosis and chemotherapeutic drugs for enhanced drug perfusion and efficacy in pancreatic cancer, *ACS Appl. Mater. Interfaces*, 2016, **8**(5), 3438–3445.
33. Y. Wan, J. Han, G. Fan, Z. Zhang, T. Gong and X. Sun, Enzyme-responsive liposomes modified adenoviral vectors for enhanced tumor cell transduction and reduced immunogenicity, *Biomaterials*, 2013, **34**(12), 3020–3030.
34. P. Xu, Q. Meng, H. Sun, Q. Yin, H. Yu, Z. Zhang, M. Cao, Y. Zhang and Y. Li, Shrapnel nanoparticles loading docetaxel inhibit metastasis and growth of breast cancer, *Biomaterials*, 2015, **64**, 10–20.
35. C. Wong, T. Stylianopoulos, J. Cui, J. Martin, V. Chauhan, W. Jiang, Z. Popovic, R. Jain, M. Bawendi and D. Fukumura, Multistage nanoparticle delivery system for deep penetration into tumor tissue, *Proc. Natl. Acad. Sci. U. S. A.*, 2011, **108**(6), 2426–2431.
36. S. Ruan, X. Cao, X. Cun, G. Hu, Y. Zhou, Y. Zhang, L. Lu, Q. He and H. Gao, Matrix metalloproteinase-sensitive size-shrinkable nanoparticles for deep tumor penetration and pH triggered doxorubicin release, *Biomaterials*, 2015, **60**, 100–110.
37. G. Hu, H. Zhang, L. Zhang, S. Ruan, Q. He and H. Gao, Integrin-mediated active tumor targeting and tumor microenvironment response

dendrimer-gelatin nanoparticles for drug delivery and tumor treatment, *Int. J. Pharm.*, 2015, **496**(2), 1057–1068.
38. H. Nagase, Substrate specificity of MMPs, in *Cancer Drug Discovery and Development: Matrix Metalloproteinase Inhibitors in Cancer Therapy*, ed, Inc, H. P., 2001, pp. 39–66.
39. B. Turk, L. Huang, E. Piro and L. Cantley, Determination of protease cleavage site motifs using mixture-based oriented peptide libraries, *Nat. Biotechnol.*, 2001, **19**(7), 661–667.
40. Y. Mi, J. Wolfram, C. Mu, X. Liu, E. Blanco, H. Shen and M. Ferrari, Enzyme-responsive multistage vector for drug delivery to tumor tissue, *Pharmacol. Res.*, 2016, **113**, 92–99.
41. M. Basel, T. Shrestha, D. Troyer and S. Bossmann, Protease-sensitive, polymer-caged liposomes: a method for making highly targeted liposomes using triggered release, *ACS Nano*, 2011, **5**(3), 2162–2175.
42. N. Li, H. Cai, L. Jiang, J. Hu, A. Bains, J. Hu, Q. Gong, K. Luo and Z. Gu, Enzyme-sensitive and amphiphilic pegylated dendrimer-paclitaxel prodrug-based nanoparticles for enhanced stability and anticancer efficacy, *ACS Appl. Mater. Interfaces*, 2017, **9**(8), 6865–6877.
43. G. Habraken, M. Peeters, P. Thornton, C. Koning and A. Heise, Selective enzymatic degradation of self-assembled particles from amphiphilic block copolymers obtained by the combination of N-carboxyanhydride and nitroxide-mediated polymerization, *Biomacromolecules*, 2011, **12**(10), 3761–3769.
44. A. Aimetti, A. Machen and K. Anseth, Poly(ethylene glycol) hydrogels formed by thiol-ene photopolymerization for enzyme-responsive protein delivery, *Biomaterials*, 2009, **30**(30), 6048–6054.
45. C. Guo, H. Kim, E. Ovadia, C. Mourafetis, M. Yang, W. Chen and A. Kloxin, Bio-orthogonal conjugation and enzymatically triggered release of proteins within multi-layered hydrogels, *Acta Biomater.*, 2017, **56**, 80–90.
46. I. Candel, E. Aznar, L. Mondragon, C. de la Torre, R. Martinez-Manez, F. Sancenon, M. Marcos, P. Amoros, C. Guillem, E. Perez-Paya, A. Costero, S. Gil and M. Parra, Amidase-responsive controlled release of antitumoral drug into intracellular media using gluconamide-capped mesoporous silica nanoparticles, *Nanoscale*, 2012, **4**(22), 7237–7245.
47. A. Harnoy, I. Rosenbaum, E. Tirosh, Y. Ebenstein, R. Shaharabani, R. Beck and R. Amir, Enzyme-responsive amphiphilic PEG-dendron hybrids and their assembly into smart micellar nanocarriers, *J. Am. Chem. Soc.*, 2014, **136**(21), 7531–7534.
48. M. Vicent, F. Greco, R. Nicholson, A. Paul, P. Griffiths and R. Duncan, Polymer therapeutics designed for a combination therapy of hormone-dependent cancer, *Angew. Chem., Int. Ed.*, 2005, **44**(26), 4061–4066.
49. L. Mondragón, N. Mas, V. Ferragud, C. de la Torre, A. Agostini, R. Martinez-Manez, F. Sancenon, P. Amoros, E. Perez-Paya and M. Orzaez, Enzyme-responsive intracellular-controlled release using silica mesoporous nanoparticles capped with e- poly- l- lysine, *Chem.–Eur. J.*, 2014, **20**(18), 5271–5281.

50. K. Patel, S. Angelos, W. Dichtel, A. Coskun, Y. Yang, J. Zink and J. Stoddart, Enzyme-responsive snap-top covered silica nanocontainers, *J. Am. Chem. Soc.*, 2008, **130**(8), 2382–2383.
51. I. Fernando, D. Ferris, M. Frasconi, D. Malin, E. Strekalova, M. Yilmaz, M. Ambrogio, M. Algaradah, M. Hong, X. Chen, M. Nassar, Y. Botros, V. Cryns and J. Stoddart, Esterase- and pH-responsive poly(beta-amino ester)-capped mesoporous silica nanoparticles for drug delivery, *Nanoscale*, 2015, **7**(16), 7178–7183.
52. M. Malvindi, R. Di Corato, A. Curcio, D. Melisi, M. Rimoli, C. Tortiglione, A. Tino, C. George, V. Brunetti, R. Cingolani, T. Pellegrino and A. Ragusa, Multiple functionalization of fluorescent nanoparticles for specific biolabeling and drug delivery of dopamine, *Nanoscale*, 2011, **3**(12), 5110–5119.
53. Y. Ding, Y. Zhou, H. Chen, D. Geng, D. Wu, J. Hong, W. Shen, T. Hang and C. Zhang, The performance of thiol-terminated PEG-paclitaxel-conjugated gold nanoparticles, *Biomaterials*, 2013, **34**(38), 10217–10227.
54. Y. Li, R. Liu, J. Yang, G. Ma, Z. Zhang and X. Zhang, Dual sensitive and temporally controlled camptothecin prodrug liposomes codelivery of siRNA for high efficiency tumor therapy, *Biomaterials*, 2014, **35**(36), 9731–9745.
55. C. Lee, W. Park, S. Park and K. Na, Endolysosomal environment-responsive photodynamic nanocarrier to enhance cytosolic drug delivery *via* photosensitizer-mediated membrane disruption, *Biomaterials*, 2013, **34**(36), 9227–9236.
56. J. Awino, S. Gudipati, A. Hartmann, J. Santiana, D. Cairns-Gibson, N. Gomez and J. Rouge, Nucleic acid nanocapsules for enzyme-triggered drug release, *J. Am. Chem. Soc.*, 2017, **139**(18), 6278–6281.
57. D. Guo, K. Wang, Y. Wang and Y. Liu, Cholinesterase-responsive supramolecular vesicle, *J. Am. Chem. Soc.*, 2012, **134**(24), 10244–10250.
58. J. Qi, J. Zhuang, Y. Lu, X. Dong, W. Zhao and W. Wu, *In vivo* fate of lipid-based nanoparticles, *Drug Discovery Today*, 2017, **22**(1), 166–172.
59. M. Schwab, G. Sax, S. Schulze and G. Winter, Studies on the lipase induced degradation of lipid based drug delivery systems, *J. Controlled Release*, 2009, **140**(1), 27–33.
60. M. Schwab, C. McGoverin, K. Gordon, G. Winter, T. Rades, J. Myschik and C. Strachan, Studies on the lipase-induced degradation of lipid-based drug delivery systems. Part II - Investigations on the mechanisms leading to collapse of the lipid structure, *Eur. J. Pharm. Biopharm.*, 2013, **84**(3), 456–463.
61. R. Mahjub, F. Dorkoosh, M. Rafiee-Tehrani and A. Schnurch, Oral self-nanoemulsifying peptide drug delivery systems: impact of lipase on drug release, *J. Microencapsulation*, 2015, **32**(4), 401–407.
62. C. Park, H. Kim, S. Kim and C. Kim, Enzyme responsive nanocontainers with cyclodextrin gatekeepers and synergistic effects in release of guests, *J. Am. Chem. Soc.*, 2009, **131**(46), 16614–16615.

63. M. Xiong, Y. Bao, X. Yang, Y. Wang, B. Sun and J. Wang, Lipase-sensitive polymeric triple-layered nanogel for "On-demand" drug delivery, *J. Am. Chem. Soc.*, 2012, **134**(9), 4355–4362.
64. R. Ostrem, L. Parhamifar, H. Pourhassan, G. Clergeaud, O. Nielsen, A. Kjaer, A. Hansen and T. Andresen, Secretory phospholipase A(2) responsive liposomes exhibit a potent anti-neoplastic effect *in vitro*, but induce unforeseen severe toxicity *in vivo*, *J. Controlled Release*, 2017, **262**, 212–221.
65. C. Pawlowski, W. Li, M. Sun, K. Ravichandran, D. Hickman, C. Kos, G. Kaur and A. Sen Gupta, Platelet microparticle-inspired clot-responsive nanomedicine for targeted fibrinolysis, *Biomaterials*, 2017, **128**, 94–108.
66. P. F. Hockl, A. Wolosiuk, J. M. Pérez-Sáez, A. V. Bordoni, D. O. Croci, Y. Toum-Terrones, G. J. A. A. Soler-Illia and G. A. Rabinovich, Glyco-nano-oncology: Novel therapeutic opportunities by combining small and sweet, *Pharmacol. Res.*, 2016, **109**(suppl. C), 45–54.
67. E. Ferguson and R. Duncan, Dextrin-phospholipase a(2): synthesis and evaluation as a bioresponsive anticancer conjugate, *Biomacromolecules*, 2009, **10**(6), 1358–1364.
68. Q. Zhao, J. Zhang, L. Song, Q. Ji, Y. Yao, Y. Cui, J. Shen, P. G. Wang and D. Kong, Polysaccharide-based biomaterials with on-demand nitric oxide releasing property regulated by enzyme catalysis, *Biomaterials*, 2013, **34**(33), 8450–8458.
69. X. Wu, X. Tang, M. Xian and P. G. Wang, Glycosylated diazeniumdiolates: a novel class of enzyme-activated nitric oxide donors, *Tetrahedron Lett.*, 2001, **42**(23), 3779–3782.
70. J. Clarhaut, S. Fraineau, J. Guilhot, E. Peraudeau, I. Tranoy-Opalinski, M. Thomas, B. Renoux, E. Randriamalala, P. Bois, A. Chatelier, A. Monvoisin, L. Cronier, S. Papot and F. Guilhot, A galactosidase-responsive doxorubicin-folate conjugate for selective targeting of acute myelogenous leukemia blasts, *Leuk. Res.*, 2013, **37**(8), 948–955.
71. G. Baier, A. Cavallaro, K. Vasilev, V. Mailander, A. Musyanovych and K. Landfester, Enzyme responsive hyaluronic acid nanocapsules containing polyhexanide and their exposure to bacteria to prevent infection, *Biomacromolecules*, 2013, **14**(4), 1103–1112.
72. A. Bernardos, L. Mondragon, E. Aznar, M. Marcos, R. Martinez-Manez, F. Sancenon, J. Soto, J. Barat, E. Perez-Paya, C. Guillem and P. Amoros, Enzyme-responsive intracellular controlled release using nanometric silica mesoporous supports capped with "Saccharides", *ACS Nano*, 2010, **4**(11), 6353–6368.

CHAPTER 10

Swelling-controlled Drug Delivery Systems

CANDACE MINHTHU DAY, THOMAS G. BARCLAY, YUNMEI SONG AND SANJAY GARG*

Centre for Pharmaceutical Innovation and Development, School of Pharmacy and Medical Science, University of South Australia, Adelaide, SA 5000, Australia
*E-mail: Sanjay.Garg@unisa.edu.au

10.1 Introduction

10.1.1 Polymers

Polymers are large molecules constructed from many smaller building blocks called monomers (Figure 10.1). The term polymer came from the Greek; "poly" meaning many and "meres" meaning parts.[1] The name polymer was first used in 1866 by Berthelot in the Bulletin of the Chemical Society of France following a scientific discovery that demonstrated the transformation of styrene into a "resinous polymer" after heating at 200 °C for a few hours.[2] Nonetheless, the polymer theory that is understood and accepted today was not established until Hermann Staudinger introduced the "macromolecular" theory in 1920. In acknowledgment of the significance of Staudinger's work establishing the field of polymer science, he was awarded the Nobel Prize in 1953.[3]

Biomaterials Science Series No. 1
Stimuli-responsive Drug Delivery Systems
Edited by Amit Singh and Mansoor M. Amiji

Published by the Royal Society of Chemistry, www.rsc.org

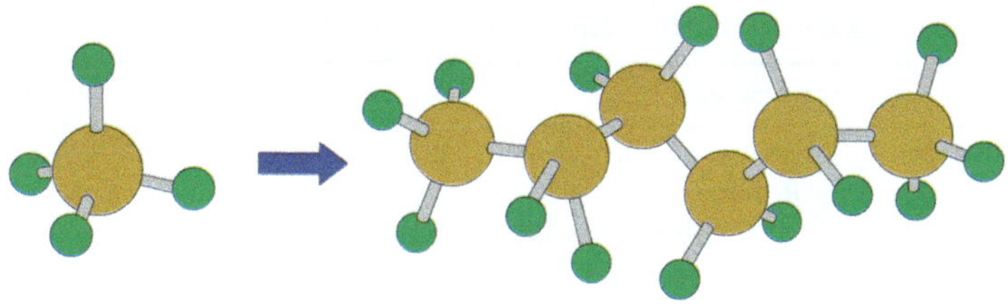

Figure 10.1 An illustration of polymer formation from individual monomers.

From the middle of the 20th century, both naturally occurring and synthesized polymers have been extensively exploited for a range of uses.[4] This includes commercial use of simple plastics as containers and fabrics, right up to the application of complex multifunctional polymeric materials in sophisticated pharmaceutical systems, including swelling controlled drug delivery.

10.1.2 The Important Roles of Polymers in Drug Delivery

- Conventional (immediate release) tablets: polymers used as binders in granulation processes, agglomerating the mixtures of powdered ingredients and excipients prior to tableting.[5]
- Capsules: polymers act as bulking agents to fill the empty volume of capsules.[5]
- Polymers have been used to prolong the gastric residence time of drugs by binding with the stomach mucus lining or floating above the gastric contents. This reduces drug waste and enhances the aqueous solubility of poorly soluble drugs.[6,7]
- Polymers have been exploited as excipients in sustained release dosage forms, maintaining systemic drug levels within the therapeutic range. This results in reduced dosing frequency, increased patient compliance and decreased costs.[8] Sustained release from gastro retentive formulations (GRFs) that remain in the stomach for an extended period of time are also able to enhance oral bioavailability.[8]
- Some water-soluble polymers, including cellulose derivatives, alginates, and starch, are used as disintegrants that assist the dispersion of tablets in the gastrointestinal tract. The rapid breakup of tablets increases the exposed surface area, leading to a more rapid release of the active ingredients.[9]
- Some polymers are also used as coating agents for solid dosage forms.[10]

Common pharmaceutical polymers are classified based on both structures and applications. A brief overview of these polymers and their uses is presented in Table 10.1.

Table 10.1 Classes of pharmaceutical polymers and their applications.

Class	Sub-class	Applications	References
Vinyl polymers	Polyvinyl alcohol Polymethacrylates	Film coating	11
	Polyvinylpyrrolidone	Binder	
	Poly(acrylic acid)	Lubricant	
		Viscosity modifying agent	
Cellulose ethers	Methylcellulose	Thickeners	
	Ethylcellulose	Emulsifiers	
	Hydroxypropylcellulose	Binders	
	Hydroxyethylcellulose	Coating agents	
	Hydroxypropylmethyl-cellulose	Stabilizers	
	Hydroxyethylmethyl-cellulose		
	Sodium carboxymethyl-cellulose		
Polyesters	Poly(lactid)	Controlled release of injectable and implant-able drugs and vaccine	
	Polycaprolacton		
Polysaccharides	Chitosan	Adsorption enhancer	
	Xanthan gum	Stabilizers Thickeners	
	Carrageenan	Emulsifiers	
	Poly(alginic acid)		
Miscellaneous polymers	Gelatin	Film coating	
	Polyanhydrides	Binder	
		Viscosity modifying agent	
	Polyethylene glycols	Thickeners	
	Polyethylene oxide	Solubilizers	
Silicones	Silicone fluids	Medical devices	12
	Silicone gels	Implants	
	Rubbers	Intravaginal delivery	
	Silicone resins		

10.1.2.1 Natural Polymers in Drug Delivery

Examples of useful naturally occurring polymers include polysaccharides (cellulose, chitin, and starch) and proteins (wool, collagen, and silk) and such polymers possess a wide spectrum of applications, ranging from oil production and clothing manufacture to the food industry (Figure 10.2). The use of naturally occurring polymers remains heavily attractive in the pharmaceutical industry, mainly due to their low cost, ready availability, high biocompatibility and excellent biodegradability.[13] Generally, natural polymers also provide functional groups for chemical modification, which are also useful for pharmaceutical applications.[13] Furthermore, natural polymers generally have higher molecular weights and they take longer to reach equilibrium swelling compared to synthetic and semi-synthetic swellants and so alter drug release characteristics.[14] Some examples of naturally occurring polymers used in drug delivery and biomedical research are listed in Table 10.2.

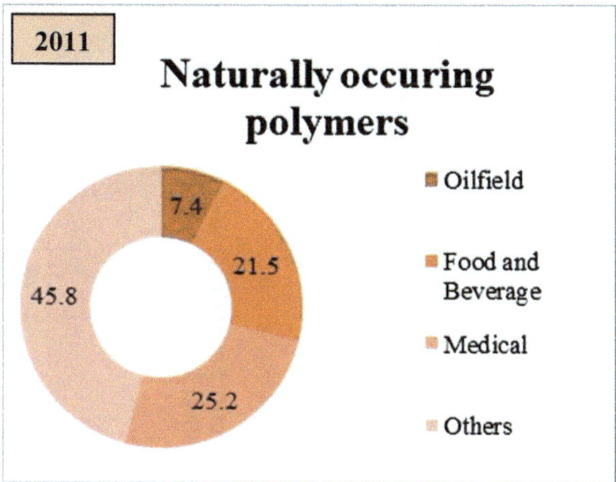

Figure 10.2 US market share of natural polymers in 2011. (Data: Transparency Market Research 2013).[15]

10.1.2.2 Synthetic Polymers in Drug Delivery

Synthetic polymers, many of which are manufactured from petroleum sources, are also heavily exploited in the pharmaceutical industry. The use of synthetic polymer bypasses disadvantages associated with natural polymers, including microbial contamination, the uncontrolled rate of hydration and reduced viscosity upon storage (Table 10.3).[28]

10.2 Introduction to Swelling Controlled Drug Delivery Systems

A swelling-controlled drug delivery system (SCDDS) is a three-dimensional network built upon polymeric chains that are chemically or physically cross-linked. Most often in drug delivery, these networks are amorphous and hydrophilic and can absorb drugs suspended in aqueous solutions. Upon drying, the network contracts into a glassy state and entraps the drug molecules, restricting them from moving out of the network. Upon further contact with aqueous solutions, such as in biological fluids, the network rehydrates and becomes rubbery at a rate determined by the chemistry of the hydrogel, relaxing the polymer chains and allowing the dense polymeric network to increase in volume. This swelling mechanism allows the release of the entrapped payload (Figure 10.3).[14]

To achieve the maximal therapeutic effects of pharmaceutical agents, SCDDSs can be tuned, not only to provide controlled release in terms of time, and therefore the location of release, but also the rate of release of biologically active compounds. The ability to support sustained release means that SCDDSs can maintain therapeutic drug concentrations for extended

Table 10.2 Biopolymers and their applications.

Name and origin	Chemical structure	Applications
Alginate (brown seaweed)	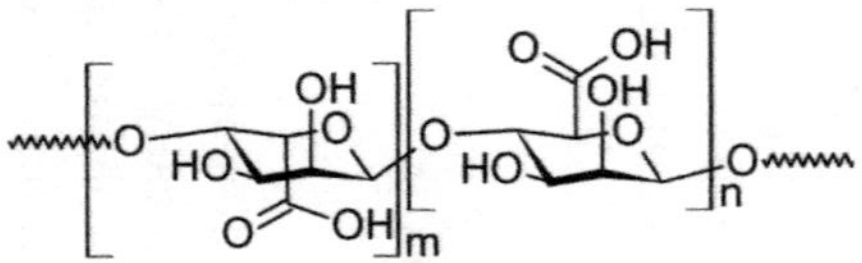	Intestinal, colon, mucosal tissue, and neurological drug delivery[16–19]
Chitosan (fungi)	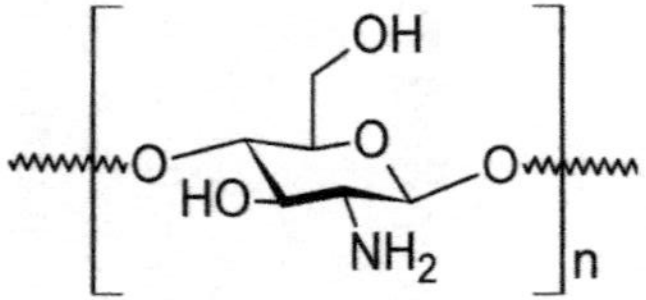	Insulin, epirubicin and protein drug oral delivery as well as wound dressings[20–22]
Gelatin (collagen from animal skin and bone)	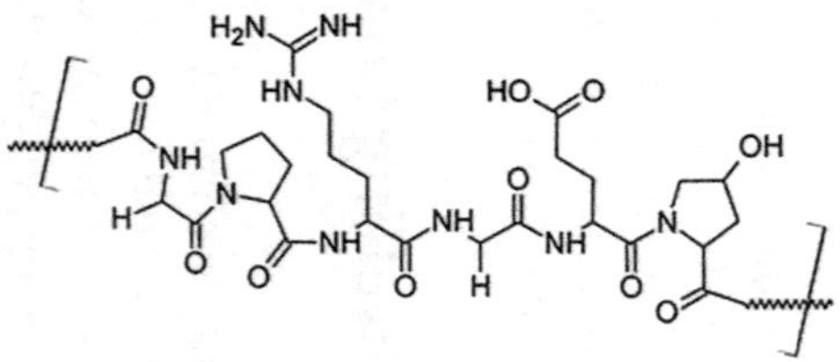	Ultrasound-triggered drug release, drug carrier system for lymphocyte uptake and vaccines[23,24]
Pectin (plants cell walls)	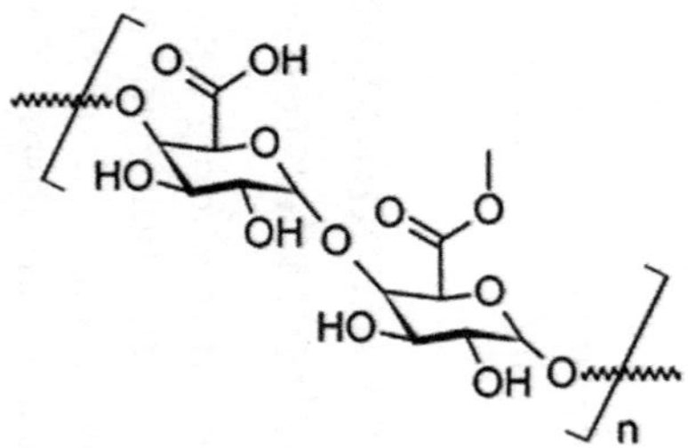	Oral anticancer drugs and vaccine delivery[25,26]
Gellan gum (bacterium, *sphingomonas elodea*)	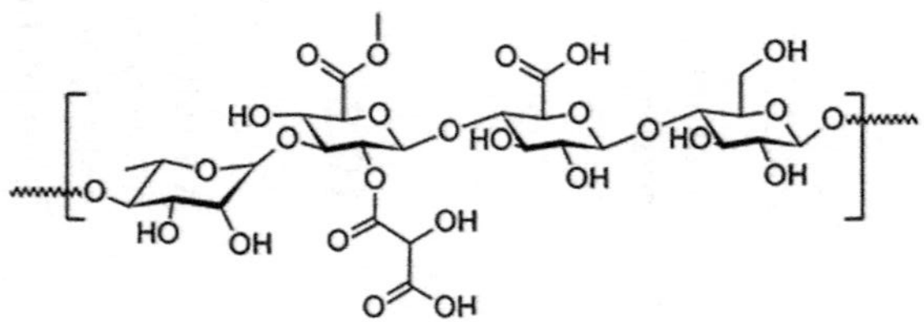	Gene delivery agent[27]

Table 10.3 Synthetic polymers and their applications.

Name	Chemical structure	Applications
Polylactic acid (PLA)	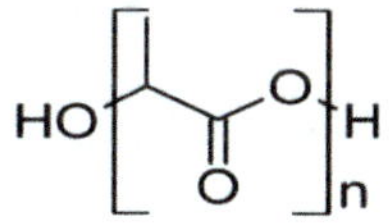	Surgical implant material; porous scaffolds for neo-tissue growth[29,30]
Polyglycolic acid (PGA)	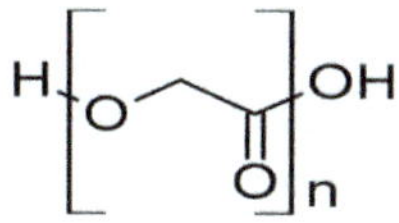	Filler materials; scaffold for spinal, bone, vaginal, intestinal and lymphatic regeneration[31]
Poly (lactide-*co*-glycolide) PLGA	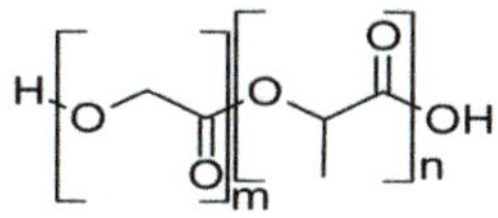	Suture materials; tissue engineering scaffolds; carriers for chemo-therapeutics, vaccines, antibiotics[31,32]
Polyethylene glycol (PEG)	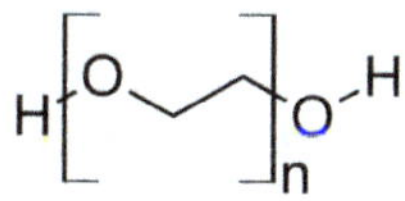	Anticancer agents carriers; targeted diagnostics agents; solubilizers; stabilizers; wound healing dressings; tissue regeneration; tissue engineering scaffolds[33–35]
Poly (phosphoester) (PPE)	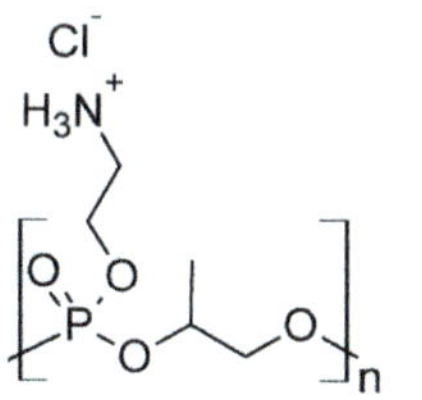	Gene and drugs nanocarriers[36,37]

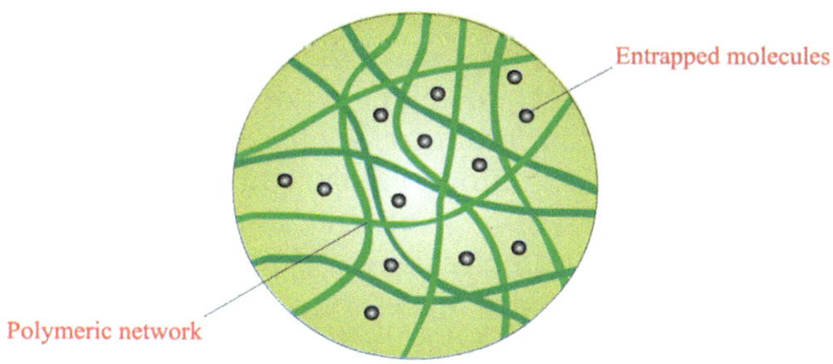

Figure 10.3 Hydrogel drug delivery system.

periods, reducing requirements for repeated dosing and increasing patient compliance. In the following section, we discuss structural properties of the polymers used in SCDDSs including their cross-linking chemistry, relative hydrophilicity, and ionic strengths. These are all aspects that can be manipulated to benefit swelling kinetics for a particular application.

10.2.1 Structuring Polymers to Control Swelling Kinetics

To understand how to use polymer structure to tune swelling, the three forces that drive swelling must also be understood. These are: (i) the osmotic force; (ii) the electrostatic force; and (iii) the polymeric entropy upon dissolution in water.[14] For example, for an ionic polymer in basic solution acid groups will be deprotonated and cations will enter the structure to balance the charge, causing a concomitant osmotic pressure that drives water into the hydrogel resulting in swelling.[38] At the same time, electrostatic interactions between charged groups in the hydrogel can influence both the ionic content of the hydrogel and the resulting supramolecular attraction and repulsion, also having an impact on swelling.[38] These osmotic and electrostatic forces must be balanced against the elastic restoring force of the system for equilibrium swelling of ionic gels to be reached.[38] The elastic restoring force of the hydrogel is related to the polymeric entropy, both of which are directly controlled by cross-link density.[14,38] Consequently, to tune swelling of a hydrogel, the osmotic force, electrostatic interactions and polymeric entropy are controlled structurally by adjusting the cross-link density, ionic content, and hydrophilicity of the incorporated polymer.

10.2.2 Cross-linking

10.2.2.1 Definition and Classification of Polymeric Cross-links

The IUPAC defines cross-linking as the bonding of polymer chains together.[39] This leads to the loss of ability of the chains to slide over each other as easily as individual chains, making polymeric materials less easily stretched, giving them higher melting points or preventing melting altogether.[40] Polymer cross-links can form either in an intermolecular or intramolecular manner. In intermolecular cross-linking functional groups from one polymer chain form linkages to a second polymer chain. This creates a larger new macromolecule. Contrastingly, intramolecular crosslinking involves the bonding between two functional groups of the same polymer chain. This does not increase molecular weight but instead leads to the formation of connective loops within that polymer, compacting the polymeric structures.[41] Intramolecular and intermolecular cross-linking tend to occur in parallel,[41] but in a concentration dependant fashion. At low polymer concentrations, intramolecular crosslinking is favoured and intermolecular cross-linking is favoured at higher concentrations (Figure 10.4).[42]

When macromolecules are highly cross-linked, three-dimensional polymeric cross-linked networks will be formed, which can be further divided into physical and chemical cross-linked networks.[14]

- *Physically cross-linked network*

A *physical* or *reversible* network is formed when polymer chains are bound *via* non-covalent physical interactions, usually hydrogen bonding, π–π interactions or simple chain entanglement.[43] Physically cross-linked networks are generally relatively easily altered by physical forces such as the addition of heat, changes in pH and shear stresses.

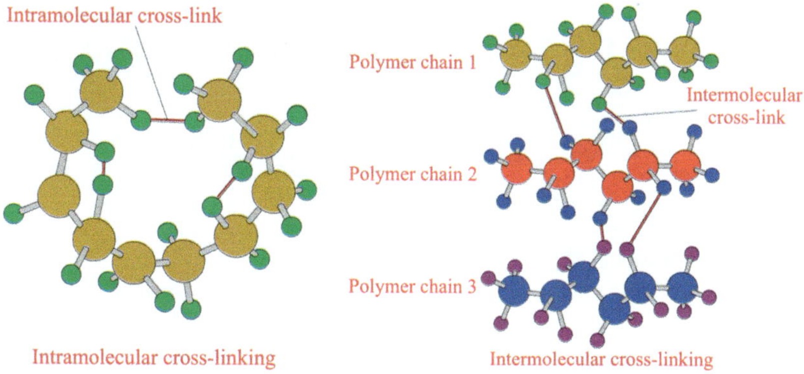

Figure 10.4 Intramolecular cross-linking *vs.* intermolecular cross-linking.

- *Chemical cross-linked network*

Chemical cross-links use covalent bonds to create the cross-links and are therefore more mechanically and thermally stable than physically cross-linked networks. Owing to their stability, chemically cross-linked polymers are widely used in the medical industry.[43]

10.2.2.2 *Effects of Polymeric Cross-linking on Swelling Controlled Kinetics*

Cross-linking changes many aspects of polymeric materials, such as their thermal stability, solubility and swellability and so cross-link density can be tailored to optimise these properties for a specific application.[41,44] This is for two main reasons. Firstly, as cross-links reduce the non-covalent intermolecular forces between polymer chains, they preventing them from collapsing onto each other.[14] This lack of interactions means that more densely cross-linked networks will swell faster than networks with fewer cross-links. Secondly, as cross-link density controls the polymeric entropy and elastic restoring force, the physical attachment between chains restricts the extent of swelling. In other words, as cross-link density increases, the elastic restoring force increases and the entropy of hydrogel molecules decrease, decreasing the swelling capacity of the hydrogel. Consequently, tailoring cross-linking provides the ability to adjust both the rate and extent of swelling of the resulting hydrogel (Figure 10.5).

10.2.2.3 *Hydrophilicity*

In addition to cross-linking, the relative hydrophilicity of the polymer also plays an essential part in establishing hydrogel swelling capacity, influencing the intermolecular interactions responsible for the diffusion water into the hydrogel and its swelling.[14] For example, all other factors being equal, a more

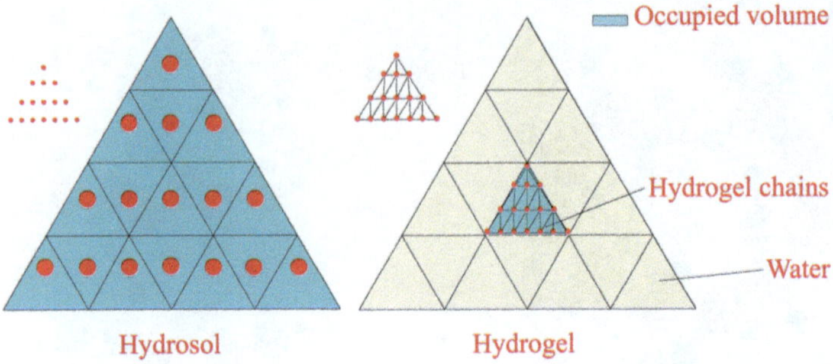

Figure 10.5 Non-cross-linked system (hydrosol) entropy *vs.* Cross-linked system (hydrogel) entropy.

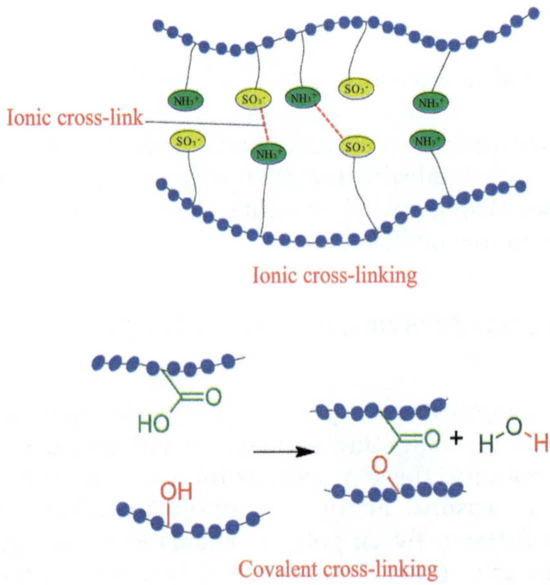

Figure 10.6 Ionic cross-linking *vs.* non-ionic (covalent) cross-linking.

hydrophilic polymer will have increased interaction with water molecules, resulting in faster diffusion into the matrix, more complete relaxation of the network and greater swelling.[44]

10.2.2.4 Ionic Strength

A significant proportion of non-covalently cross-linked SCDDSs rely on electrostatic interactions between ionic groups as the cross-link (Figure 10.6) and many covalently linked SCDDSs also contain ionic groups. As described above, the swelling process is a result of a balance of osmotic pressure, electrostatic forces and polymeric entropy and the ionic content of a hydrogel is intrinsically linked to all three of these factors. This means that the swelling rate and capacity of a hydrogel is very sensitive to salt concentration and the pH of absorbed solutions and ionic content of the swelling polymer is often used to control SCDDS swelling and drug release properties.

10.3 Drug Release Mechanism from Swelling Controlled Delivery Systems

Drug release is defined as the migration of biologically active compounds entrapped inside their carriers from their original location to the outer surface of the carriers, and then to the release medium.[45] This process is controlled by various elements, including the physicochemical characteristics of the drugs, the material structures of delivery systems, the properties of the release medium and/or a combination of all these factors.[46]

There are two approaches by which swelling controlled delivery systems are combined with the drug, both applicable for both water-soluble and water-insoluble drugs. Firstly, the swellable hydrogel can be compressed into a tablet after being combined with the drug and excipients and secondly, a drug-containing tablet can be coated with the swellable hydrogel.[14] For both types of tablet, the underlying mechanism by which drug molecules are released is through the swelling of polymer matrices upon contact with the aqueous medium. This results in the formation of a gel layer (on the surface of the system), meaning all subsequent movement of water into the tablet and drugs out of the tablet is mediated through the gel layer. As water progresses into the tablet it relaxes the swellable polymer, which triggers the escape of entrapped drug molecules by dissolution and diffusion of the drug out of the system and/or degradation of the polymeric drug delivery system. Consequently, when glassy hydrophilic polymeric systems are in contact with body fluid, the mechanisms controlling drug release include diffusion, polymer relaxation, and degradation, which are each described in detail in the following sections.[46,47] If swelling, polymer dissolution, and drug release can occur simultaneously under suitable conditions with appropriate initial drug distribution and polymer composition, then zero-order drug release profiles, in which the rate of drug release is only proportional to time, can be produced.[46,48,49]

Zero-order drug release profiles are important because one of the major preferences in drug release profiles is to maintain the drug plasma concentration between its minimum effective concentration (MEC) and maximum toxic concentration (MTC), or within its therapeutic window (Figure 10.7). When a single large dose of a drug is administrated the drug plasma level can rise above MTC, giving a toxic effect, and then readily falls below the MEC providing little therapeutic benefit. Meanwhile, if multiple smaller doses are administrated in regular intervals, the fluctuation of drug plasma concentration will be reduced, optimising the therapeutic outcomes. However, high

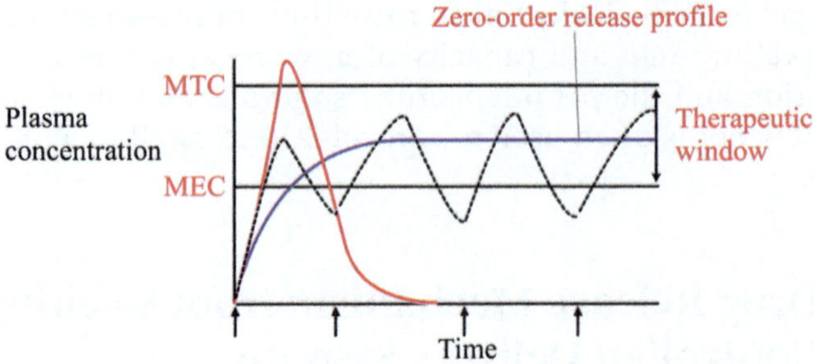

Figure 10.7 Drug plasma concentration profile observed in a single large dose (red line), multiple doses (dash black line), and zero order (blue line). The therapeutic window presented inbetween MTC and MEC is the region where the drug proposes therapeutic effects without toxicity. MTC: maximum toxicity concentration, MEC: minimum effective concentration.

dosing frequency is associated with patient compliance problems (Figure 10.7). Therefore, it is desirable to minimize drug plasma fluctuations and improve patient compliance by developing swelling controlled delivery systems that have a zero-order drug release pattern, producing a constant rate of release of the drug.[47,50]

10.3.1 Drug Diffusion-controlled Release

Drug diffusion is a fundamental aspect of controlled-release. For hydrogel systems in which swelling is rapid and drug diffusion out of the hydrogel is slow, then the drug release is diffusion controlled. Diffusion-controlled drug delivery systems can be classified as monolithic devices and reservoir systems, based on their structural properties and initial drug loading patterns.[51]

10.3.1.1 Monolithic Devices

Monolithic devices are those in which there is uniform drug loading into the matrix and the release patterns are governed by the diffusion of drug molecules through the monolith's matrix material (Figure 10.8).[46] Typically, in such systems, the release patterns present an initial burst of drug release, due to the absence of diffusion barriers/membranes at the surface. Following the initial release, release rate declines overtime with increasing diffusion distance, since drug molecules located deeper inside the monolith matrix must travel farther to the surface for the diffusion to take place.[46,51] This relationship between time and distance to release is especially relevant to cylindrical and spherical monolithic systems[47] and in this case, the desirable zero-order release pattern is unlikely to be achieved due to these geometric effects. However, near-zero-order release patterns can be approached

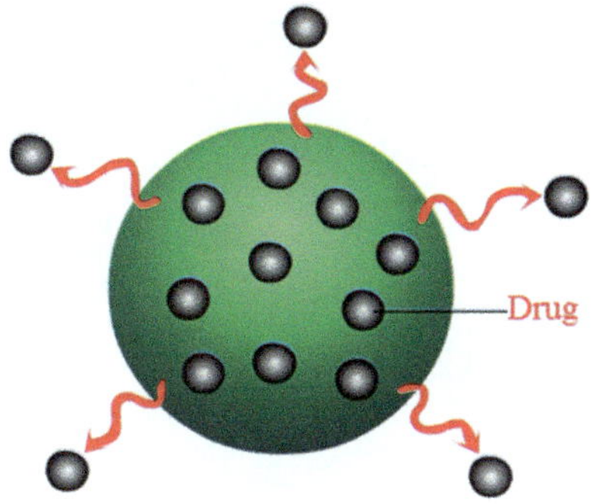

Figure 10.8 Illustration of a diffusion-controlled monolithic device.

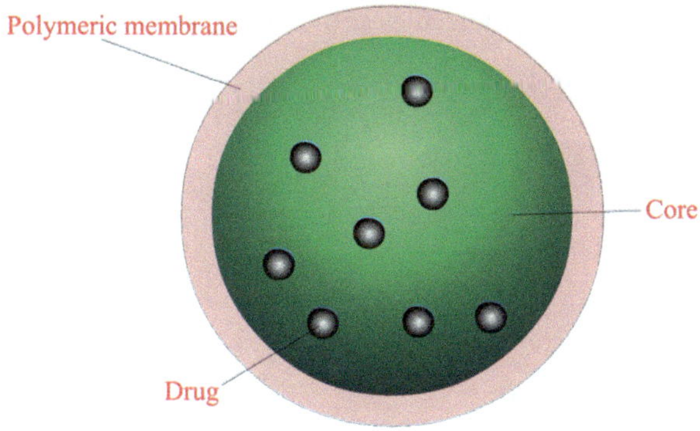

Figure 10.9 Illustration of a diffusion-controlled reservoir system.

if the monolith has a wedge, hemisphere or cone shape, that avoid the geometric issues with more regular shapes.[47] However, the difficulty in fabricating these shapes remains a challenging obstacle in engineering such materials.

10.3.1.2 Reservoir Systems

Diffusion-controlled drug release occurs in reservoir systems in which drugs molecules are contained in a core and have a surrounding polymeric membrane[46] (Figure 10.9). Upon contact with body fluids, the difference in drug concentration across the membrane drives the diffusion of the drug through the polymeric membrane and into the body.[46]

- *Non-constant activity reservoir system:*

A non-constant activity reservoir system refers to a diffusion controlled drug delivery system in which the drug concentration is below its solubility in body fluid at body temperature, and the drug molecules are all dissolved in

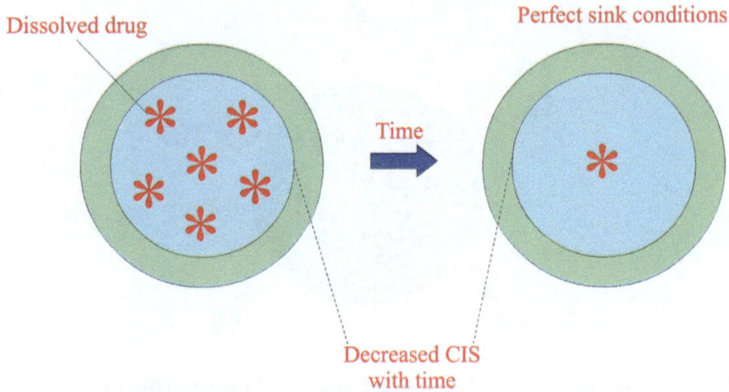

Figure 10.10 Non-constant activity reservoir system. CIS represents the concentration of drug molecules at the membrane inner surface.[14]

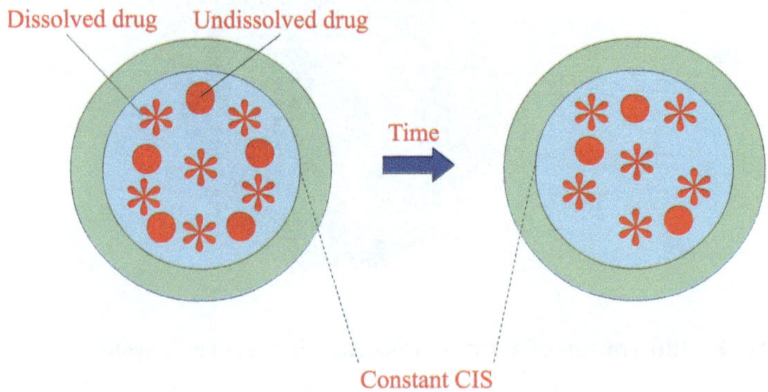

Figure 10.11 Constant activity reservoir system. CIS represents the saturation concentration of drug molecules at the membrane inner surface, provided drug excess is always present.[14]

a system wet with body fluid (Figure 10.10). As such, drugs released through the membrane are not replaced and so the concentration of drug molecules at the membrane decreases with time, decreasing the rate of release.[46] In perfect sink conditions, in which the properties of the membrane remain constant, drug release kinetics will be first-order.[51]

Constant activity sources are produced when the concentration of the drug is greater than its solubility in body fluid and part of the drug remains undissolved when wet with body fluids. As drug is released from the reservoir, more drug dissolves maintaining the concentration of soluble drug inside the reservoir[46,51] and as drug diffusion is slower than water penetration and drug dissolution in these systems, the release is diffusion controlled (Figure 10.11).[51] As such, the release of the drug remains constant providing zero-order kinetics until all of the drug dissolves, converting the system into a non-constant activity reservoir for the remainder of the drug release. Of course, these systems are complex and in practice it should be noted that the dissolution of other components in the core might significantly influence the solubility of the drug compounds in the core and so accurate predication of release kinetics at 37 °C is difficult.[52]

10.3.2 Polymer Relaxation-controlled Swelling

In contrast to diffusion controlled systems, in which swelling is relatively rapid, if the polymer is slow to swell and the drug could diffuse more rapidly out of the system than the rate water permeates into the hydrogel, then the drug release process is swelling or polymer relaxation controlled.[47] For these systems, as water molecules migrate into the network, both water diffusion and polymer chain relaxation phenomena will be achieved in sequence. As such, the rates at which drug molecules are liberated are determined by both water diffusion rate and the chain relaxation rate of polymers and so either property can be used to control water penetration and subsequent drug release.[51] Most often in pharmaceutically relevant systems, water diffusion is rapid compared with polymer chain relaxation. In this case, as the polymer swells there are three distinct regions: (i) swollen polymer, in which the water concentration is equal and at a maximum; (ii) a narrow swelling region of rubbery polymer, in which water concentration rapidly reduces; and (iii) the dry region of glassy polymer.[51] For homogenous systems, because polymer swelling is time independent, there is a constant rate for the uptake of water and for drugs with relatively rapid dissolution and diffusion zero-order kinetics of drug release results (Figure 10.12).[51]

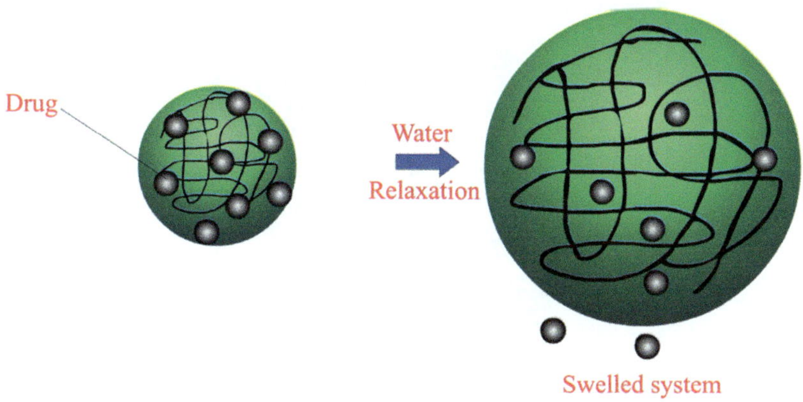

Figure 10.12 Upon contact with water, both water diffusion and polymer chain relaxation take place and the polymeric system becomes swollen, leading to the escape of entrapped drug molecules.

10.3.3 Degradation-controlled Release

Another parameter of the hydrogel affecting release rate is the dissolution or degradation of the polymer. For a polymer that isn't cross-linked, or has a low cross-linking, dissolution it is usually slow compared to the swelling due to the requirement for the polymer chains to disentangle.[47] Polymer dissolution rate is controlled by many factors, including polymer molecular weight, monomer composition and crystallinity of the dry hydrogel.[53] Polymer degradation can occur simultaneously with dissolution for systems comprised of biodegradable polymers including poly(amino acids), polysaccharides and polyesters. The degradation by-products of these polymers should be biocompatible and readily eliminated from the body without long-term adverse effects.[46,54,55,56] An illustration of degradation swelling controlled release is demonstrated in Figure 10.13 below.

The process of drug release from biodegradable SCDDSs is dependant on the component polymer. For example, polymer matrices comprising of poly(esters) including poly(lactic-*co*-glycolic acid) (PLGA), polylactic acid (PLA), and polycaprolactone (PCL) involve both diffusion and bulk degradation/erosion.[46,57] In the first instance, concentration gradients that induce water penetration into the systems occur more rapidly than polymer degradation and diffusion dominates the drug release process. In the later stages of release, polymer degradation becomes more important as the polymer chains undergo bulk erosion owing to fast hydration and the dissolution of the polymer into the bulk.[46,57] Eventually, a simultaneous degradation of the entire matrix takes place, inducing the rapid escape of any remaining entrapped drug molecules.[46] In contrast, more reactive polyanhydrides and poly(orthoesters) only degrade at the surface of the polymer matrices and so water diffusion into the network is much slower than polymer erosion.[46,58,59]

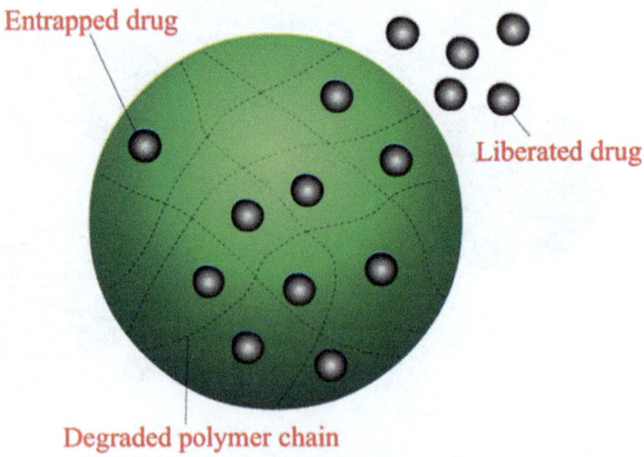

Figure 10.13 Upon contact with water, the polymer chains undergo bulk erosion due to rapid hydration, dissolution, and hydrolysis resulting in the disintegration of the entire matrix and the release of entrapped drug molecules.

10.4 Swellable Products

In the pharmaceutical industry, swellable materials have been used to develop three classes of swellable products dependant on their swelling characteristics. These include swellable matrices, superdisintegrants, and swellable devices. Some examples of commercial swelling control products are mentioned in Table 10.4 below.

10.4.1 Swellable Matrices

Swellable matrices involve loading of drug particles into monolithic devices and releasing them by diffusion through the monolith's matrices[46] and are used when slow and low swelling in the GI tract is required. These systems are fabricated by compressing a powdered combination of hydrophilic polymer particles and discrete drugs[67] and advantages include ease of production using standard pharmaceutical tableting procedures. The compression force applied to prepare swellable matrices plays a dominant role in controlling their porosity, which directly affects the release patterns of entrapped drug molecules.[68] Due to the desired low and slow swelling characteristics, swellable matrices are usually constructed from naturally derived polymers such as cellulose derivatives, chitosan, alginate or pectin, but can also use synthetic polymers such as cross-linked polyacrylic acid (carbopol) and poly (vinyl alcohol) (PVA).[68]

10.4.1.1 Swelling Mechanism of Swellable Matrices

Swellable matrices are differentiated from true swelling-controlled delivery systems that are nonporous matrices.[67] In contrast, the construction of swellable matrices means they are porous dosage forms in which a rapid

Table 10.4 Commercialized swellable products and their applications.

Drug name	Application	Manufacturers	References
PS Olanzapine ODT	Schizophrenia and psychoses symptoms treatment	Ascent Pharma Pty Ltd (Australia)	60
Apo-ondansetron	Cancer chemotherapy induced nausea prevention	Amneal pharmaceuticals Pty Ltd (Australia)	61
Mirtazapine AN ODT	Antidepressant	Sandoz Pty Ltd (Australia)	62
APO-Lansoprazole ODT	Stomach acid inhibitor	Apotex Pty Ltd (Australia)	63
Rizatriptan AN ODT	Migraine headaches, treatment	Amneal pharmaceuticals Pty Ltd (Australia)	64
Geomatrix®	Drug delivery devices	Jago Pharma AG (Switzerland)	65
Hydrofiber®	Wound dressing	ConvaTec (England)	66

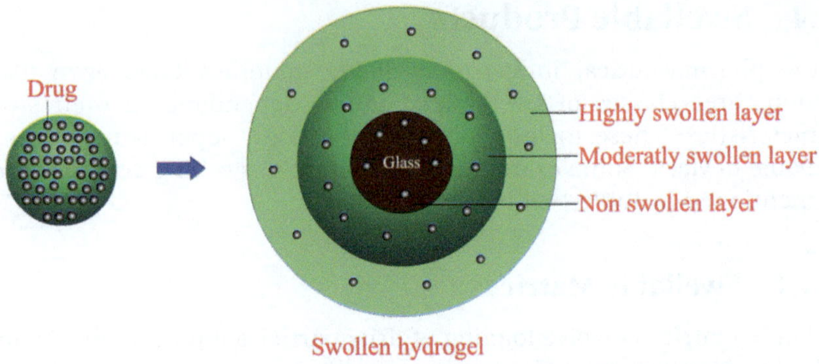

Figure 10.14 Swelling process of a hydrogel matrix, clearly indicating three layers formed within the matrix following water penetration: From the centre: non-swollen glassy core (brown colour), moderately swollen rubbery layer (dark green colour) and highly swollen region that serves as a diffusion barrier (light green colour).

transition into a gel state at the surface of the matrix serves as a diffusion barrier to water and drug transportation.[14,67] As water further diffuses into the dry, glassy hydrogel, its movement is controlled by the rate of relaxation of the polymer chains and their transition to the gel state. This results in three distinct regions in the hydrogel determined by the level of hydration and swelling (Figure 10.14). As the process progresses, in the highly swollen layer at the outer surface the increasing hydration leads to the eventual disentanglement and dissolution of the polymer chains and erosion of the matrix. Consequently, there are many aspects that control drug release from swellable matrices (including polymer swelling, matrix erosion, water penetration, drug dissolution and diffusion properties), but the dynamics of gel layer formation and progression is key to drug movement. As such, the kinetics are often diffusion controlled and zero-order initially and progressing to first-order over the course of the drug release as erosion becomes more significant.[67] Even though the characteristics of the gel transition from glass to rubber states is the key factor controlling drug release from swellable matrices, additional factors can also influence the rate of drug release.[14,68] These factors include drug size, solubility, polymeric hydrophilicity and viscosity, drug/polymer interaction, matrix pore size and distribution, water permeation pressure.[14,68,69]

10.4.1.2 Advantages of Swellable Matrices

- Simplify the formulation process.
- Provide continuous release of drug.
- Improve patient compliance.
- Provide high drug loading.

- Minimize drug plasma fluctuation.
- Reduce drug administration variations.
- Reduce dosing frequency.
- Decrease adverse effects associated with high drug plasma level.
- Diminish health care cost burden.[68,69]

10.4.2 Superdisintegrants

Superdisintegrants are excipients incorporated into solid dosage forms to allow rapid moisture penetration into the dosage matrix, weakening the structure of the formulation form for dissolution within 30 seconds of contact with water.[14,70] Therefore, superdisintegrants require polymers that swell very quickly in a size-dependent manner to enable the desired rapid drug release.[14,71] Figure 10.15 represents the mechanism by which superdisintegrants can activate drug release.

One of the major applications of superdisintegrants in the pharmaceutical industry is to make orally disintegrating tablets (ODTs), these are solid dosage forms containing active ingredients, which upon contact with saliva when placed on the tongue, will rapidly disintegrate within a matter of seconds.[72] Therefore, the use ODTs is more attractive than other conventional oral, sublingual and buccal tablets, especially in elderly patients with chewing impairment and young children.[73] Examples of technologies used for the formulation of ODTs include lyophilisation or freeze-drying, moulding, spray drying, melt granulation and direct compression; of which direct compression that uses commercially available excipients and equipment, is the most cost-saving method.[74,75]

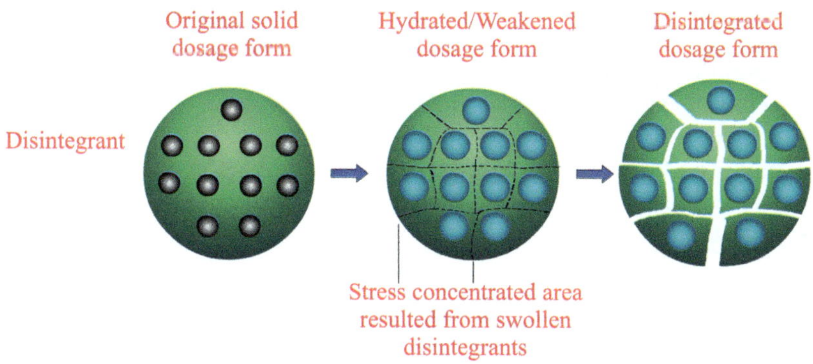

Figure 10.15 Upon contact with water, both water diffusion and polymer chain relaxation take place; disintegrants become significantly swollen, weakening the structure and creating areas of concentrated stress. The erosion of the whole solid dosage form occurs rapidly, breaking it into smaller fragments.

Table 10.5 Swellable superdisintegrants and properties.

Name	Properties	Reference
Cellulose	Highly crosslinked porous swelling in two dimensions by 4–8 fold in less than 10 seconds.	14,70,71
Starch	Cross-linked sustained release matrix swells in three dimensions by 7–12 fold in less than 30 seconds.	70 and 71
Alginic acid	Rapid swelling *or* capillary action suitable for both dry and wet granulation.	70 and 71
Poly (vinyl pyrrolidone) (PVP)	Highly cross-linked porous capillary action and diffusion	14

To ensure a desirable rate at which the solid dosage formulation will come apart, each gram of superdisintegrants must absorb 10–40 g of water and they need to be evenly dispersed inside the solid matrix.[14] Some examples of swellable superdisintegrants and their properties are presented in Table 10.5.

10.4.2.1 Ideal Superdisintegrants

Recently, there has been a considerable increase in popularity of orally swelling controlled disintegrants on the market due to their known convenience and patient compliance.[14] As superdisintegrants are excipients added into solid dosage forms, there are some requirements that they must meet, including:

- good mouth-feel;
- low solubility;
- poor gel formation;
- good flow properties;
- generally small porous particles;
- high compressibility;
- have good hydration capacity;
- lack of complex formation with active ingredients;
- non toxic.[14,71,76]

10.4.2.2 Method of Incorporation

The are three methods of incorporation of superdisintergrants. Firstly, intragranular refers to the combination of superdisintegrants with other powdery excipients during the granulation process.[71] This method is very well suited to direct compression. Secondly, extragranular refers to the incorporation of superdisintegrants blended with prepared granules before the compression process takes place. This method is suitable for wet granulation.[71] Thirdly, internal and external mixing of disintegrants refers to the incorporation of superdisintegrants by combining intra and extra granulation.[71]

10.4.3 Swellable Devices

Swelling devices are developed to orally deliver the drug into the gastrointestinal tract (GIT) through rapid, size independent swelling.[14] As such, the swelling component is generally a synthetic or semi-synthetic polymer[14] that can be the major component of the matrix or incorporated just into the tablet core to enhance the drug solubility increasing the rate of drug release.[11] The latter type is more popular with manufacturers and they can be formulated using all tableting techniques.[14] Structurally, swelling agents used in swellable devices are highly cross-linked, or arrange themselves as interconnected open cells, giving these devices the ability to reach the swelling capacity of 5–40 g g^{-1} by wicking/capillary action (Figure 10.16).[14]

Despite the advantage of often being relatively easy to manufacture, the release rates of drugs from swelling devices are not necessarily steady. This is because after being administrated into GIT, gastric factors such as pH, the quantity of water and other stomach contents can alter the properties of swelling devices, applying stress that induces the contraction and de-swelling of the device. Therefore, for GIT delivery purposes ideal swelling devices are mechanically strong in their fully swollen state to withstand the effect of excessive gastric pressures. They must also be compatible with the incorporated drugs, stomach contents and be able to cope with dynamic gastric changes in pH or temperature.[14] Furthermore, since swelling devices are fabricated from synthetic polymers, the production also faces the challenge of ensuring safety with pure, non-toxic formulations that are biocompatible and biodegradable. Finally, the complexity of some swelling devices means that scale-up for manufacturing can be difficult.[14]

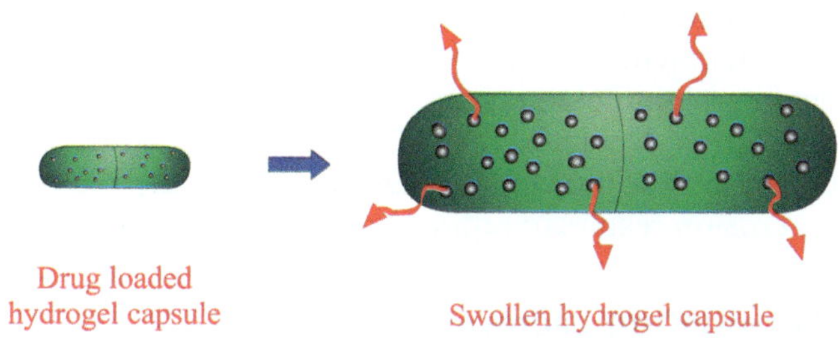

Figure 10.16 An oral hydrogel capsule as an example of swellable devices.

10.5 Factors Influencing Drug Release in Swelling Controlled Systems

The feasibility of taking a swelling controlled drug delivery system to market heavily depends on the release profile of drugs molecules, which can be directly affected by many factors derived from both the physical processes within the swellable devices and the physicochemical properties of drug molecules and/or aggregated particles. This complex system means that swelling controlled drug delivery systems offer a wide spectrum of pharmaceutical applications.[77] Nonetheless, in order to assist swelling device fabrication, related elements influencing the drug release profile must be well studied to produce functional, safe and reliable swellable products.

10.5.1 Physical Properties of the Swelling Device

The physical properties of swelling devices are fundamental elements controlling the onset of polymer chain erosion, especially in non-biodegradable swellable hydrogels.[14] When high hydration leads to a highly swollen matrix, the resulting lack of interactions between polymer chains in the hydrogel means it can be eroded. As this occurs, the hydrogel matrix is broken into smaller particles providing a larger surface exposed to the swelling medium and therefore triggering more drug release.[14]

The level of hydration of an SCDDS is controlled by the rate at which the device matrix absorbs aqueous fluid, which also directly affects the physical properties of the devices glassy, tough rubber and soft rubber layers. If swelling happens rapidly, water absorption across the hydrogel structure makes the layers less distinguishable and erosion of the hydrogel will be rapid, leading to the rapid release of the incorporated drugs.[14] In contrast, if the kinetics of swelling are slow, the glass, tough rubber, and soft rubber layers remain distinct and polymer chains still partially incorporated into the tough rubber and glass layers cannot be released, leading to erosion occurring at a much slower rate.[14]

The physical characteristics of the gel layers are also controlled by the polymer molecular weight. For polyethylene oxide (PEO) polymers, hydration rates of low and high molecular weight polymers are similar, but higher molecular weight polymers tended to swell more and erode more slowly than lower molecular weight polymers. As such, swelling and drug diffusion rates are the key to high molecular weight PEO drug release, while erosion is much more important for low molecular weight polymers and the combination of swelling and erosion results in rapid, constant release kinetics.[78,79]

10.5.2 Formulation Factors

Other components in the hydrogel such as drugs and excipients have a strong influence on the properties of the drug delivery system and hence drug release. Formulation factors such as the concentration of drugs or other additives that can either dilute the polymer chain to chain interactions[79–81] or serve as osmolytes to draw more water into the polymer networks and so can often accelerate swelling kinetics of the system[51,82] and even increase its swelling capacity.[93] In contrast, the incorporation of binding agents into the hydrogel structures can cause the retardation in release rates.[83] Other additives that interact with the drug and reduce its solubility can also ultimately retard drug release.[82]

Formulation technique can also play a significant role in the drug release behaviours from swelling controlled delivery carriers. For instance, the type and the amount of solvent used will affect the degree of swelling and gel forming ability of a polymer directly.[83,84]

10.5.3 Drug-related Factors

Several parameters of the drug incorporated in the hydrogel affect the drug release properties. For example, swellable hydrogels tendency to erode increases in proportion with drug solubility since the presence of undissolved drugs and drug particles in the matrix reduces the swelling and polymer chain relaxation, influencing the resistance of the hydrogels to erosion.[83] Furthermore, while soluble drugs are transported through the drug delivery system as it hydrates and has swelling and diffusion controlled release, sparingly soluble or non-soluble drugs usually have drug release profiles determined by the erosion of the system.[14] The size of the drug molecules and drug particles are also important factors affecting release as large drug molecules with a molecular weight greater than 500 g mol^{-1} tend to have very poor diffusivity in hydrogel structures because of the constraints of the gel matrix.[83] Furthermore, drug particle shape and size can also affect drug release kinetics through controlling the drug dissolution rate owing to the difference in effective surface area.[83]

10.5.4 Drug Delivery System Geometry

The geometry (shape and size) of the drug delivery system also plays a significant role in the drug release kinetics from swelling controlled delivery systems as drug release rate is proportional to the surface area of the tablet.[85] This is not only because a larger surface area provides more access to the external water to convert the system from glass to gels that can release drugs, but also because larger tablets have longer diffusion pathways for the drugs to travel to be released.[83,86]

10.5.5 Degree of Swelling Agent Saturation

The greatest volume of aqueous medium that hydrogels can absorb is defined as the equilibrium swelling capacity. The equilibrium swelling capacity is influenced by several factors, all of which directly influence drug release

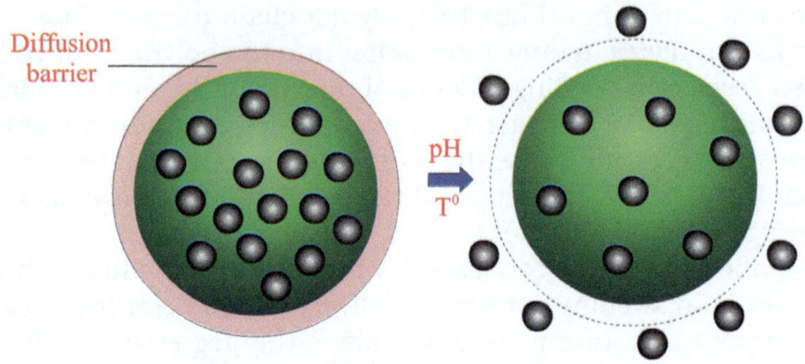

Figure 10.17 External parameters such as pH or temperature could induce the destruction of the diffusion barrier surrounding the hydrogel matrix, leading to the escape of entrapped drug molecules.

behaviours.[87] These include the ionic character and hydrophilicity of the monomers incorporated into the polymer chains,[88] the polymer interactions with the incorporated drugs,[89] the molecular weight of the polymers;[79,90] as well as the degree of cross-linking in the hydrogel.[91,92] Additional components in the drug delivery system such as drugs and excipients can also influence equilibrium swelling capacity.[93]

10.5.6 Stimuli Responsive Release

Swelling kinetics and drug release profiles can be altered by stimuli derived from the changes in external factors such as temperature and pH[50,94] that occur upon ingestion.[46] The influence of these factors is highly dependent on the chemical and physical properties of the delivery device. For instance, for stomach release, pH influences swelling when the hydrogel contains ionic groups and erosion in the stomach is directly influenced when linking chemistries include acid degradable groups such as esters.[50] For colon release, the formulation must be chosen to minimise swelling and erosion in the stomach but promoting erosion in the colon using polysaccharides such as inulin that are indigestible to human enzymes, but readily broken up by microbes in the gut.[95] Finally, external factors such as pH and temperature can induce the destruction of the diffusion layer surrounding the SCDDS, accelerating the release of entrapped drugs (Figure 10.17).

10.6 Mechanistic Mathematical Models for Predicting Drug Release

Mechanistic mathematical models are essential in the pharmaceutical industry as they provide good estimations of drug release for many drug delivery systems, reducing the time and costs required for laboratory investigations during product development.[96] Moreover, quantitative analysis using mechanistic mathematical models can contribute to building an understanding of the drug release mechanisms based on phenomena including diffusion, swelling, degradation and drug transport.[96,97] This knowledge is required for efficient optimisation and troubleshooting of the drug delivery system.[96] To use mechanistic mathematical models effectively in the development of swellable drug delivery systems a number of factors need to be taken into account.[96]

- The more phenomena that are taken into consideration in the model, the more accurate the model generally is. However, considering more aspects also increases the complexity of the mathematical theory, creating issues with computation. As such, if dominant physiochemical and biological processes are identified, only these should be taken into account during the development of these models.[96]
- In correlative theoretical models, each phenomenon is considered as independent functions. Considering these phenomena in a coupled manner is much better for accurate prediction of drug release.[97]
- Theoretical data obtained using mechanistic mathematical models must always be compared to independent experimental results. This enables evaluation of the validity of the mathematical model and allows optimization of that theory by carefully and systematically fitting the model to the experimental data.[96]
- Mechanistic mathematical models used to describe drug delivery are simplifications of real systems and consequently their suitability can be limited. As such, once experimental evidence suggests issues with the model, the model should either be rectified based on the new knowledge or abandoned.[96]

There is a range of drug release models that represent a large number of drug release systems available. Initially, in this section, we will describe some simple models of drug delivery including models based on one-dimensional transport that has nonetheless been very important in the development of drug delivery models for more complex systems. We will then highlight some examples of predictive mechanistic mathematical models that involve coupled processes of drug transportation and other phenomena for the prediction of drug release from swellable carriers. Generally, the first 60% of drug release can be fitted to these models[98] and release becomes less predictable as the system is depleted of drugs.

The zero-order release model is regarded as ideal drug release behaviour for many systems and is particularly important in swelling controlled delivery systems. Zero order release kinetics can be represented through eqn (10.1) below:

$$M_t = M_\infty + k_0 t \tag{10.1}$$

where M_t is the amount of dissolved/released drug, M_∞ is the original amount of drug, k_0 is the zero order release constant and t is time.[99]

The Higuchi model was developed for the study of the release of water soluble and insoluble drugs uniformly suspended in thin ointment films into a perfect sink condition.[100] This model can also be relevant to swellable drug release systems when Fickian diffusion controls drug release.[78] The Higuchi theoretical model can be expressed by eqn (10.2) below and simplified to eqn (10.3) when the drug loading is in excess of the solubility of the drug in the matrix:[11,98]

$$M_t = A[DC_s(2C - C_s)t]^{0.5} \tag{10.2}$$

$$M_t = A[2DC_sCt]^{0.5} \tag{10.3}$$

Where M_t is the total amount of drug released at time t, A is the surface area of the device, D is the diffusion coefficient of the drug, C is the initial total drug loading and C_s is the saturation solubility of the drug in the system. The assumptions made for the Higuchi model mean that modifications are required for many drug delivery systems.[98] For example, polymer matrices with pores require modification to the Higuchi model, as do those in which degradation is a significant factor in drug release.[11]

Diffusion models are also often exploited to describe the processes relating to the movement of drug molecules in one dimension through a concentration gradient. These are based on Fick's laws of diffusion.[101] Fick's first law of diffusion connects diffusive flux of a substance to its concentration under steady state conditions and proposes that the magnitude of the diffusive flux of the drug is proportional to the concentration gradient as it moves from a highly concentrated region to a region of lower concentration. This law is mathematically represented by eqn (10.4) in one dimension:

$$J = -D(\mathrm{d}c/\mathrm{d}x) \tag{10.4}$$

where J indicates the amount of substance passing through a unit surface area per time unit, D is the diffusion coefficient and $\mathrm{d}c/\mathrm{d}x$ represents concentration gradient where c is the concentration and x the position of the substance.[102]

Fick's second law of diffusion predicts how diffusion changes the concentration (c) with time (t).[98,102] For a system having a diffusion constant of D, and diffusing in one-dimension in direction x, Fick's second law can be written as eqn (10.5) below:

$$\mathrm{d}c/\mathrm{d}t = D(\mathrm{d}^2c/\mathrm{d}x^2) \tag{10.5}$$

In which: $t = 0 \leftrightarrow \frac{-L}{2} < x < \frac{L}{2} \leftrightarrow c = c_0$ while $t > 0 \leftrightarrow x = \pm\frac{L}{2} \leftrightarrow c = 0$.[98,102]

When the diffusion of water into swellable hydrogels is the rate limiting step, known as case 1 diffusion, Fick's Law is used to define the relationship between the amount of water absorbed by the hydrogel and time.

For chemically cross-linked hydrogels in which an equilibrium fluid uptake is observed, that uptake and subsequent drug transport can be defined by eqn (10.6):

$$M_t/M_\infty = 4(Dt/\pi L^2)^{0.5} \tag{10.6}$$

Where: M_t is the mass substance transported at time t, M_∞ indicates the mass of substance absorbed at equilibrium, L is the thickness of the material, and D is the diffusion coefficient.[11]

Similarly to the Higuchi equation, Fickian diffusion is based on a number of assumptions. As such, care must be taken when applying either model to different shapes of drug delivery device, porous devices and swellable systems that change size due to swelling and erosion.[98] Furthermore, commonly drug release from swellable delivery systems does not agree with either the Higuchi Model or Fickian behaviour in which swelling must be negligible.[98] This is because often for swellable systems, relaxation of polymer chains from the glassy to rubbery states, rather than diffusion, governs the rate of water penetration into hydrogels, and consequently is the rate limiting step in drug release.[11] Many chemically and physically cross-linked hydrogels follow this mechanism, known as case 2 diffusion, and so the modelling of drug release involves both the polymer chain relaxation time and the release of drug molecules at the interface of rubbery and glassy phases of swollen hydrogels.[11,98] As such, mathematical modelling of drug release from these systems becomes extremely complicated[11] and semi-empirical equations simplified with constants for both diffusion (k_1) and polymer chain relaxation (k_2) have been used to describe the release of drugs (eqn (10.7)). This has been further generalised to power law models, initially for water transport into glassy polymers[103] and then for drug release from those systems (eqn (10.8)), known as the Peppas equation:[98,104]

$$M/M_\infty = k_1\sqrt{t} + k_2 t \tag{10.7}$$

$$M/M_\infty = kt^n \tag{10.8}$$

where M is the mass of drug released at time t, M_∞, is the total initial drug loading, k_1 is a constant related to diffusion of the drug, k_2 is a constant related to polymer relaxation and k is a general constant and n is the release exponent.

The release exponent in the Peppas equation can be used to characterise the drug release with $n = 0.5$ indicative of Fickian drug release with a case 1 mechanism and $n = 1$ indicative of the case 2 mechanisms having constant or zero-order drug release. In between these two values, the drug release is described as anomalous.[11,98] For the Peppas equation, there are several coupled processes presented, including the water diffusion into the tablet, the swelling of the tablet, gel formation and diffusion of the drug out of the tablet and polymer matrix dissolution.[102] Therefore, if the constants can be well defined, Peppas models are more suitable to be used for the prediction of drug release behaviours from swellable delivery systems, rather than Higuchi or Diffusional models.

Another well-known empirical drug release model is the Weibull model (eqn (10.9)), which has been used to describe dissolution processes and it is useful for the comparison of the release patterns of various matrix type drug delivery systems.[102,105] However, investigations into the effect of the cross-linking degree on carbohydrate hydrophilic swellable matrix tablets that fitted the resultant data to both the Weibull model and the Peppas model found that two out of three samples fit better to the Peppas model.[106] This demonstrates the greater role of the use of the Power law to predict drug release profiles from swellable systems.

$$C = C_0[1 - \exp(-(t - T)^b/a)] \qquad (10.9)$$

Where C is the amount of drug dissolved as a function of time t, C_0 is the total amount of drug being released, T accounts for the lag time measured as a result of dissolution, a indicates a scale parameter describing the time dependence while b represents the shape of dissolution curve progression.[102,105]

A sequential layer model was developed by Siepmeann and Peppas to determine the swelling and drug release profiles from hydrophilic matrix tablets taking a large number of complicated drug release factors into account.[106] This model was built upon Fick's Second law of diffusion in cylinders and includes consideration of both radial and axial diffusion, the concentration dependence of drug diffusivity, matrix swelling and erosion.[86] To this model was added the concept of layer by layer swelling in which the tablet is broken into layers and each layer of a tablet considered sequentially to address factors such as inhomogeneous swelling, drug solubility and high initial drug concentration.[107] While the experiments in which the model was developed were based upon hydroxypropyl methylcellulose, its applicability has been demonstrated for a wide range of polymers, drugs, tablet size and shape and drug release media.[106] As such, the sequential layer model possesses an incredible predictive ability on drug release patterns.[106] Other models have been developed to specifically deal with drug release from swelling and dissolving polymer tablets, addressing issues of tablet shape[108,109] and moving boundary conditions.[110]

One important factor that needs to be taken into account during the use of these models is the fact that these mechanistic mathematical models can only be used for *in vitro* release studies. *In vitro* data has been one of the keys to optimize the dosage forms as they provide a relationship between the structure and the function of the material matrices, and provide further insights into the *in vivo* release data of formulations of interest. Nonetheless, the actual physiological conditions are more complicated than all the buffers that are used for *in vitro* studies, meaning that the effects of various proteins, cells, and enzymes on drug release behaviours are not fully understood and usually not estimated through current mechanistic mathematical models. That being said, if mathematical models that can reflect the actual physiological conditions such as pH, enzyme concentration and osmotic pressure can be developed, it would produce predictive models with extremely accurate estimations regarding drug release profiles for SCDDSs.

10.7 Disadvantages of Swelling Controlled Systems

Swelling controlled release systems are not only applicable in oral delivery, but have also found use in more complicated bioadhesive nasal, buccal and rectal administration.[14] In fact, it is unarguable that swelling controlled drug delivery systems, similar to other controlled drug vehicles, possess great benefits over conventional dosage forms. This is due to controlled drug delivery offering decreased frequency of dosing and reduced side effects, improving patient compliance. These factors are of benefit to all treatments, but particularly for patients with chronic illnesses.[77,111] However, there are still several disadvantages of these formulations which the manufacturers need to be aware of:

- Cost burdens: SCDDS can require expensive specialised equipment and ingredients (such as polymers) during the formulation development process.
- Effects of food and gastric content: the drug release rate between doses can be greatly affected by food consumption and gastric transit time.
- Formulation related problems: swellable controlled release products generally cannot be crushed or made into chewable products as this will cause the loss of their desirable slow release properties and unwanted toxicity effects may arise.
- Drug release is complicated by many factors: that there are so many variables that influence drug release in terms of ingredients and formulation and the interaction of the product with biological systems means that optimisation is difficult.[77,112,113]

To work effectively, SCDDSs require design such that there is local or systemic delivery of the active drug and that delivery occurs at a predetermined rate for a specified period of time to achieve therapeutic plasma levels,[77,114,115] these are goals that are very difficult to achieve. Consequently, both pros and cons of possible swelling controlled delivery systems must be carefully evaluated prior to the development of such systems in order to minimize the costs that arise from the problems that may occur later during the manufacturing process and to produce an effective product.

10.8 Conclusion and Future Perspectives

Swelling controlled drug delivery systems have the potential to be highly effective drug delivery tools. The advantage of these systems is the ability to offer oral administration (as well as other routes) while providing tuneable release of both hydrophilic and hydrophobic therapeutic agents over selectable periods of time. Drug release can be adjusted by a range of factors discussed in this chapter including the swellable polymer and cross-linking characteristics, formulation ingredients and techniques and the form of the drug used. Besides the needs to consider all these variables, optimisation is further complicated, because of multiple mechanisms controlling the drug release, including drug diffusion, polymer swelling, and polymer erosion. The complexity in the design of SCDDs results in the requirement for expensive specialised equipment and difficulties in scale-up. Nonetheless, the complex mechanisms of drug release from swelling systems are becoming better understood, leading to the development of effective predictive mechanistic models. Furthermore, advances in polymer technology and manufacturing are increasing the availability of safe, biocompatible and biodegradable polymers suitable for use in swelling systems. This developing knowledge needs to be combined with the understanding of disease progression as well as drug pharmacokinetic and pharmacodynamic properties for truly and practically optimised systems.

Once the complexity inherent in swelling drug delivery systems is fully overcome, there are significant opportunities for pharmaceutical formulations that treat a large range of diseases. This is because the very complexity that makes design and manufacture difficult provides diverse and versatile systems that can be fabricated into a wide range of designs in which various mechanisms can be adapted to achieve a drug release for the desired therapeutic outcome.

References

1. R. Ebewele, *Polymer Science and Technology*, CRC Press, New York, 2000, pp. 3–5.
2. M. C. R. Berthelot, *Acad. Sci.*, 1866, **62**, 905–909.
3. Y. Gnanou and M. Fontanille, *Organic and Physical Chemistry of Polymers*, Wiley-Interscience, Hoboken, US, 2008.
4. M. Fontanille and Y. Gnanou, *Mechanical Properties of Polymers*, John Wiley & Sons, Inc., 2007, pp. 427–465.
5. R. Kaur and S. Kaur, *JDDT*, 2014, **4**(3), 32–36.
6. S. Arora, J. Ali, A. Ahuja, R. Khar and S. Baboota, *AAPS PharmSciTech*, 2005, **6**(3), E372–E390.
7. A. Y. Kaushik, A. K. Tiwari and A. Gaur, *Int. J. Pharm. Invest.*, 2015, **5**(1), 1–12.
8. K. Gandhi, S. Deshmane and K. Biyani, *Int. J. Pharm. Sci. Rev. Res.*, 2012, **14**(2), 57–66.
9. A. Haywood and B. Glass, *Aust. Prescr.*, 2011, **34**, 112–114.
10. D. Sharma, S. Kalasare and R. Kamble, *Asian J. Pharm. Technol. Innovation*, 2013, **1**(2), 1–16.
11. D. Jones, *Pharmaceutical Applications of Polymers for Drug Delivery*, iSmithers Rapra Publishing, United Kindom, 2004, vol. 15, pp. 3–34.
12. M. F. Maitz, *Biosurf. Biotribol.*, 2015, **1**(3), 161–176.

13. J. Singh, *WJPPS*, 2016, **5**(4), 805–816.
14. H. Omidian and K. Park, *J. Drug Delivery Sci. Technol.*, 2008, **18**(2), 83–93.
15. Transparency Market Research, *Natural Polymers Market (Cellulose Ethers, Starch and Fermentation Products, Protein Based Polymers and Others) for Medical, Food and Beverage and Oilfield Applications*, U.S. Industry Analysis, Size, Share, Growth, Trends and Forecast, 2012–2018.
16. F. Tuğcu-Demiröz, F. Acartürk, S. Takka and Ö. Konuş-Boyunağa, *Eur. J. Pharm. Biopharm.*, 2007, **67**(2), 491–497.
17. G. Ciofani, V. Raffa, T. Pizzorusso, A. Menciassi and P. Dario, *Med. Eng. Phys.*, 2008, **30**(7), 848–855.
18. K. Moebus, J. Siepmann and R. Bodmeier, *Eur. J. Pharm. Biopharm.*, 2009, **72**(1), 42–53.
19. N. Mennini, S. Furlanetto, M. Cirri and P. Mura, *Eur. J. Pharm. Biopharm.*, 2012, **80**(1), 67–75.
20. L. Chen, Z. Tian and Y. Du, *Biomaterials*, 2004, **25**(17), 3725–3732.
21. Y. S. Wang, L. R. Liu, Q. Jiang and Q. Q. Zhang, *Eur. Polym. J.*, 2007, **43**(1), 43–51.
22. U. Ubaidulla, R. K. Khar, F. J. Ahmad and P. Tripathi, *Pharm. Dev. Technol.*, 2009, **14**(1), 99–108.
23. M. Santoro, A. M. Tatara and A. G. Mikos, *J. Controlled Release*, 2014, 210–218.
24. N. Sahoo, R. K. Biswas, A. Guha and K. Kuotsu, *Int. J. Biol. Macromol.*, 2015, **81**, 317–331.
25. T. W. Wong, G. Colombo and F. Sonvico, *AAPS PharmSciTech*, 2011, **12**(1), 201–214.
26. C. Sandolo, S. Péchiné, A. Le Monnier, S. Hoys, C. Janoir, T. Coviello and N. Tsapis, *Eur. J. Pharm. Biopharm.*, 2011, **79**(3), 566–573.
27. R. Goyal, S. K. Tripathi, S. Tyagi, K. Ravi Ram, K. M. Ansari, Y. Shukla and K. Gupta, *et al.*, *Eur. J. Pharm. Biopharm.*, 2011, **79**(1), 3–14.
28. A. Gavasane and H. Pawar, *Clin. Pharmacol. Biopharm.*, 2014, **3**(2), 121.
29. T. Mekonnen, P. Mussone, H. Khalil and D. Bressler, *J. Mater. Chem. A*, 2013, **1**(43), 13379–13398.
30. R. Alsaheb, R. Aladdin, N. Othman, R. Malek, O. Leng, O. R. Aziz and H. El Enshasy, *J. Chem. Pharm. Res.*, 2015, 7(12), 51–63.
31. B. D. Ulery, L. S. Nair and C. T. Laurencin, *J. Polym. Sci., Part B: Polym. Phys.*, 2011, **49**(12), 832–864.
32. J. Conn, R. Oyasu, M. Welsh and J. M. Beal, *Am. J. Surg.*, 1974, **128**(1), 19–23.
33. L. T. Dang, N. T. Feric, C. Laschinger, W. Y. Chang, B. Zhang, G. A. Wood and M. Radisic, *et al.*, *Biomaterials*, 2014, **35**(27), 7786–7799.
34. O. Jeon, J. E. Samorezov and E. Alsberg, *Acta Biomater.*, 2014, **10**(1), 47–55.
35. M. K. Nguyen, O. Jeon, M. D. Krebs, D. Schapira and E. Alsberg, *Biomaterials*, 2014, **35**(24), 6278–6286.
36. J. Wang, H. Q. Mao and K. W. Leong, *J. Am. Chem. Soc.*, 2001, **123**(38), 9480–9481.

37. J. Wen, H. Q. Mao, W. Li, K. Y. Lin and K. W. Leong, *J. Pharm. Sci.*, 2004, **93**(8), 2142–2157.
38. S. K. De, N. R. Aluru, B. Johnson, W. C. Crone, D. J. Beebe and J. Moore, *J. Microelectromech. Syst.*, 2002, **11**(5), 544–555.
39. PAC, *Definitions of terms relating to the structure and processing of sols, gels, networks, and inorganic-organic hybrid materials*, IUPAC Recommendations, 2007, vol. 79, p. 1801.
40. A. Schmidt, *Handbook of Polymer Synthesis, Characterization and Processing*, ed. E. Saldívar-Guerra and E. Vivaldo-Lima, Angew. Chem. Int. Ed., John Wiley and Sons, Hoboken, 2013, vol. 35, p. 358.
41. J. C. Hernández-Ortiz and E. Vivaldo-Lima, *Handbook of Polymer Synthesis, Characterization, and Processing*, John Wiley & Sons Inc, 2013, pp. 187–204.
42. D. Mecerreyes, V. Lee, C. J. Hawker, J. L. Hedrick, A. Wursch, W. Volksen and R. D. Miller, *et al.*, *Adv. Mater.*, 2001, **13**(3), 204–208.
43. F. Billmeyer, *Textbook of Polymer Science*, Wiley-lntuluenm, New York, 2nd edn, 1971.
44. R. Hirlekar, P. Phansalkar and O. Mainkar, *WJPR*, 2016, **4**(12), 2017–2024.
45. Y. Fu and W. J. Kao, *Expert Opin. Drug Delivery*, 2010, **7**(4), 429–444.
46. J. H. Lee and Y. Yeo, *Chem. Eng. Sci.*, 2015, **125**, 75–84.
47. A. K. Bajpai, S. K. Shukla, S. Bhanu and S. Kankane, *Prog. Polym. Sci.*, 2008, **33**(11), 1088–1118.
48. P. I. Lee, *J. Pharm. Sci.*, 1984, **73**(10), 1344–1347.
49. S. Kaity, J. Isaac and A. Ghosh, *Carbohydr. Polym.*, 2013, **94**(1), 456–467.
50. A. K. Bajpai, J. Bajpai, S. Shukla and R. A. Kulkarni, *J. Polym. Sci., Part A: Polym. Chem.*, 2004, **41**(2), 211–230.
51. J. Siepmann and F. Siepmann, *Fundamentals and Applications of Controlled Release Drug Delivery*, Springer, Boston, MA, US, 2012, pp. 153–170.
52. S. I. F. Badawy and M. A. Hussain, *J. Pharm. Sci.*, 2007, **96**(5), 948–959.
53. S. Fredenberg, M. Wahlgren, M. Reslow and A. Axelsson, *Int. J. Pharm.*, 2011, **415**(1–2), 34.
54. H. S. Yoo and T. G. Park, *J. Controlled Release*, 2001, **70**(1–2), 63–70.
55. M. Prabaharan, J. J. Grailer, S. Pilla, D. A. Steeber and S. Gong, *Biomaterials*, 2009, **30**(29), 5757–5766.
56. S. H. Lee, H. Mok, Y. Lee and T. G. Park, *J. Controlled Release*, 2011, **152**(1), 152–158.
57. N. Kamaly, B. Yameen, J. Wu and O. C. Farokhzad, *Chem. Rev.*, 2016, **116**(4), 2602–2663.
58. J. C. Middleton and A. J. Tipton, *Biomaterials*, 2000, **21**(23), 2335–2346.
59. F. Burkersroda, L. Schedl and A. Göpferich, *Biomaterials*, 2002, **23**(21), 4221–4231.
60. Pharmaceutical Benefits Scheme (PBS), *OLANZAPINE*. Department of Health Australia government, 2017.
61. Pharmaceutical Benefits Scheme (PBS), *ONDANSETRON*. Department of Health Australia government, 2017.

62. Pharmaceutical Benefits Scheme (PBS), *MIRTAZAPINE*. Department of Health Australia government, 2017.
63. Pharmaceutical Benefits Scheme (PBS), *LANSOPRAZOLE*. Department of Health Australia government, 2017.
64. Pharmaceutical Benefits Scheme (PBS), *RIZATRIPTAN*. Department of Health Australia government, 2017.
65. L. Augsburger and S. W. Hoag, *Pharmaceutical Dosage Forms - Tablet*, CRC Press, Baton Rouge, US, 3rd edn, 2008, vol. 3.
66. E. Caló and V. V. Khutoryanskiy, *Eur. Polym. J.*, 2015, **65**, 252–267.
67. P. Colombo, R. Bettini, P. Santi and N. A. Peppas, *Pharm. Sci. Technol. Today*, 2000, **3**(6), 198–204.
68. B. V. Kadri, PhD Thesis, University of Toronto, Canada, 2001.
69. M. U. Ghori and B. R. Conway, *Am. J. Pharmacol. Sci.*, 2015, **3**(5), 103–109.
70. P. Mohanachandran, P. Sindhumol and T. Kiran, *Int. J. Pharm. Sci. Rev. Res.*, 2011, **6**(1), 105–109.
71. S. Rawat, D. Derle, S. Fukte, P. Shinde and B. Parve, *WJPPS*, 2014, **3**(5), 263–278.
72. Food and Drug Administration (FDA), *Guidance for Industry Orally Disintegrating Tablets*, U.S. Department of Health and Human Services, 2008.
73. F. Baldi and P. Malfertheiner, *Digestion*, 2003, **67**(1), 1–5.
74. S. Singh and M. Jaimini, *Int. J. Pharm. Invest.*, 2014, **2**(3), 815–830.
75. A. Wrzesinski, H. Bertrand, M. Moses and Y. Zhang, *Pharmtech*, 2010, **34**(7), 54–65.
76. I. Rashid, M. Al-Remawi, A. Eftaiha and A. Badwan, *J. Pharm. Sci.*, 2008, **97**(11), 4955–4969.
77. A. Nokhodchi, S. Raja, P. Patel and K. Asare-Addo, *BioImpacts*, 2012, **2**(4), 175–187.
78. J. Sujja-areevath, D. L. Munday, P. J. Cox and K. A. Khan, *Eur. J. Pharm. Sci.*, 1998, **6**(3), 207–217.
79. L. Maggi, L. Segale, M. L. Torre, E. Ochoa Machiste and U. Conte, *Biomaterials*, 2002, **23**(4), 1113–1119.
80. M. Efentakis, H. Al-Hmoud and G. Buchton, *Int. J. Pharm.*, 1991, **70**, 153–158.
81. A. Nokhodchi, S. Norouzi-Sani, M. R. Siahi-Shadbad, F. Lotfipoor and M. Saeedi, *Eur. J. Pharm. Biopharm.*, 2002, **54**(3), 349–356.
82. T. Kristmundsdottir, K. Ingvarsdóttir and G. Sémundsdóttir, *Drug Dev. Ind. Pharm.*, 1995, **21**(13), 1591–1598.
83. M. V. S. Varma, A. M. Kaushal, A. Garg and S. Garg, *Am. J. Drug Delivery*, 2004, **2**(1), 43–57.
84. Y. Kawashima, H. Takeuchi, T. Hino, T. Niwa, T. L. Lin, F. Sekigawa and K. Kawahara, *Pharm. Res.*, 1993, **10**(3), 351–355.
85. J. L. Ford, M. H. Rubinstein, F. McCaul, J. E. Hogan and P. J. Edgar, *Int. J. Pharm.*, 1987, **40**(3), 223–234.

86. J. Siepmann, H. Kranz, R. Bodmeier and N. A. Peppas, *Pharm. Res.*, 1999, **16**, 1748–1756.
87. E. Vallés, D. Durando, I. Katime, E. Mendizábal and J. E. Puig, *Polym. Bull.*, 2000, **44**(1), 109–114.
88. J. Chern, W. Lee and M. Hsieh, *Ind. Eng. Chem. Res.*, 2004, **43**, 6150–6156.
89. P. M. Torre, Y. Enobakhare, G. Torrado and S. Torrado, *Biomaterials*, 2013, **24**(8), 1499–1506.
90. W. F. Lee and Y. H. Lin, *J. Appl. Polym. Sci.*, 2003, **90**(6), 1683–1691.
91. X. Huang and C. S. Brazel, *Chem. Eng. Commun.*, 2003, **190**(4), 519–532.
92. G. G. Buonocore, M. A. Del Nobile, A. Panizza, M. R. Corbo and L. Nicolais, *J. Controlled Release*, 2003, **90**(1), 97–107.
93. Y. M. Mohan, J. P. Dickson and K. E. Geckeler, *Polym. Int.*, 2007, **56**(2), 175–185.
94. Z. Mohamadnia, M. J. Zohuriaan-Mehr, K. Kabiri, A. Jamshidi and H. Mobedi, *J. Bioact. Compat. Polym.*, 2007, **22**(3), 342–356.
95. D. Mandracchia, N. Denora, M. Franco, G. Pitarresi, G. Giammona and G. Trapani, *J. Biomater. Sci., Polym. Ed.*, 2011, **22**(1–3), 313–328.
96. J. Siepmann and F. Siepmann, *Int. J. Pharm.*, 2008, **364**(2), 328–343.
97. A. N. Ford Versypt, D. W. Pack and R. D. Braatz, *J. Controlled Release*, 2013, **165**(1), 29–37.
98. J. Siepmann and N. A. Peppas, *Int. J. Pharm.*, 2011, **418**(1), 6–12.
99. G. Singhvi and M. Singh, *Int. J. Pharma Sci. Res.*, 2011, **2**(1), 77–84.
100. T. Higuchi, *J. Pharm. Sci.*, 1961, **50**, 874–875.
101. A. Fick, *J. Membr. Sci.*, 1995, 33–38.
102. H. Shaikh, R. Kshirsagar and S. Patil, *WJPPS*, 2015, **4**(4), 324–338.
103. G. Sinclair and N. Peppas, *J. Membr. Sci.*, 1984, **17**(3), 329–331.
104. N. A. Peppas, *Pharm. Acta Helv.*, 1985, **60**(4), 110–111.
105. S. Dash, P. N. Murthy, L. Nath and P. Chowdhury, *Acta Pol. Pharm.*, 2010, **67**(3), 217–223.
106. F. M. Carbinatto, A. D. de Castro, R. C. Evangelista and B. S. F. Cury, *Afr. J. Plant Sci.*, 2014, **9**(1), 27–34.
107. J. Siepmann and N. A. Peppas, *Pharm. Res.*, 2000, **17**, 1290–1298.
108. P. Borgquist, PhD Thesis, Lund University, Sweden, 2005.
109. P. Borgquist, A. Körner, L. Piculell, A. Larsson and A. Axelsson, *J. Controlled Release*, 2006, **113**(3), 216–225.
110. N. Wu, L. S. Wang, D. C. W. Tan, S. M. Moochhala and Y. Y. Yang, *J. Controlled Release*, 2005, **102**(3), 569–581.
111. C. Maderuelo, A. Zarzuelo and J. M. Lanao, *J. Controlled Release*, 2011, **154**(1), 2–19.
112. O. Kayser, A. Lemke and N. Hernandez-Trejo, *Curr. Pharm. Biotechnol.*, 2005, **6**(1), 3–5.
113. S. Kumar, A. Kumar, V. Gupta, K. Malodia and P. Rakha, *Asian J. Pharm. Technol.*, 2012, **2**(2), 38–43.
114. X. Chen, H. Wen and K. Park, *Oral Controlled Release Formulation Design and Drug Delivery*, John Wiley & Sons, Inc, 2010, pp. 257–277.
115. A. B. Nair, H. Vyas and A. Kumar, *J. Basic Clin. Pharm.*, 2010, **1**(2), 71–75.

CHAPTER 11

Biologically-inspired Stimuli-responsive DDS

JEONG-HWAN KIM

Yokohama City University, Graduate School of Medicine, 3-9 Fukuura, Kanazawa-ku, Yokohama, 236-0004, Japan
*E-mail: biohalo@gmail.com

11.1 Introduction

Nature creates smart composite interfaces (SCIs) with complex structures and various properties *via* evolution-optimized innovation.[1–9] Whereas existing functional material interfaces limit their usages due to their simplicity, SCIs, on the basis of inspiration by natural machines, have wider applications due to their smart functionalities.[6–11] Bio-inspired SCIs offer an exceptional opportunity to lay the foundations for a novel design of biomedical systems through the combination of diverse scientific disciplines that permit exploiting of the lessons from nature-based synthetic processes.[6,9,10] Mimicking the way biological SCIs are formulated and how they interact with other physiological components or trigger specific responses may markedly enhance the efficacy and safety of drug delivery system (DDS).

Selective shape-morphing systems are a hallmark of natural SCIs, which can respond or adapt efficiently to a wide range of external stimuli from the surrounding environment (*e.g.*, humidity, light, or touch).[1–9] These systems are mainly exemplified by botanical behaviors/motions, where a variety of organs such as tendrils, bracts, leaves, and flowers respond to the stimuli,

Biomaterials Science Series No. 1
Stimuli-responsive Drug Delivery Systems
Edited by Amit Singh and Mansoor M. Amiji

Published by the Royal Society of Chemistry, www.rsc.org

which leads to dynamic conformations governed by the organ composition and microstructural anisotropy of the cell walls.[1–6] For example, the touch-me-not-plant (*Mimosa pudica*) has leaves that fold inward and bend when touched or shaken. The Venus flytrap (*Dionaea muscipula*) has wide-open leaves with sensitive hair-like sensors on the surface. Once triggered by an insect, the leaves close and trap the insect. Some flowers are sensitive to light or temperature, such as the Moss rose (*Portuclaca grandiflora*), which opens only in bright sunlight and closes at night.[1–6] Scientists have been trying for the last few decades to learn from such biosystems to design and create smart polymer materials that can mimic such behavior and utilize them in technological applications.[6–17] A specific example is polymer brush technology, a layer of polymer chains with one end attached or grafted onto a thin film surface, which has attracted extensive research studies in the past decade due to its potential application in sensors, actuators and drug delivery systems.[12,14,16,17] The variations in polymer chain conformations in response to external stimuli lead to a transition between the shrunk and expanded state. The huge difference between these two states generates large stress and causes strong bending of the surface/brush.[12,14,16,17] The response mechanism of the brush depends on the polymer type and molecular weight.[12] For example, poly(*N*-isopropylacrylamide) brushes are sensitive to temperature, while triblock poly(styreneblock-2-vinylpyridine-*block*-ethylene oxide) copolymer brushes are sensitive to pH.[12,14,16] Despite all these interesting studies and applications, the research in this area is far from mature. Most of the previously reported systems respond to external stimuli by simple movements such as limited bending, longitudinal shrinkage or expansion.[12,14–17] More complicated and controlled movements are required to expand the applications of these systems, especially in the biomedical field.[15–17]

Hydrogels offer structural scaffolding to universal tissue constructs,[16–18] as well as to surface coat medical implants for enhanced biocompatibility by tethering appropriate therapeutic biomolecules such as growth factors, peptides or specific tissue-type drugs[18–21] However, the standalone hydrogel-based approaches are limited by short half-life, low binding efficiency of the drugs or peptides, and inability to induce long-term therapeutic effects.[21] Using synthetic DDS platforms involving polymers, lipids or nanoparticles (NPs), it is usually difficult to meet clinical prospects. Consequently, natural counterparts range from biomolecules to cells are worth exploring in more depth, as they are favorably optimized for their exclusive functions *in vivo* and hold features that are often required in DDS platforms.[16] With a better understanding of these biological systems, together with the availability of advanced therapeutic tools that are useful for re-engineering the diverse tissue systems, scientists have been able to exploit natural constituents for multiple applications in the delivery of drugs, proteins, small interfering RNA, and other therapeutic agents. Here, we review the natural-synthetic hybrid DD nano-platforms that could provide the foundation of active bio-inspiration for new therapeutic applications.

11.2 Bio-inspired Synthetic Designs of DDS

Bio-inspired surfaces and interfaces may enable the DDS nanocarriers to deal with the physiological barriers and selectively target diseased tissues. In particular, stimuli-responsive composites offer novel mechanisms for the control of drug release, which can also be integrated into tissue systems as scaffolds, networks, or coating agents, and exploited for attaining *in vivo* disease targeting in a controlled manner.[15–20] Although there are still few in number, some bio-inspired systems have already confirmed their potential in the clinical field.[16–20]

Recent advances in the synthesis of novel DDS materials and understanding of biological systems have paved the way towards bridging the gap between artificial (synthetic) and natural (biological) systems that has conventionally been quite large. Merging outlooks from the synthetic and biological arenas will offer a paradigm shift for the design of clinically translatable bio-inspired systems toward biomedical applications, such as smart DDS, biomaterials, and tissue engineering (Figure 11.1).

For instance, incorporation of nanostructured materials into a smart polymer matrix and controlling their shape, size, and distribution can give subsequent composite enhanced properties for more controlled movements in response to external stimulus. Hence, the adaptable behavior mimicked from biological systems can contribute to DD applications through *in vitro* and *in vivo* experimentations.

11.2.1 Bio-inspired Nanoscale Composites Architecture

Anisotropic nanocomposites evolved recently as a new composite system for which it is capable to create architecture controlled on a sizable scale, possessing intimate similarity to biological matrices. It involves diverse

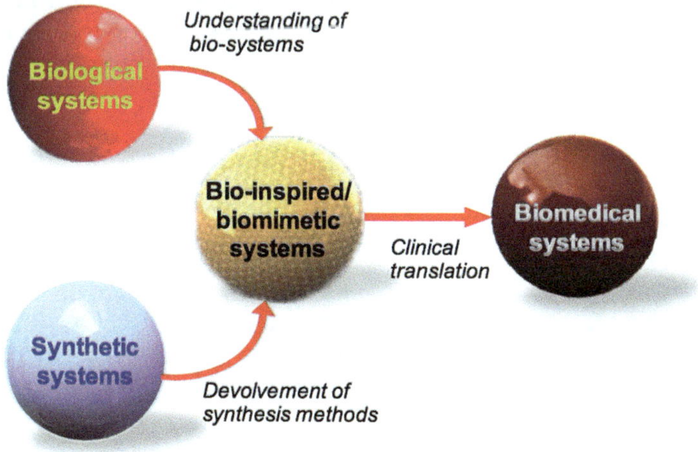

Figure 11.1 Bio-inspired/biomimetic systems for biomedical systems *via* emerging bio-systems and synthetic systems.

methodologies that can be used to organize nanoscale structural units into a desired configuration, for example, atom/molecules manipulation at the building block level, controlled ordering/orientation, and commended self-assembly.[22–27] The grand challenging question is how to assemble nanoparticular structures with designed architectures without explicitly prescribing particle sites. The modulation of the template and surface geometry play a key role by incorporating topographically distinct active surface domains, for example, patch/valence, and offer a convenient tool in assisting programmable assembly.[28–30] Simple natural molecules, such as DNA,[30–32] proteins,[33] and saccharides,[34,35] can provide us with the template for precise assembly of those components interacting on an extremely limited surface/valence, enabling a promising approach for the controlled design of DDS.

11.2.1.1 Layer-by-layer-NC-based Bioinspired DDS

Heterogeneous supramolecular DDSs can be built from template-guided nanosized molecular assemblies, amongst which most building blocks are self-assembled monolayers, bio-surface functionalizations, and layer-by-layer (LbL) assemblies.[16,26,27,36,37] The latter (LbL) method is the most versatile due to the required mild conditions, as well as the variety of materials that can be organized in an ordered manner.[26,27,36,37]

In this section, numerous features of LbL assembly on bioinspired DDSs are discussed, including their formulation, the tuning of their shell absorbency, their pharmaceutical purposes, and the advantages and challenges as presented in Figure 11.2. The wide-variety of hydrophobic solid drugs can be loaded in a multilayered (LbL) capsule, which can coated with distinct types of polymeric materials, a ranging from synthetic to naturally derived polymers, to colloids and other active composites.[36,37] These coatings allow high loading of the drug cores, tuning of the core dissolution rate *in vitro*, and shielding against the degradation of the loaded drug molecules, which can target the tissue of interest. In spite of the versatility of the obtainable

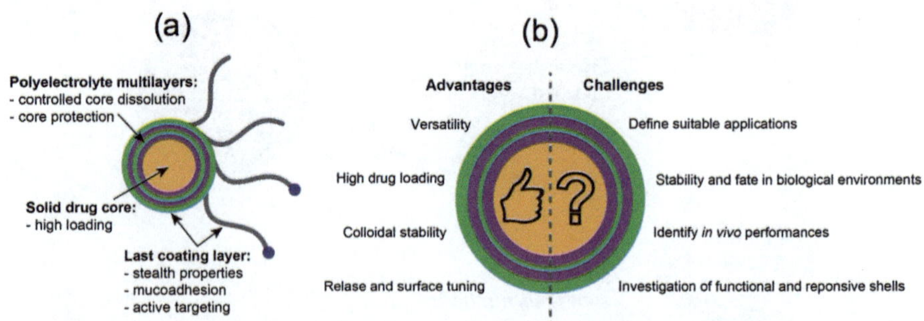

Figure 11.2 Schematic structure of layer-by-layer encapsulated solid drug cores (a) and the advantages and challenges (b). Reproduced with permission.[36] © 2017 WILEY-VCH Verlag GmbH & Co. KGaA, Weinheim.

LbL shells designs, there are still a very limited number of findings dedicated to these smart and responsive multilayer shells on the solid drug cores.[36,37]

Compared with previously developed DD systems, the role in which the LbL-coated solid drug cores might have enhanced importance has not yet been evidently revealed and the performances in biomedical settings have been rarely explored in appropriate depth. Many of the studies reported up-to-date on LbL-coated DDSs have focused on well-established orally administered compounds (*e.g.*, steroid drugs or nonsteroidal anti-inflammatory drugs).[38] On the other hand, LbL-encapsulation of cytotoxic drugs could provide great advantages over existing formulations, particularly for targeted tumor therapy[36–38] in terms of the improvement of colloidal stability, controlled release, reduction of side effects, and long-circulating injectable nanoformulations based on the enhanced permeability and retention (EPR) effect.[38] Nevertheless, biodistribution and pharmacokinetics studies should be complemented, which are mandatory to determining the efficacy of these designs in anticancer DDS. With complementing of the passive targeting, the functionalization of NP surfaces with active targeting moieties could potentially generate DDS with tumor selectivity, this concept has also yet to be comprehensively investigated in combination with LbL-DDSs.

Furthermore, there are many aspects concerning the behavior of LbL-coated drug cores in biological systems that remain unknown or poorly comprehended, which hinders the translation of these DDSs into clinical stages. For instance, the understanding of their cellular uptake mechanism and intracellular fate is very limited. Another critical characteristic lacking comprehensive studies is the reliability of the different polymer coatings and their impact on drug release *in vivo*. Additional chief issues that need to be clarified are the potential toxicity and removal paths of the engaged polyelectrolytes, especially polycations and non-biodegradable ones. Accordingly, investigating the influence of the physicochemical properties and structure of LbL-coated drug cores on their biological performances is an urgent prerequisite for the rational design of these classes of composites and their future translation into the clinic.

11.2.1.1.1 2-D Metal-polymer Nanocomposites as LbL-DDS. Metal-polymer nanocomposites can lead to innovative 2-D hybrid structures, which are of particular interest for a variety of reasons, such as ease of processability, prospects for large-scale manufacturing, and considerably lower density than pure metals.[26,27] Furthermore, they demonstrate combinatory physical, chemical, and optical properties largely distinctive from other materials, which can be realized by altering their geometry and composition, and thus offer various multifunctional platforms that have attracted interest as disease targeting tools in contrast imaging, drug delivery, hyperthermia, and radiotherapy.[26,27] Due to the plasmonic nanostructure with light-triggered heating function, the use of plasmonic gold nanoparticles (AuNPs), for example Au nanorods (AuNRs), combined with organic components, for example polymers, liposomes, or micelles, in light-responsive photothermal therapies and drug delivery is largely prevalent in eliminating tumours in a controlled and minimally-invasive manner.[38,39] However, the AuNRs are known to be melted by laser photothermal heating,[40] which are unstable, and more photothermally stable and complex Au nanostructures in this regime with a higher surface area for advanced drug loading and release performance are rarely studied.[41]

Layered structures from 2-D sheets are often discovered in nature and capable of versatile functionalities of living systems.[42–44] Although the dynamic morphology-control of micro-/nano-scale entities is of importance in nature, the process is not simple in artificial structures and very difficult to control/fine-tune the shape due to a spontaneous (uncontrollable) rolling-up mechanism.[45,46] To synthesize the layered structure in a simple and controllable manner, a heterogeneous LbL process using stepwise deposition *via* a gas phase sputtering technique would be suitable in terms of hierarchical ordering of the building blocks including metal NPs (structural and functional template/scaffold) and different type of polymers (stabilizing and responsive surface), as illustrated in Figure 11.3. Moreover, the exfoliation of free standing and stable dispersions of 2-D structured artificial materials remains a difficult task[46] and its biomedical application is limited due to a lack of biocompatibility and water-solubility.

The simple and effective synthesis method of fabricating individual metal-polymer based 2-D materials, for example, micro-/nanosheets with appropriate optical properties such as localized surface plasmon resonance (LSPR),

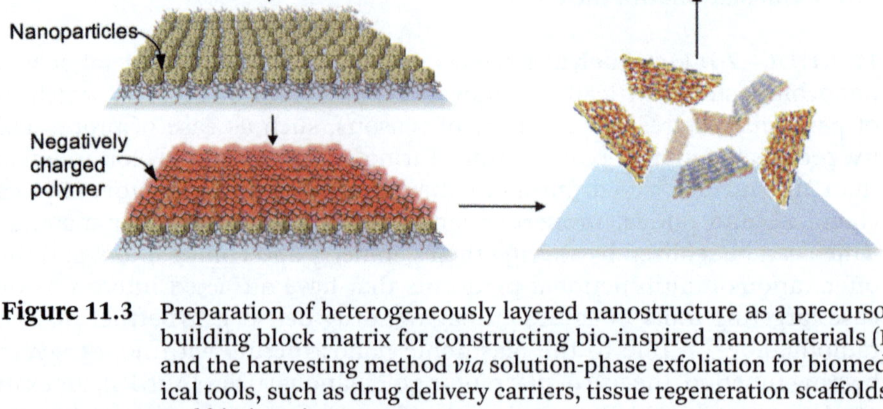

Figure 11.3 Preparation of heterogeneously layered nanostructure as a precursor building block matrix for constructing bio-inspired nanomaterials (I) and the harvesting method *via* solution-phase exfoliation for biomedical tools, such as drug delivery carriers, tissue regeneration scaffolds, and bio-imaging agents.

have not only not been fully studied, but also, how to obtain a stable colloidal dispersion remains an urgent challenging issue due to their aggregation. Therefore, 2-D metal-polymer composites are urgently demanded, as a reasonable precursor for designing novel shape-controllable dispersions, and the sophisticated synthetic techniques to prepare such a shape-controlled nanosheet-based drug carrier platform remains relatively rarely explored and challenging. Recently, Kim *et al.* demonstrated that 2-D gold-polymer composite nanosheets can be fabricated by a simple LbL-based technique *via* a physical vapor deposition and liquid exfoliation.[47] Branched polyethylenimine (bPEI) was chosen as a surface functionalization material because it is a smart polymer "proton sponge", which expands at protonated states (lower pH values) and shrinks at non-protonated states (higher pH values).[48,49] The shape transformations at different protonation states, which correspond to different pH values, are monitored by transmission electron microscopy (TEM), atomic force microscopy (AFM), UV-visible spectroscopy, and zeta-potential measurements. They thus seem promising for applications that necessitate controlled and gradual two-way shape-changing materials with magnificent surface area and loading capacity. For example, a large number of diverse NPs could be loaded onto the nanosheets by sputtering, followed by rolling by pH, for a customizable function. Furthermore, due to their LSPR- and SERS-adaptable plasmonic properties, nanosheet/nanoscroll colloids hold attractive potential for further applications in smart DD vehicles, biosensing, catalysis, and nanotechnology contexts, from basic bio-inspired materials research to commercially available smart products. Finally, the synthesis method combines physical vapor deposition and liquid exfoliation and thus can be applied to other NP building blocks and smart polymers.

11.2.1.1.2 Bio-inspired Janus Nanostructures for Smart DDS. More dynamic and complex material properties can result when the LbL-metal-polymer nanocomposites are realized as shape-transformable nanostructures in response to physicochemical stimuli permitting precisely tuned regulation over favorable functional activities, mainly in plants.[42,50–56] A natural example is the discovery of fibrous motors in plants that behave as the opening gear of seed/spore containers, such as legume pods or fern sporangia, to release seeds/spores.[50–56] The pod walls are 'Janus' 2-D LbL structures with two distinct fibrous hydration layers.[53,54] Dehydration of the maturing pod walls leads to a torsional kink that breaks the pod and blows out the seeds in different directions.[53–56] The asymmetric bilayer model is complex and still mysterious, but is being investigated as it is exceptionally useful to mimic multifunctional imitative designs where fibers can function both as structural scaffolds and as responsive gears.

By taking symmetry-breaking approaches to convey different properties within a single 2-D structure, as described in Figure 11.3, recent research focus lies on 'Janus' nanostructures,[57,58] for example, Janus particles, Janus cylinders, and Janus platelets and sheets. Particularly, polymer-based 2-D nanosheets[59,60] that have an outsized surface volume-to-area with extensive components from different polymers that respond differently to an external stimulus, in example, pH, temperature, or wettability, and act like a natural fibrous system. Polyelectrolytes have been used for the large-scale preparation of Janus nanosheets that can generate emulsions with immiscible liquids.[59–61] In contrast to the colloidal stability of existing emulsion-based drugs, for example micelles and liposomes, Janus nanosheets are more difficult to turn over at an emulsion interface because of their highly asymmetric shape, and make an emulsion more stable.[59–61] The underlying mechanism of the drug (seed) loading and triggered-release activities include the entrapment of the drug (seed) payloads inside the nanosheet and a stimuli-induced phase change, respectively, analogous to natural seed dispersal systems. This will offer us bio-inspiration to design smart DDS platforms, although heterogeneous nanosheets merged with responsive materials in this regime are rarely reported.[59–61] In 2016, Kim and Lu demonstrated an efficient design and facile LbL synthesis of Janus composite nanosheets (JCNs), which are comprised of Au islands embedded in two oppositely pH-sensitive natural polymer chains (Figure 11.4).[41] The scrolling of JCNs is pH-dependent because two different pH-responsive types of natural polymers, for example chitosan (Chi) and gum arabic (GA), were employed as the smart polymers that expand/shrink in protonated states (lower pH values) and shrink/expand at non-protonated states (higher pH values), respectively.[47] In particular, the drug release appears like a spore launcher in fern sporangium (Figure 11.4a). While the simple synthetic polymer-based Au nanosheets often produce randomly scrolled structures, such as double-directionally rolled/folded structures,[47] the rolling of JCNs is uni-directional due to the oppositely charged polymer coating (Figure 11.4b), which can produce a more controlled and higher quality 1-D scroll structure with improved pH-sensitivity. Furthermore, the thermo-sensitive PEG-*b*-PCL was functionalized at the Chi-layer on the JCN, allowing the controlled loading of a drug (Doxorubicin) and the release of drug molecules by unrolling in response to NIR light-triggered heating (Figure 11.4c): selective coating of the PEG-*b*-PCL site creates a JCN with a hydrophobic interior layer and a hydrophilic exterior surface that mimics biological components, for example proteins and biomembranes. Additionally, the shape changes at different protonation/drug loading/photothermal states, which correspond to different pH/hydrophobicity/temperature values, were monitored by TEM, AFM, UV-Vis spectroscopy, and zeta-potential analysis. As a proof-of-concept of the efficacy of bio-inspired smart LbL nanostructures that are response governable to specific stimuli, either endogenous (scrolling by changes in pH or hydrophobic drug molecules) or exogenous (un-scrolling by heating *via* NIR light), the NIR-guided *in vitro* photothermal treatments of JCNs on HeLa cells were demonstrated.[41]

Triggered release could be achieved upon remote NIR-activated heating, which leads to a phase change of the crystalline PCL domains into the amorphous state. The use of these JCNs as an insoluble DD vehicle and an effective photosensitizer for PTA or TP-PTA *in vitro* were also demonstrated on

HeLa cell treatment under a local two-photon irradiation, by employing the cytotoxicity assay (indirect method), and fluorescence microscopy (direct method) (Figure 11.4f). In particular, the scrolled JCN geometry incorporating PCL moieties inside the wall became an advantage in inhibiting the drug release before targeting the desirable sites. The collective result of the synergistic contribution of phototherapy and chemotherapy, under guidance from imaging, and the stable complex formulation make JCNs an attractive agent for targeted cancer treatment, cancer-cell ablation, smart molecular probes, DD vehicles or smart tissue scaffolds, in a minimally-invasive fashion.

11.2.1.1.3 Bioinspired LbL-DDS-based Scaffolds. As one of the most important tools for tissue engineering, scaffolds are forged extracellular lattices which are employed for supporting 3-D tissue arrangement, and are used to convey cells and drugs into the body which are either embed or permeated.[16,17,36] Scaffold networks can be utilized to achieve DD with high stacking and efficiency to a specific site, and have been created to enhance bone in growth and recovery in the treatment of bone deformities.[16,17,36] To meet the challenges of recovering bone damage to ailment or injury, biodegradable scaffolds are being investigated as an approach to recover bone without the requirement for an auto-or allograft.[62,63] For example, drugs, anti-toxins, antimicrobials, and development components should be delivered to the affected site because the blood circulation in the rigid deformity locales is poor. A few DD scaffolds have been created to exemplify clinical medications, for example biodegradable polymers (engineered or natural), as particulates and a permeable lattice.[62,63] Taking into account utilization of biodegradable material, for example PLA or PLGA, is the most worthwhile medication delivery scaffold as it supports basic implantation of the scaffold and then the biodegradable material is bioresorbable into the body. Looking into feasibility, there is good scope in developing injectable sol–gel scaffolds due to their ease of use, adaptability, and involve the utilization of a safe adjuvant; at present some of them have been approved by the Food and Drug Administration.[62,63]

Simple structured 2-D scaffolds have been widely explored in the area of skin engineering. Several synthetic skin grafts are already available in the market, providing replacement of epidermal layers, dermal layers, or a combination of the two.[64] The majority of these grafts are comprised of scaffolds seeded with fibroblasts and/or keratinocytes. However, these grafts have negatively influenced the vascularization of these skin grafts, which requires more attention to incorporate more complex geometry into the grafts to accommodate better blood supply to the transplanted sites. Heterogeneous scaffolds based on smart materials have been studied in diverse tissue engineering fields, such as cardiovascular and endothelium tissue engineering, skin growth or wound healing, neural regeneration, dental engineering, and corneal and retina applications.[65] Natural polymer scaffolds are advantageous in terms of biomimeticity and structural complexity. These naturally occurring materials, however, lack controlled degradation and have the potential to be immunogenic. Numerous factors of synthetic materials including functionality, mechanical,

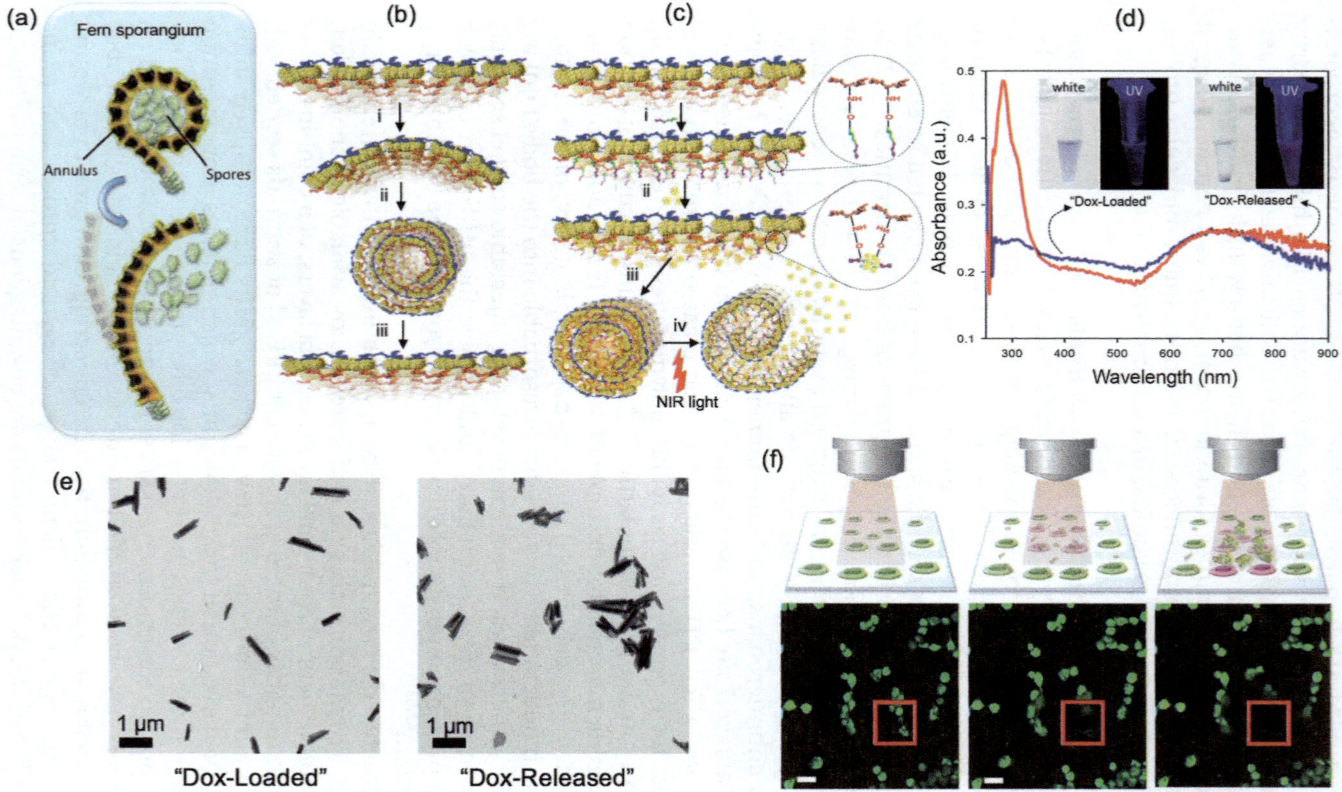

Figure 11.4 (a) Illustration of mimicking real spore launching systems in fern sporangia. (b) The sequential asymmetric preparation of Janus composite nanosheets: (I) chitosan matrix was spin-coated onto a glass wafer, (II) Au was sputtered onto the polymer-coated surface until a near-percolating Au thin film was obtained, (III) the Au surface was functionalized with gum arabic and (IV) dipped in acetate solution and exfoliated by ultra-sonication, followed by purification and re-suspension of the nanosheets in deionized water. (c) The functionalization of PEG-b-PCL block copolymers on the chitosan side of nanosheet (I), Dox loading (II), triggered scrolling (III), and NIR-induced unscrolling of a nanosheet resulting in drug release (IV).

strength, and degradation can be coordinated by manipulating the chemical and physical properties during manufacturing. Generally, both natural and synthetic polymer-based scaffolds are more efficacious when composited or copolymerized with another material. Resultant composite scaffolds comprising of two or more polymers convey the properties of each material, which permits for the compensation of drawbacks from coexisting polymers.[65]

More complex two-layered scaffolds based on the LbL technique have been developed for the purpose of engineering vasculature and more tubular structures.[65] For example, scaffold synthesis with an inner layer of epithelial cells and an outer layer of muscle cells for urethral engineering. Similar techniques have been employed in vascular engineering with fibroblasts and endothelial cell layers.[66]

Carbon/Hydrogel LbL Nano-scaffolds for Smart DDS. The interacting forces at the nanoscale carbon/polymer interfaces play a critical role in ordering the structures and dynamics of the resulting hybrid composite systems to create enhanced or even new functions that are originally lacking in their individual components. Although it currently remains greatly challenging to address the interacting forces at interfaces through experimental measurements, it is of great significance to comprehend and clarify how the interacting forces at interfaces modify the hybrid nanostructures and their bio-physico-chemical properties for practical applications. In contrast to carbon nanotubes (CNTs), carbon nanoscrolls (CNSs) have a core size that can be tuned by surface functionalization of graphene/graphite before rolling.[67,68] With these novel and expanded functionalities, CNSs are highly promising for cutting-edge multifunctional biological applications, such as biosensors, molecular probes, and medical therapeutic agents.[69]

Recently, Cheng *et al.* reported a CNS-protein-based hybrid structure,[70] a CNS-silk crystallite composite with a hydration level governable by an interlayer interaction in the CNS, where the size-controlled CNS core not only shields the crystallite from a weakening effect of water, but also distinctly strengthens the crystallite as shown in Figure 11.5: As the CNS interlayer interaction energy changes, the core size of CNS become altered, which in turn changes the hydration level of the silk crystallite, the interaction of CNS and silk consequently modifies the hybrid structure and mechanical behavior. Their smart hybrids exhibit promising features for potential functional applications in tissue engineering, biosensing, DDS, and technical textiles.

The optical characterization (d) and TEM images (e) of the Dox-loaded/released nanosheets. The two-photon (TP) photo thermal effect of JCNs loaded with Dox against HeLa cells. The cells were exposed to JCNs before therapy (left), with JCNs after 30 min two-photon laser irradiation (center) Dox release after the therapy (right), mimicking the release under physiological conditions after the TP-treatment. Scale bar: 30 mm. Reproduced from ref. 41 with permission from The Royal Society of Chemistry.

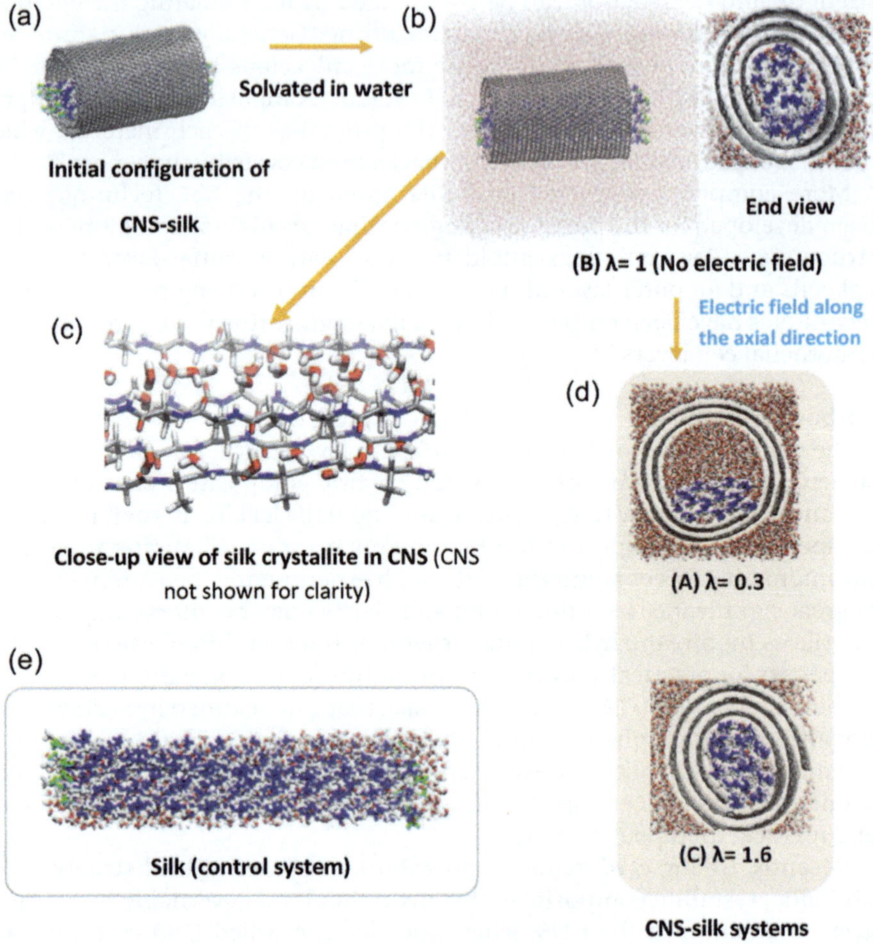

Figure 11.5 CNS-silk hybrid structure (a) before and (b) after beingsolvated in water. (c) Close-up view of equilibrated structure of silk crystallite and water molecules. (d) Silk crystallite in different sized CNS core. (e) The control structure of silk crystallite. Reproduced from ref. 70 with permission from The Royal Society of Chemistry.

Paul *et al.* demonstrated a novel DNA-activated graphene oxide (GO)/methacrylated gelatin (GelMA) hydrogel,[71] which can take advantage of the unique features coupled with a GelMA hydrogel and GO-based angiogenic injectable gene delivery system to display enhanced acute myocardial infarction (AMI) therapeutic efficacy without any substantial side effects (Figure 11.6). In particular, the hybrid composite hydrogels not only provide the *in vivo* biocompatibility, but also induced tissue revascularization at the injured region and improved contractile performance, as validated by vasculo-genesis, scar area analysis, and echocardiogram analysis. This approach is specifically beneficial compared to commonly considered stem cell therapy whose application is restricted mostly by recurrent immune rejections by patients,

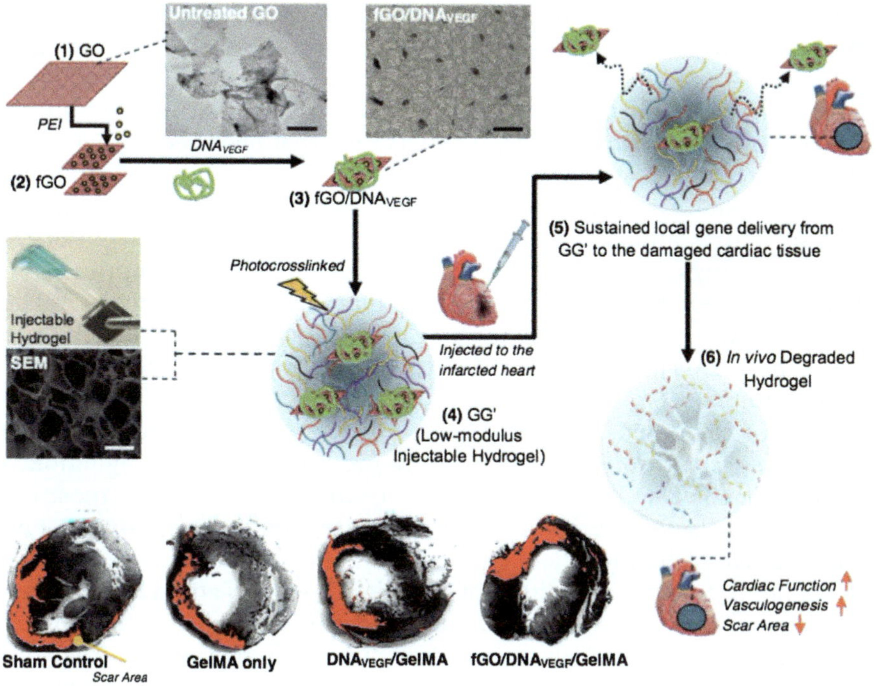

Figure 11.6 Injectable hybrid nanoscaffolds based on graphene oxide (GO)-GelMA (GG′) composites. First, GO (1) is coated with branched polyelectrolyte, polyethylenimine (PEI) to form fGO (2), which is further functionalized with anionic plasmids (DNAVEGF) to form fGO/DNAVEGF (3) (Scale bar: 1 μm). These composites are incorporated within a prepolymer matrix of GG′ (4), following by exposure to UV to obtain a cross-linked network with a porous network (5). Injected into rat hearts with AMI for local gene delivery (6). The results imply notably reduced scar site of infarcted hearts treated with GO-GG′. In the bottom panel, sham treated and untreated (control) infarcted groups are presented. The red area represents ECM deposition in the scar tissue while the gray area represents the myocardium, showing the promising therapeutic efficacy of GO-GG′. Reproduced with permission from A. Paul, A. Hasan, H. A. Kindi, A. K. Gaharwar, V. T. S. Rao, M. Nikkhah, S. R. Shin, D. Krafft, M. R. Dokmeci, D. Shum-Tim, and A. Khademhosseini, *ACS Nano* 2014, **8**, 8050. Copyright 2014, American Chemical Society.[71]

conservation of cell viability and retention at the target location, probabilities of teratoma formation, along with a plethora of ethical, logistical and technical challenges for cell isolation and culturing.

11.2.1.1.4 Bio-inspired Printing Techniques Using LbL-DDS

3-D Bioprinting. 3-D printing is a technology compatible for biomedical applications due to its capability to fabricate greatly complex and predesigned structures from personalized designs with a rapid turnaround.[72] As a cutting-edge 3-D printing technique, laser-based 3-D printing, for example stereolithography (SLA)-directed 3-D printing based on digital light processing (DLP) allows high-speed processing of scaffolds with higher resolution (1–30 μm with SLA; <100 nm with two-photon polymerization) and higher mechanical strength, compared with existing various fast-prototyping techniques.[73] Hydrogels are 3-D networks that have been widely used as bio-inspired tissue scaffolds and templates in recent tissue engineering applications because of their tailorable high-water content, cell-compatibility, and structural similarities to the natural extracellular matrix.[74–77] Generally, the processing of hydrogels using SLA-based 3-D printing comprises of reactive monomers, a photoinitiator (PI), and/or a cross-linking agent.[73] The PI plays a critical role in defining the rate of polymerization and the subsequent functions of the printed object, along with the time necessary for 3-D fabrication.

For live cell-based fabrication with hydrogels, cell-compatible and mild processing conditions are required by preventing heating, agitation, use of organic solvents, or ultraviolet (UV) irradiation.[78] However, most up-to-date protocols for fabricating hydrogels use poorly water-soluble PIs, in the absence of highly efficient water-borne PIs, which need extensive stirring and/or heating or addition of organic solvents to attain clear precursor solutions.[79] To overcome this limitation, Pawar *et al.* investigated a rapid 3D printing technique of hydrogels in aqueous solutions based on the conversion of water-insoluble PIs into highly efficient water-dispersible NPs.[80]

Smart Hydrogels for Bioinspired 4-D Printing. Precise architectural reconfiguration and surface functionalization of the printing materials are the two major post-3-D printing modification processes and potential techniques for post-3-D printing to attain preferred biofunctionality, towards improved biomedical applications. To address the challenge, smart fluids have been studied as a post-3-D printing material by self-assembly, or "4-D" printing material because it enables creation of the dimension of transformation over time, which could make something without relying on chips/sensors, wires, or motors.[81–83] By programming the printer with a precise geometric code based on the materials' configuration (*e.g.*, compositions, rheology, angles) and dimensions, the shape can be tailorable when encountered with external stimuli such as water, movement, or a change in temperature, making it possible for 4-D. With a minimal theoretical framework of plant-inspired architectures, precise smart shape-morphing objects can be programmed to be fabricated that change morphology on dispersion in water, yielding complex 3-D morphologies.[81] For instance, one can use smart filler ink fluid through the printhead, that rapidly solidifies once printed. The composite hydrogel architectures can be encoded with localized, anisotropic swelling behavior controlled by the alignment of cellulose fibrils along with 4-D printing pathways.[81] A variety of hydrogel materials can be used interchangeably resulting in different stimuli-responsive behavior, while the cellulose fibrils can be exchanged with other anisotropic fillers of choice, including conductive fillers.

Due to the limited availability of printable materials, for example PIs, the biofunctionality of many 3-D-printed components has not been intensively studied. In this regard, post-3-D printing modification can be the solution that could solve the limit by linking the gap between 3-D printing technology and requirements of bio-safety. Therefore, blending the potential of 4-D printing and smart biofunctional hydrogels is an excellent novel paradigm in the creation of a functional 3-D tissue.

Shida Miao *et al.* reported a new type of 4-D printable PI resin,[84] which is synthesized by a polymer oil polymer, soybean oil epoxidized acrylate compound. By mixing with a small amount of 3-D printable resins, the resin mixture presents shape transformation in response to temperature variations from −18 °C to 37 °C. Printed scaffolds of this compound demonstrated successful human bone marrow mesenchymal stem cell (hMSCs) growth with outstanding biocompatibility. The exceptional shape memory effect inspired by a botanical system and biocompatibility could allow researchers to advance the development of natural resin-based stereolithography for advanced DD scaffolds.

11.3 Challenges and Future Insights

The surface-engineered (LbL) DDSs is a cutting-edge emerging field, which takes advantage of the heterogeneous structure with high drug loading efficiency, and exceptional smart DD mechanisms, such as stimuli-responsive selective targeting, and high stability and biocompatibility. Very recently, this study has involved an innovative modular approach to create bioactive functionalization/coating on perfused 3-D or 4-D bioprinted scaffolds that release drugs, small molecules, or growth factors, and displays their stimuli-responsive profiles toward improving tissue regeneration/repairing, for example vascularized bone regeneration. The field of biologically inspired LbL DD scaffolds is, however, still in its infancy and there are several challenges that need to be overcome. Firstly, processes that are mandatory for the development of such scaffolds, for example *ex vivo* treatments to produce or incorporate/release therapeutic drugs, make it difficult to maintain the interfacial integrity of natural counterparts in the manner of biomimetic hierarchical architecture and dynamic bio-signaling events, particularly those of living tissues/organ systems. Secondly, major desirable properties of different cells/tissues should be effectively combined into a single synthetic composite so that tailorable DD LbL scaffolds could be designed and improved for specific aims. In this regard, further optimization of synthetic LbL nanocomposite systems that mimic cells/tissues will boost up controlled drug release and minimize structural alterations and possible side effects. This study will have great potential for a wide-variety of innovative clinical applications such as on-demand DD carriers, remote DD evaluation platforms of therapeutic drugs, transplantable scaffold systems, complex tissue/organ regenerations, and smart biosensors/lab-on-a-chips.

References

1. Y. Forterre, J. Skotheim, J. Dumais and L. Mahadevan, *Nature*, 2005, **433**, 421–425.
2. B. J. Awell, P. E. Kriedemann and C. G. N. Turnbull, *Plants in Action*, Macmillan Education AU, 1999.
3. E. Reyssat and L. Mahadevan, *J. R. Soc., Interface*, 2009, **6**, 951–957.
4. S. Armon, E. Efrati, R. Kupferman and E. Sharon, *Science*, 2011, **333**, 1726–1730.
5. P. Fratzl and I. Burgert, *Philos. Trans. R. Soc., A*, 2009, **6**, 1541–1557.
6. C. Sanchez, H. Arribart and M. M. Giraud-Guille, *Nat. Mater.*, 2005, **4**, 277–288.
7. K. Liu, Y. Tian and L. Jiang, *Prog. Mater. Sci.*, 2013, **58**, 503–564.
8. K. Liu and L. Jiang, *Nanotoday*, 2011, **6**, 155–175.
9. F. Xia and L. Jiang, *Adv. Mater.*, 2008, **20**, 2842–2858.
10. H. D. Espinosa, J. E. Rim, F. Barthelat and M. J. Buehler, *Prog. Mater. Sci.*, 2009, **54**, 1059–1100.
11. A. Finnemore, P. Cunha, T. Shean, S. Vignolini, S. Guldin, M. Oyen and U. Steiner, *Nat. Commun.*, 2012, **3**, 966–972.
12. A. Martien, *et al.*, *Nat. Mater.*, 2010, **9**, 101–113.
13. R. Lupitskyy, Y. Roiter, C. Tsitsilianis and S. Minko, *Langmuir*, 2005, **21**, 8591–8593.
14. S. T. Milner, *Science*, 1991, **251**, 905–914.
15. P. M. Mendes, *Chem. Soc. Rev.*, 2008, **37**, 2512–2529.
16. L. Rodrigues and M. Mota, *Bioinspired Materials for Medical Applications*, Woodhead Publishing, 2016.
17. M. R. Aguilar, C. Elvira, A. Gallardo, B. Vasquez and J. S. Romań, Smart polymers and their applications as biomaterials, in *Topics in Tissue Engineering*, ed. N. Ashammakhi, R. L. Reis and E. Chiellini, University of Oulu, Oulu, Finland, 2007, vol. 3, pp. 2–27.
18. A. Khademhosseini, J. P. Vacanti and R. Langer, *Sci. Am.*, 2009, **300**, 64–71.
19. L. Schukur, P. Zorlutuna, J. M. Cha, J. H. Bae and A. Khademhosseini, *Adv. Healthcare Mater.*, 2013, **2**, 195–205.
20. S. Khetan, M. Guvendiren, W. R. Legant, D. M. Cohen, C. S. Chen and J. A. Burdick, *Nat. Mater.*, 2013, **12**, 458–465.
21. G. W. Ashley, J. Henise, R. Reid and D. V. Santi, *Proc. Natl. Acad. Sci. U. S. A.*, 2013, **110**, 2318–2323.
22. L. Xu, *et al.*, *Chem. Soc. Rev.*, 2013, **42**, 3114–3126.
23. G. R. Yi, D. J. Pine and S. Sacanna, *J. Phys.: Condens. Matter*, 2013, **25**, 193101.
24. B. G. P. van Ravensteijn, M. Kamp, A. van Blaaderen and W. K. Kegel, *Chem. Mater.*, 2013, **25**, 4348–4353.
25. Y. F. Wang, *et al.*, *Nature*, 2012, **491**, 51–61.
26. K. Ariga, Y. Yamauchi, G. Rydzek, Q. Ji, Y. Yonamine, K. C. W. Wu and J. P. Hill, *Chem. Lett.*, 2014, **43**, 36–68.
27. K. Ariga, Q. Ji, J. P. Hill, Y. Bando and M. Aono, *NPG Asia Mater.*, 2012, **4**, e17.

28. Z. Zhang and S. C. Glotzer, *Nano Lett.*, 2004, **4**, 1407–1413.
29. R. M. Choueiri, E. Galati, H. Thérien-Aubin, A. Klinkova, E. M. Larin, A. Querejeta-Fernández, L. Han, H. L. Xin, O. Gang, E. B. Zhulina, M. Rubinstein and E. Kumacheva, *Nature*, 2016, **538**, 79–83.
30. Q. Li, *Anisotropic Nanomaterials: Preparation, Properties, and Applications*, Springer, 2015.
31. W. Liu, J. Halverson, Y. Tian, A. V. Tkachenko and O. Gang, *Nat. Chem.*, 2016, **8**, 867–873.
32. J.-W. Kim, J.-H. Kim and R. Deaton, *Angew. Chem., Int. Ed.*, 2011, **50**, 9185–9190.
33. K. L. Gurunatha, A. C. Fournier, A. Urvoas, M. Valerio-Lepiniec, V. Marchi, P. Minard and E. Dujardin, *ACS Nano*, 2016, **10**, 3176–3185.
34. Z. Wang, K. Wang, X. Lu, C. Li, L. Han, C. Xie, Y. Liu, S. Qu and G. Zhen, *Adv. Healthcare Mater.*, 2015, **4**, 927–937.
35. E. Rodriguez-Velazquez, M. Alatorre-Meda and J. F. Mano, *Curr. Pharm. Des.*, 2015, **21**, 4837–4850.
36. A. Polomska, J.-C. Leroux and D. Brambilla, *Macromol. Biosci.*, 2017, 1600228.
37. J.-H. Kim, H. Eguchi and M. Umemura, *et al.*, *NPG Asia Mater.*, 2017, **9**, e367.
38. Z. Dai, *Advances in Nanotheranostics I: Design and Fabrication of Theranosic*, Springer, 2015.
39. E. C. Dreaden, L. A. Austin, M. A. Mackey and M. A. El-Sayed, *Ther. Delivery*, 2012, **3**, 457–478.
40. L. Nie, M. Chen, X. Sun, P. Rong, N. Zheng and X. Chen, *Nanoscale*, 2014, **6**, 1271–1276.
41. J.-H. Kim and T.-M. Lu, *RSC Adv.*, 2016, **6**, 17179–17187.
42. Y. Bar-Cohen, *Biomimetics: Biologically Inspired Technologies*, CRC Press, Boca Raton, FL, 2005, pp. 1–552.
43. R. Elbaum, L. Zaltzman, I. Burgert and P. Fratzl, *Science*, 2007, **316**, 884.
44. P. Fratzl, R. Elbaum and I. Burgert, *Faraday Discuss.*, 2008, **139**, 275.
45. G. Wan, A. A. Solovev, G. S. Huang, M. F. Maitz, N. Huang and Y. F. Mei, *J. Mater. Chem.*, 2012, **22**, 12983.
46. V. Nicolosi, M. Chhowalla, M. G. Kanatzidis, M. S. Strano and J. N. Coleman, *Science*, 2013, **340**, 1226419–1226437.
47. J.-H. Kim, M. Bohra, V. Singh, C. Cassidy and M. Sowwan, *ACS Appl. Mater. Interfaces*, 2014, **6**, 13339–13343.
48. C. K. Choudhury and S. Roy, *Soft Matter*, 2013, **9**, 2269–2281.
49. X. Gong and T. Ngai, *Langmuir*, 2013, **29**, 5974–5981.
50. M. Lattuada and T. A. Hatton, *Nano Today*, 2011, **6**, 286.
51. I. Burgert and P. Fratzl, *Philos. Trans. R. Soc., A*, 2009, **367**, 1541.
52. S. Armon, E. Efrati, R. Kupferman and E. Sharon, *Science*, 2011, **333**, 1726.
53. R. Elbaum, L. Zaltzman, I. Burgert and P. Fratzl, *Science*, 2007, **316**, 884.
54. P. Fratzl, R. Elbaum and I. Burgert, *Faraday Discuss.*, 2008, **139**, 275.
55. E. Reyssat and L. Mahadevan, *J. R. Soc., Interface*, 2009, **6**, 951.

56. X. Noblin, N. O. Rojas, J. Westbrook, C. Llorens, M. Argentina and J. Dumais, *Science*, 2012, **335**, 1322.
57. A. Walther and A. H. E. Müller, *Chem. Rev.*, 2013, **113**, 5194.
58. F. Liang, K. Shen, X. Qu, C. Zhang, Q. Wang, J. Li, J. Liu and Z. Yang, *Angew. Chem., Int. Ed.*, 2011, **50**, 2379.
59. X. Ji, Q. Zhang, F. Liang, Q. Chen, X. Qu, C. Zhang, Q. Wang, J. Li, X. Song and Z. Yang, *Chem. Commun.*, 2014, **50**, 5706.
60. H. Yang, F. Liang, X. Wang, Y. Chen, C. Zhang, Q. Wang, X. Qu, J. Li, D. Wu and Z. Yang, *Macromolecules*, 2013, **46**, 2754.
61. Z. Zhao, F. Liang, G. Zhang, X. Ji, Q. Wang, X. Qu, X. Song and Z. Yang, *Macromolecules*, 2015, **48**, 3598.
62. A. B. Allen, L. B. Priddy, M. T. Li and R. E. Guldberg, *Ann. Biomed. Eng.*, 2015, **43**, 555–567.
63. H. F. Ko, C. Sfeir and P. N. Kumta, *Philos. Trans. R. Soc., A*, 2010, **368**, 1981–1997.
64. S. Böttcher-Haberzeth, T. Biedermann and E. Reichmann, *Burns*, 2010, **36**, 450–460.
65. L. Edgar, K. McNamara, T. Wong, R. Tamburrini, R. Katari and G. Orlando, *Materials*, 2016, **9**, 332.
66. H. Orabi, S. Bouhout, A. Morissette, A. Rousseau, S. Chabaud and S. Bolduc, *Sci. World J.*, 2013, **2013**, 154564.
67. X. Xie, L. Ju, X. Feng, Y. Sun, R. Zhou, K. Liu, S. Fan, Q. Li and K. Jiang, *Nano Lett.*, 2009, **9**, 2565–2570.
68. P. Yin, S. Shah, M. Chhowalla and K. Lee, *Chem. Rev.*, 2015, **115**, 2483–2531.
69. J.-H. Kim and M. Benelmekki, *ACS Appl. Mater. Inter.*, 2016, **8**, 33121–33130.
70. Y. Cheng, L.-D. Koh, F. Wang, D. Li, B. Ji, J. Yeo, G. Guan, M.-Y. Han and Y.-W. Zhang, *Nanoscale*, 2017, **9**, 9181–9189.
71. A. Paul, A. Hasan, H. A. Kindi, A. K. Gaharwar, V. T. S. Rao, M. Nikkhah, S. R. Shin, D. Krafft, M. R. Dokmeci, D. Shum-Tim and A. Khademhosseini, *ACS Nano*, 2014, **8**, 8050.
72. S. Wang, J. M. Lee and W. Y. Yeong, *Int. J. Bioprint.*, 2015, **1**, 3.
73. T. Billiet, M. Vandenhaute, J. Schelfhout, S. Van Vlierberghe and P. Dubruel, *Biomaterials*, 2012, **33**, 6020–6041.
74. S. Hong, D. Sycks, H. F. Chan, S. Lin, G. P. Lopez, F. Guilak, K. W. Leong and X. Zhao, *Adv. Mater.*, 2015, **27**, 4035–4040.
75. A. L. Rutz, K. E. Hyland, A. E. Jakus, W. R. Burghardt and R. N. Shah, *Adv. Mater.*, 2015, **27**, 1607–1614.
76. J. Wei, J. Wang, S. Su, S. Wang, J. Qiu, Z. Zhang, G. Christopher, F. Ning and W. Cong, *RSC Adv.*, 2015, **5**, 81324–81329.
77. C. B. Highley, C. B. Rodell and J. A. Burdick, *Adv. Mater.*, 2015, **27**, 5075–5079.
78. C. G. Williams, A. N. Malik, T. K. Kim, P. N. Manson and J. H. Elisseeff, *Biomaterials*, 2005, **26**, 1211–1218.

79. B. D. Fairbanks, M. P. Schwartz, C. N. Bowman and K. S. Anseth, *Biomaterials*, 2009, **30**, 6702–6707.
80. A. A. Pawar, G. Saada, I. Cooperstein, L. Larush, J. A. Jackman, S. R. Tabaei, N.-J. Cho and S. Magdassi, *Sci. Adv.*, 2016, **2**, e1501381.
81. A. S. Gladman, E. A. Matsumoto, R. G. Nuzzo, L. Mahadevan and J. A. Lewis, *Nat. Mater.*, 2016, **15**, 413–418.
82. Q. Ge, H. J. Qi and M. L. Dunn, *Appl. Phys. Lett.*, 2013, **103**, 131901.
83. S. Tibbits, *Archit. Des.*, 2014, **84**, 116–121.
84. S. Miao, W. Zhu, N. J. Castro and M. Nowicki, *Sci. Rep.*, 2016, **6**, 27226.

CHAPTER 12

Stimuli-responsive Materials in Theranostics

HAO WU[†a], LU ZHANG[†a], YUANPEI LI*[a] AND KIT S. LAM*[a,b]

[a]Department of Biochemistry and Molecular Medicine, UC Davis Comprehensive Cancer Center, University of California Davis, Sacramento, CA 95817, USA; [b]Department of Internal Medicine, Division of Hematology/Oncology, University of California Davis, Sacramento, CA 95817, USA
*E-mail: kslam@ucdavis.edu, lypli@ucdavis.edu

12.1 Introduction

The development of multifunctional materials such as nanoparticles as therapeutic and diagnostic (theranostic) agents has great impact in the management of cancer and other diseases.[1–4] Several advantages can be offered by these theranostic agents: (i) sensitive and accurate diagnosis of diseases, (ii) non-invasively monitoring of drug distribution and pharmacokinetics, and (iii) real-time assessment of treatment outcomes, and therefore timely adjustment of individualized therapy.[5–7] These unique features suggest that theranostic agents may have great potential to be applied towards personalized nanomedicine for disease- and patient-specific diagnosis and treatment.

Stimuli-responsive materials represent ideal building blocks to construct smart theranostic systems to improve diagnostic and therapeutic functions

[†]Hao Wu and Lu Zhang contributed equally to this book chapter.

Biomaterials Science Series No. 1
Stimuli-responsive Drug Delivery Systems
Edited by Amit Singh and Mansoor M. Amiji

Published by the Royal Society of Chemistry, www.rsc.org

for the treatment of cancer and other diseases.[5,8–10] For instance, nanoprobes constructed with linkers that are sensitive to the stimuli in the tumor microenvironment, could be utilized to significantly increase the sensitivity of cancer detection.[11,12] The introduction of stimuli-responsive assembly units could make nanoparticles responsive to stimuli present in the tumor local microenvironment or to exogenously applied stimuli. Therefore, the encapsulated drugs could be released predominately in cancerous tissue or inside cancer cells, resulting in the enhancement of therapeutic efficacy and lessening of systemic side effects.[6–8] In this book chapter, we will first focus on theranostic systems that are responsive to chemical and biological stimuli, such as pH, redox potential and enzymes. After that, we will introduce physical based theranostic systems, triggered by ultrasound, heat, light, magnetic or electric fields.

12.2 Chemical and Biological Stimuli-sensitive Theranostic Systems

Among various chemical and biological stimuli, acidic environments, reducing conditions and enzymes have been widely utilized to design responsive theranostic systems for smart drug delivery and disease diagnosis.[13,14] Tumors are generally more acidic (pH 6.5) than normal tissues (pH 7.4) because of hypoxia, high cell turnover, cell death, improper removal of debris, and high rate of glycolysis, leading to the accumulation of lactic acid or other type of organic acids under both aerobic and anaerobic conditions.[15] Tumor cells have a higher redox potential than normal cells because of the higher intracellular concentration (2–10 mM) of glutathione tripeptide (GSH) compared to that of normal cells (0.2–1 mM).[16] Tumors also express more hydrolytic enzymes such as proteases, which can be exploited for stimuli-sensitive theranostic systems.[17]

12.2.1 pH-responsive Theranostic Systems

As a responsive factor, the pH value varies in different tissues or organs (Figure 12.1), such as the stomach and liver, and in some special disease states (ischemia, infection, inflammation, and tumorigenesis).[18] Due to the high rate of glycolysis in tumor cells, the pH level of tumors is lower than that of normal tissues. Specifically, it has been demonstrated that the tumor microenvironment is acidic, with pH values ranging from 5.7 to 6.5, while the pH of normal tissue is about 7.4.[19] An even greater pH difference can be found at the subcellular level, such as late endosomes and lysosomes with an acidic pH, in the range of 4.5–5.5.[8,9] This pH gradient has been exploited in smart theranostic systems, in which the nanoparticles are internalized through endocytosis and trapped within endosomal and lysosomal compartments.[20,21] The two common strategies to elicit pH-triggered responses are protonation and acid labile bond cleavage.[22,23]

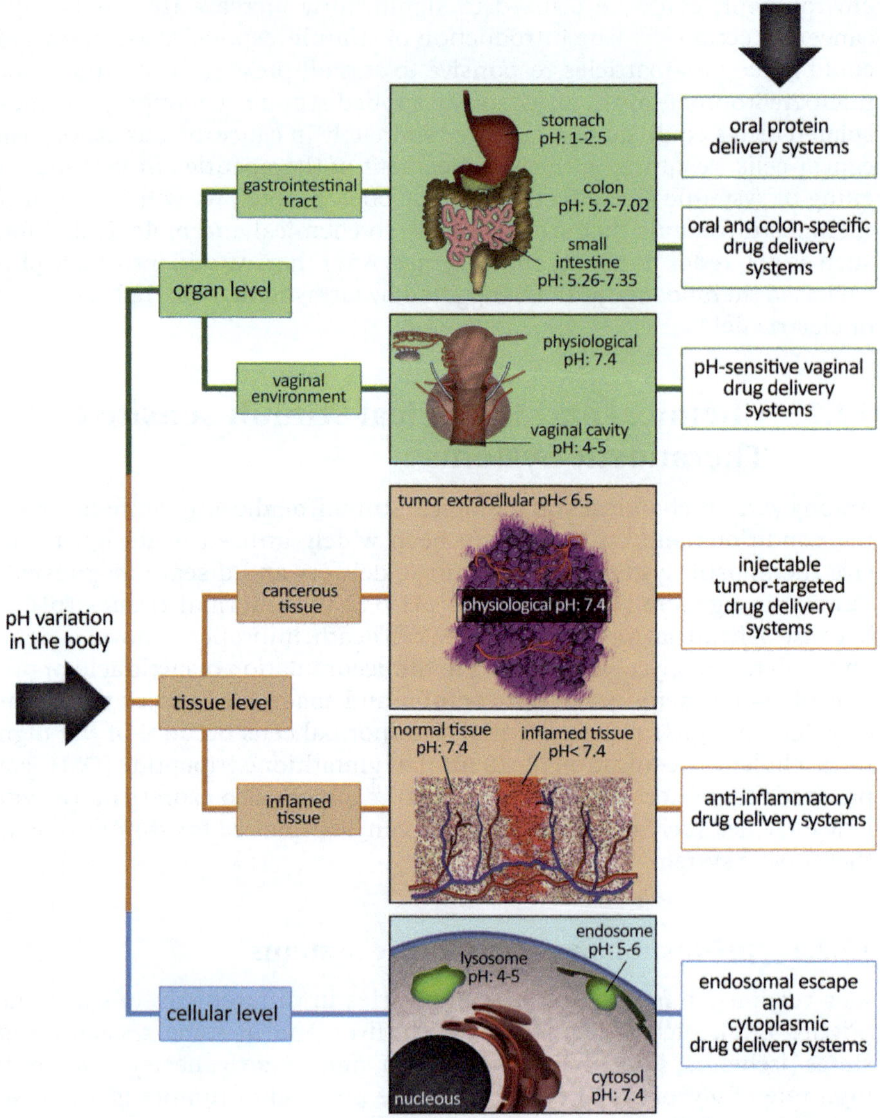

Figure 12.1 pH variation in the body, in the organism, tissue, and cellular levels. (Reproduced with permission from ref. 18, *Journal of Controlled Release*, **253**, S. Bazban-Shotorbani, M. M. Hasani-Sadrabadi, A. Karkhaneh, V. Serpooshan, K. I. Jacob, A. Moshaverinia, M. Mahmoudi, Revisiting structure-property relationship of pH-responsive polymersfor drug delivery applications, 46–63, Copyright 2017, with permission from Elsevier).[18]

12.2.1.1 pH Triggered Protonation

Protonation is one kind of commonly utilized mechanism to establish pH-triggered theranostic systems. In this strategy, the ionizable chemical groups of smart materials are utilized as responsive structure to construct the carriers. At physiological pH, these ionizable groups remain deprotonated or deionized. However, acidic pH allows protonation or charge reversal of the responsive materials, causing structural transformation or disassembly of the theranostic systems, and leads to specific drug release at desirable extracellular or intracellular sites.[24,25] Additionally, lower pH induced protonation brings forth a hydrophilic–hydrophobic phase transition for anionic and cationic polymers. The commonly used ionizable groups include amino, imidazolyl, sulfonates and carboxyl groups.[26]

Leveraging the protonation mechanism, the pH-responsive theranostic systems can release drugs and trigger a sensitive reporting signal under an acidic microenvironment.[27–29] For instance, one kind of intracellularly acid-switchable multifunctional micelle was designed by Wang *et al.*[29] for combination photo- and chemotherapy against drug-resistant tumors. This micelle displayed silent fluorescence, magnetic resonance imaging (MRI), and photodynamic therapy (PDT) activities in the blood, which could be dramatically activated upon cellular uptake and intracellular dissociation inside cancer cells. As shown in Figure 12.2, this micellar nanoplatform exhibited good potential for triple-modal tumor imaging (fluorescence imaging, MRI, and photoacoustic imaging (PAI)) and a combination of photo/chemotherapy against drug-resistant tumors.[29] In another system, Yu *et al.* reported a novel triple-layered pH-responsive micelle,[30] co-loaded with alkylated cisplatin prodrug and siRNA against NF-Kappa B, used for targeted treatment of metastatic breast cancer. This nanoparticle responded well to the weakly acidic microenvironment of early endosomes and was found to be taken up by orthotopic 4T1 breast tumor efficiently.

Anionic polymers displaying carboxylic groups are hydrophilic at neutral pH because they are deprotonated, but upon reaching an acidic environment, protonation occurs, resulting in an increase in hydrophobicity and precipitation of nanocarriers and therefore drug release.[31] Deng *et al.*[32] reported the development of a series of multifunctional PEG-b-PCL copolymers bearing an acid-labile β-carboxylic amides-modified PCL chain, with the ability of acid-triggered charge-reversal from negative to positive. They used these copolymers to construct novel micelles for doxorubicin (DOX) delivery and pH-responsive release. These pH-induced charge-reversal micelles were able to deliver the drugs into tumor cells efficiently, resulting in cell death.

12.2.1.2 Acid Labile Bond Cleavage

Acid labile bond cleavage is another common strategy to achieve tumor-responsive therapy and diagnosis, by a smart extracellular drug release or a sensitive endosomal escape mechanism.[33] This approach uses acid labile chemical bonds, either to covalently link drug molecules to the nanocarriers, or to construct a new nanostructure. These acid labile chemical bonds, shown

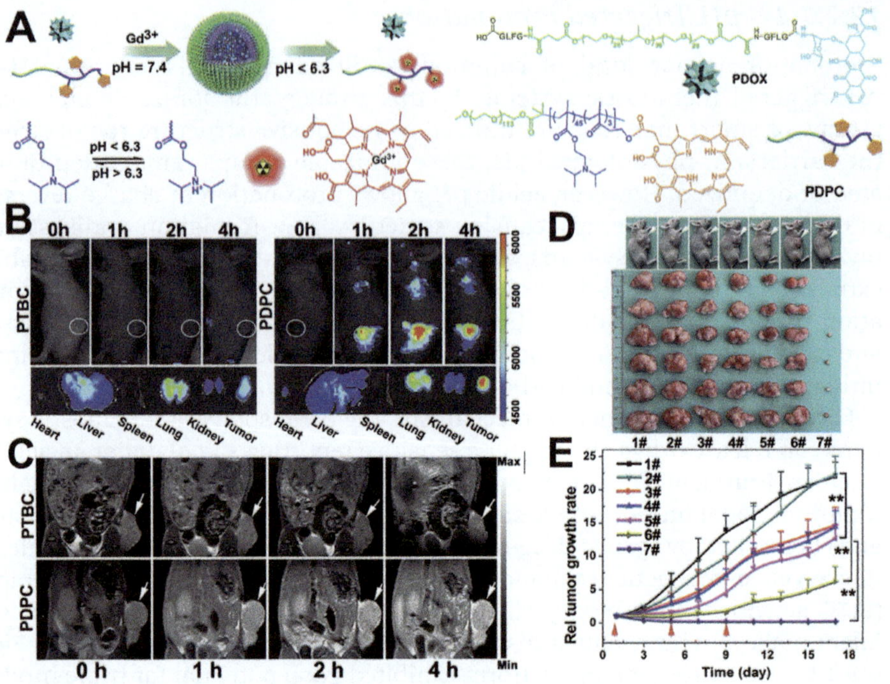

Figure 12.2 (A) Self-assembly and structural composition of the acid-switchable micelles. (B) Fluorescence images of MCF-7/ADR tumor-bearing mice post i.v. injection of the PDPC or PTBC micelles. (C) T1-weighted MR images of MCF-7/ADR tumor-bearing mice pi of the PDPC or PTBC micelles. (D) Photographs of MCF-7/ADR tumor-bearing nude mice and the tumors taken at the end of the antitumor studies; the white arrows indicate the tumor location. (E) Tumor growth inhibition profiles of PDOX/PDPC micelles in combination with NIR laser irradiation. (Reproduced with permission from T. T. Wang, D. G. Wang, H. J. Yu, M. W. Wang, J. P. Liu, B. Feng, F. Y. Zhou, Q. Yin, Z. W. Zhang, Y. Z. Huang and Y. P. Li, *ACS Nano*, 2016, **10**, 3496–3508. Copyright 2016 American Chemical Society.)[29]

in Table 12.1, are stable at neutral pH but are degraded or hydrolyzed under an acidic microenvironment.[34] This unique property makes these bonds promising candidates for the construction of pH-sensitive theranostic systems.[26]

Zhong *et al.*[35] reported the development of hyaluronic acid (HA)-shelled acid-activatable paclitaxel (PTX) prodrug micelles, using an hyaluronic acid-β-dendritic oligo-glycerol block copolymer as the backbone. The HA shell functions as native ligands against CD44 receptors present on breast cancer cells, promoting rapid internalization of the nanoparticles into CD44-positive tumor cells, followed by PTX release inside the acidic environment of the endosomes, resulting in a superior antitumor effect against CD44-positive human breast tumor xenografts.

The applications of acid cleavable hydrazine linkages have been exploited in various drug carriers,[36] including polymer-drug conjugates, dendrimers and dendritic polymer-drug conjugates, star shaped polymers, amphiphilic

Table 12.1 Examples of acid-labile chemical bonds and their degradation products.

Name	Acid-labile chemical bonds	Degradation product
Orthoester	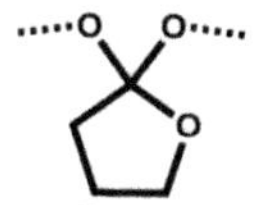	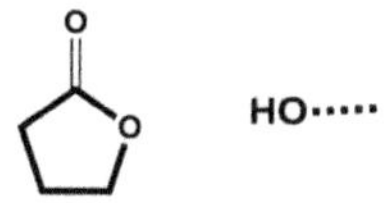
Ester	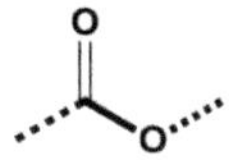	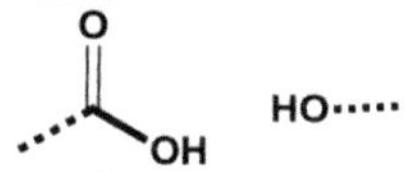
Hydrazone	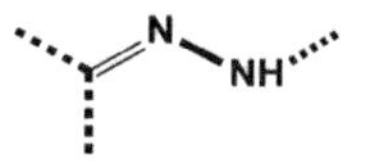	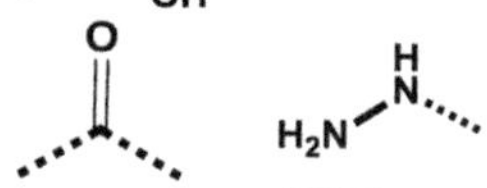
Imine	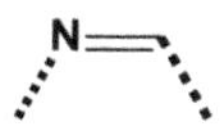	
Cis-aconityl	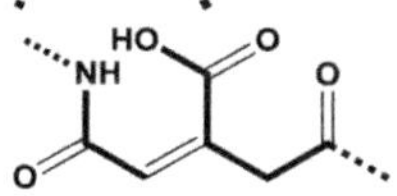	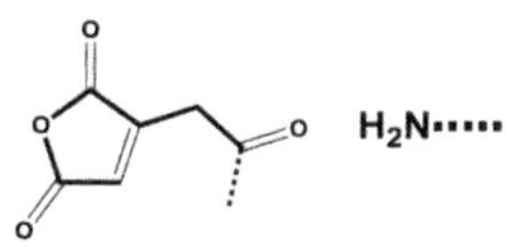
Acetal/ketal	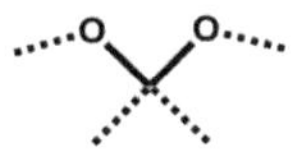	

polymer-drug conjugates, amphiphilic dendritic polymer-drug conjugates, liposomes and inorganic materials. Ye *et al.*[13] reported the development of an alendronate-monoethyl adipate-(hydrazone)-doxorubicin conjugate (ALN-MA-hyd-DOX), that could be used for site-specific delivery of DOX to bone tumor tissue. ALN-MA-hyd-DOX released DOX in a pH-dependent manner by the cleavable hydrazine linkages, causing more apoptosis and higher cytotoxicity against both wild-type tumor cells and DOX-resistant tumor cells. This bone-targeting conjugate holds promise for the treatment of bone neoplasms.

Hou *et al.*[37] described the development of a well-defined acid-sensitive *cis*-aconitic anhydride-modified doxorubicin (CAD), which could self-assemble together with the photosensitizer chlorin e6 (Ce6) to form TCAD@Ce6 nanoparticles (NPs), a novel near-infrared fluorescence imaging and chemo-photodynamic combination therapy theranostic system against cancer. The accelerated release of doxorubicin (DOX) and chlorin e6 (Ce6) from the TCAD@Ce6 NPs could be achieved through hydrolysis of the acid-sensitive amide linker under mild acidic conditions (pH = 5.5). The *in vitro* and *in vivo* experiments of TCAD@Ce6 NPs exhibited a significantly higher uptake of DOX and Ce6 in lung cancer.

We[1] reported a novel class of dual pH and diol-responsive boronate cross-linked micelles (BCM), formed by self-assembly of two versions of well-defined cholic-acid based telodendrimers, decorated with either boronic acid or catechol. The stability of these micelles was improved by pH- and diol-responsive crosslinking with boronate esters. This novel nanocarrier shows great promise for drug delivery with minimal premature drug release at a physiological glucose level and physiological pH (pH 7.4) in blood circulation, but can be activated to release the drug on demand, under an acidic tumor microenvironment or in the acidic sub-cellular compartments upon uptake by target tumor cells, or *via* administration of mannitol, on demand (Figure 12.3).

12.2.2 Reduction-responsive Theranostic Systems

Redox-sensitive theranostic systems usually contain disulfide bonds, which can be rapidly cleaved by high levels of GSH (~2–10 mM) inside the cell, and in tumor tissues.[26] The redox-sensitive system has been exploited in capped mesoporous materials,[38] dendrimer,[39] liposomes[40] and other nanocarriers. The oxidation responsiveness has also been explored for triggered drug release in some inflammatory tissues, with high level of reactive oxygen species.

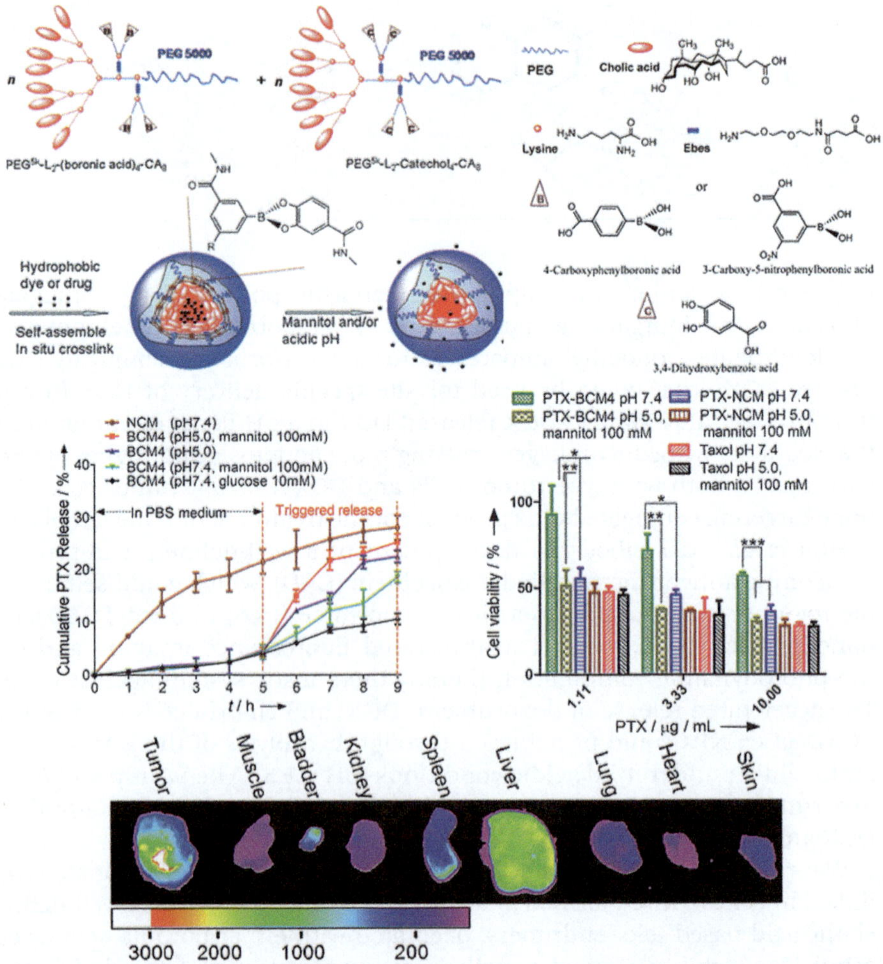

Figure 12.3 Schematic representation of BCM and its applications *in vivo*.

Luo *et al.*[41] described two novel redox sensitive PTX-oleic acid conjugates (PTX-S-OA and PTX-2S-OA), prepared by ligating OA to PTX through a single thioether bond or dithioether bond, respectively. PTX-S-OA was found to be more redox sensitive than PTX-2S-OA. Micelles assembled from PEG-PTX-S-OA exhibited high drug loading (57.4%) and potent antitumor activity against a human epidermoid carcinoma xenograft. Combination of PEGyated versions of PTX-S-OA and PTX-2S-OA is expected to yield a dual-redox sensitive nanotaxane with differential release of PTX.

We reported[42] a robust, smart and highly versatile 'all-in-one' porphyrin and cholic acid-based amphiphilic telodendrimer nanoconstruct by integrating a variety of imaging, and therapeutic functions with redox-sensitive drug release properties. The theranostic system can be reversibly stabilized by disulfide bonds and be used as amplifiable multimodality nanoprobes for near-infrared fluorescence imaging (NIRFI), MRI, positron emission tomography (PET), and photothermal therapy (PTT) and PDT (Figure 12.4).

12.2.3 Enzyme-responsive Theranostic Systems

The altered expression profile of specific hydrolytic enzymes[26] (such as proteases, phospoholipases or glycosidases) observed in pathological conditions (such as cancer or inflammation), has been exploited to achieve enzyme-mediated drug release from nanocarriers at the tumor sites or pathological tissues.

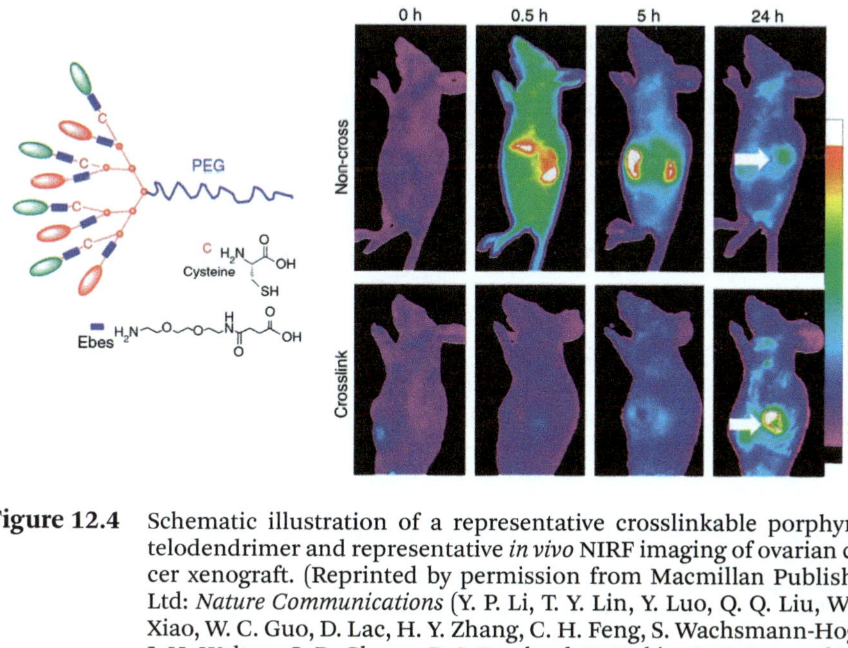

Figure 12.4 Schematic illustration of a representative crosslinkable porphyrin–telodendrimer and representative *in vivo* NIRF imaging of ovarian cancer xenograft. (Reprinted by permission from Macmillan Publishers Ltd: *Nature Communications* (Y. P. Li, T. Y. Lin, Y. Luo, Q. Q. Liu, W. W. Xiao, W. C. Guo, D. Lac, H. Y. Zhang, C. H. Feng, S. Wachsmann-Hogiu, J. H. Walton, S. R. Cherry, D. J. Rowland, D. Kukis, C. X. Pan and K. S. Lam, A smart and versatile theranostic nanomedicine platform based on nanoporphyrin, *Nat. Commun.*, 2014, **5**). Copyright 2014).[42]

Most enzyme-responsive DDSs utilize enzyme-mediated drug delivery in the extracellular environment. Van Rijt *et al.*[43] reported the development of avidin-capped multifunctional mesoporous silica nanoparticles (MSNs), functionalized with heptapeptide linkers that could be specifically cleaved by MMP9. The resulting MSNs enabled the controlled release of chemotherapeutics in tumor areas with high expression levels of MMP9, as demonstrated in a lung cancer xenograft model.

Hu *et al.*[44] reported a "transformable" core–shell based nanocarrier (designated CS-NG) system, in which the components can be enzymatically assembled into microsized extracellular depots at the tumor site by hyaluronidase, an overexpressed enzyme at the tumor microenvironment (Figure 12.5). The enhanced cytotoxicity against MDA-MB-231 cells and improved antitumor efficacy with CS-NG could be attributed to the inhibition of cellular internalization and prolonged retention of the drug at the tumor sites in a *in vivo* xenograft model.

12.2.4 Multi-responsive Theranostic Systems

Multi-responsive theranostic systems are capable of prolonging the circulation time of the drug in the blood, increasing the accumulation of the drug in tumor tissue, and promoting the drug release in response to multiple stimuli. Tang *et al.* reported the development of a tumor microenvironment-adaptive nanoparticle co-loaded with PTX and the anti-metastasis siRNA targeting twist.[45] The nanoplatform comprised of a pH-sensitive core, a cationic shell, and a matrix metalloproteinase (MMP)-cleavable polyethylene glycol (PEG) corona conjugated by a peptide linker. This pH and MMP dual-responsive system led to more efficient cellular uptake in tumor cells, increased intra-tumor accumulation of both PTX and siRNA, and significantly inhibited tumor growth and pulmonary metastasis in the breast cancer xenograft model.

12.3 Physically Stimuli-responsive Theranostic Systems

In this section, we will discuss theranostic materials that take advantage of physically applied stimuli to achieve diagnosis and treatment of disease. These stimuli include ultrasound, temperature, light, magnetic field, and electric fields.

12.3.1 Ultrasound-triggered Theranostic Systems

Ultrasound (US) represents an effective, unique and exciting theranostic modality for tracking drug carriers, attaining spatiotemporal control of drug release at the desired site and improving drug deposition.[5,46–49] In addition, US modalities are appealing because of their non-invasiveness, the absence of ionizing radiations, and the facile regulation of tissue penetration depth[50]

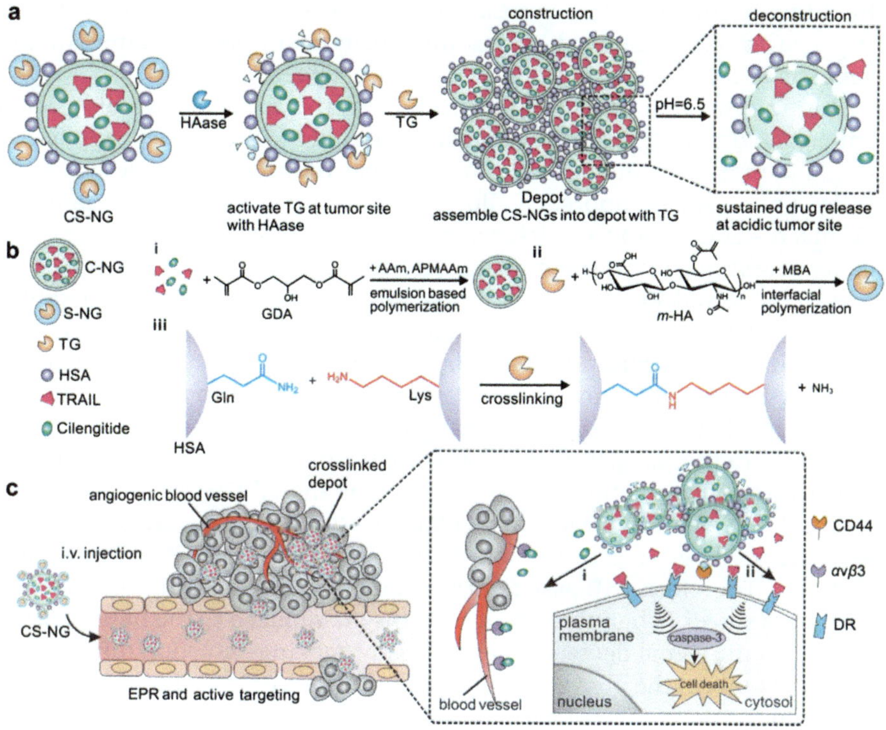

Figure 12.5 Schematic of tumor microenvironment-mediated construction of combination drug-delivery depots for sustained drug release using CSNG. (a) A typical CS-NG, consisting of core-NG (C-NG) loaded with therapeutics, shell-NG (S-NG) with encapsulated transglutaminase (TG), and human serum albumin (HSA), coated on the surface of C-NG. The polymeric matrix of C-NG is acid-degradable, and the polymeric matrix of SNG is made from hyaluronic acid (HA) that can be degraded by HAase. At the tumor site, the overexpressed HAase digests the HA and releases TG, which catalyzes the cross-linking of HSA and assembles CS-NG into the "drug-delivery depots". The matrix of C-NG can be gradually degraded and subsequently release the encapsulated therapeutics at the acidic tumor microenvironment. (b) Construction of C-NG, S-NG and cross-linked depots. (i) TRAIL and cilengitide are loaded into C-NG using emulsion based polymerization method. (ii) TG is encapsulated into S-NG using interfacial polymerization method. (iii) TG catalyzes the cross-linking of Gln and Lys on HSA to form depots at the tumor site. (c) Co-delivery of TRAIL and cilengitide by CS-NG for combination cancer treatment through *i.v.* injection. CS-NGs can accumulate in the tumor site *via* the EPR effect and active targeting effect of HA. Both therapeutics are released from the cross-linked CS-NGs depots. (i) Cilengitide targets the integrins of endothelial cells and inhibits angiogenesis. (ii) TRAIL targets the death receptors (DRs) on the cell membrane and triggers the apoptosis. (Reproduced with permission from Q. Y. Hu, W. J. Sun, Y. Lu, H. N. Bomba, Y. Q. Ye, T. Y. Jiang, A. J. Isaacson and Z. Gu, *Nano Lett.*, 2016, **16**, 1118–1126. Copyright 2016 American Chemical Society.)[44]

by tuning frequency, duty cycles, acoustic pressure and exposure time.[51,52] Physical effects elicited by ultrasound include simple pressure variation, acoustic fluid streaming, cavitation and local hyperthermia.

Highly echogenic microbubbles (MBs) in the form of gas-filled microspheres, administered intravenously, have been extensively applied for fundamental scientific research and clinical US imaging for decades. This unique non-invasive approach allows dynamic detection of tissue perfusion by macro- and micro-vasculature.[53,54] Although MBs have achieved significant success in clinical US imaging, they still facing some critical issues that need to be solved. First, the large size of MBs limits the imaging to blood-flow and vasculature. Second, the low stability of commercial lipid MBs makes them difficult for continuous US imaging and long-duration observations. Third, MB-assisted US imaging is typically non-specific, making it difficult to distinguish between benign and malignant lesions such as tumors. Last but not least, traditional MBs display only single functionality, making it difficult for integration with other diagnostic imaging modalities for precision diagnosis. When circulating within the blood vessels, the large size of traditional MBs are too large to escape from the vasculature and penetrate into tumor tissues.[55] Therefore, MB assisted US imaging or therapy only occurs within the blood vessel. When the size of the MBs are reduced to the nanoscale, the US echogenic signal decreases substantially, making detection difficult.[54]

To circumvent the above-mentioned limitations of MBs, Emelianov *et al.*[56] reported the development of a "small-to-big" strategy based on phase-transformable small NPs with intelligent responsiveness for US-based theranostics. These small-sized NPs efficiently escape from the vascular circulation and penetrate into tumor tissue. Upon external triggering, the small NPs transform *in situ* into larger microparticles/MBs to exert US-based theranostic functions. Using an opposite approach, Zheng *et al.*[57] reported the design of large-particulate MBs that can also be transformed *in situ* into small NPs by using various intrinsic or external triggers ("big-to-small" strategy). The *in vivo* generation of NPs from administered MBs can penetrate tumors *via* enhanced permeability and retention (EPR) effects and therefore can enhance theranostic applications.[54]

Several strategies have been developed to achieve the "small-to-big" transformation for US-based cancer theranostics.[58,59] Acoustic droplet vaporization (ADV) is a typical strategy to induce phase transformation of nanodroplets to form MBs for extravascular US imaging. The concept of ADV refers to the phase transformation from liquid droplets to gas bubbles by US irradiation. MBs could also be generated from some gas-generating polymer NPs. Wang *et al.*[60] reported an optical droplet vaporization (ODV) process for near infrared (NIR)-triggered vaporization of a perfluorocarbon (PFC) liquid droplet. ODV can be achieved by combing two components with distinct performances, for example, NIR absorbance (Fe_3O_4 NPs) and a phase shifting droplet (perfluorohexane, PFH, boiling point: 56 °C). Fe_3O_4 NPs (embedded in the shell) and perfluorohexane (PFH) (within the core) were concurrently encapsulated into poly(lactic-coglycolic acid) (PLGA) nanocapsules (designated as PFH@PLGA/Fe_3O_4) *via* a modified emulsion-evaporation strategy

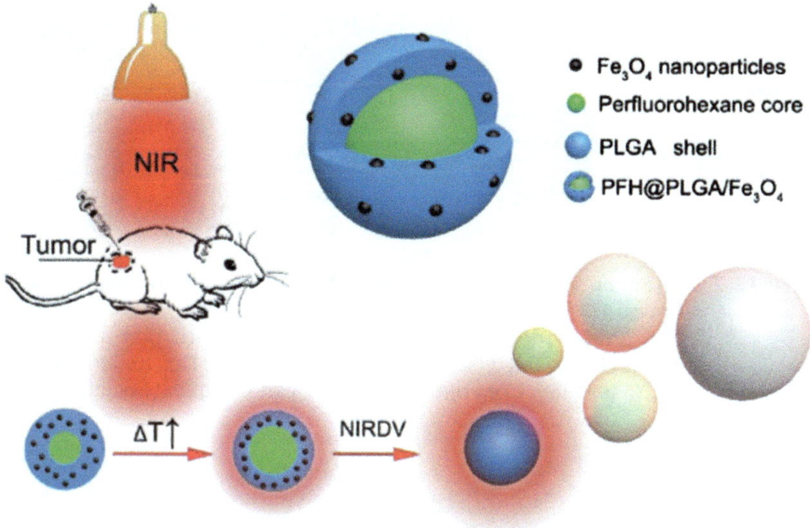

Figure 12.6 Schematic of the Structure of PFH@PLGA/Fe3O4 nanocapsules and a schematic of the phase-transition process by the method of near-infrared radiation droplet vaporization (NIRDV). Reproduced with permission from Y. Zhao, W. Song, D. Wang, H. Ran, R. Wang, Y. Yao, Z. Wang, Y. Zheng and P. Li, *ACS Applied Materials & Interfaces*, 2015, 7, 14231–14242. Copyright 2015 American Chemical Society.

(Figure 12.6).[61] Typically, the droplet vaporization is triggered by the thermal effect of the outer stimulus to increase the temperature exceeding the boiling point of the droplet. It has been well demonstrated that magnetic NPs could effectively respond to an external alternative current (a.c.) magnetic field to elicit the thermal effect.[62] In addition, radiofrequency (RF) ablation has been widely used for minimally invasive and localized tumor hyperthermia,[63] which is also considered as an irradiation source to trigger the phase transformation of droplets to generate MBs within tumor tissues. For instance, solid DL-menthol (DLM, melting point: 32–36 °C) was encapsulated into PLGA nanocapsules. After exposure to RF, the encapsulated DLM exhibited a continuous solid–liquid–gas (SLG) triphase-transformation behavior, yielding echogenic MBs (Figure 12.7a). Based on this observation, a new RF droplet vaporization (RDV) strategy for contrast-enhanced US imaging (Figure 12.7b and c) and continuous cavitation for cancer therapy was established.

To solve the critical issue of the large particle size of MBs, Huynh *et al.*[57] developed the "big-to-small" strategy, in which MBs comprised of a bacteriochlorophyII-lipid shell and perfluoropropane gas core (1–10 μm) were converted into small NPs (5–500 nm) upon exposure to US irradiation (Figure 12.8). Their research demonstrates that the "big-to-small" strategy can accumulate and retain the delivery nanosystems within tumor tissue, and exert specific theranostic functions for subsequent cancer treatment. However, this strategy is not beneficial for intra-tumor US imaging because the post-formed nanobubbles are too small to generate enough contrast signals. However, it does have

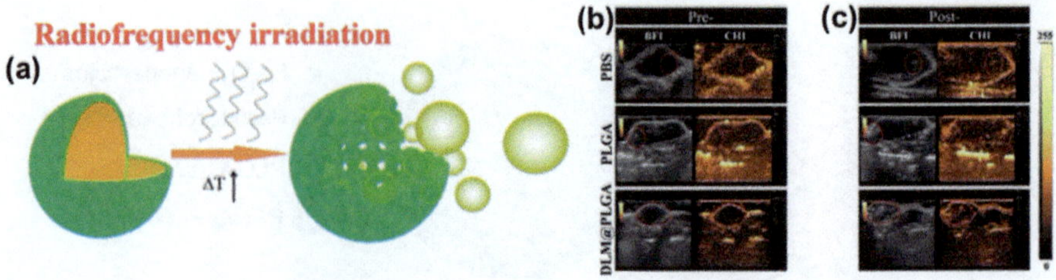

Figure 12.7 (a) Schematic of the RSV strategy for MBs generation by RF irradiation. *In vivo* US imaging (b) before and (c) after RSV process post receiving PBS, PLGA and DLM@PLGA. Reproduced with permission from K. Zhang, P. Li, H. R. Chen, X. W. Bo, X. L. Li and H. X. Xu, *ACS Nano*, 2016, **10**, 2549–2558. Copyright 2016 American Chemical Society.

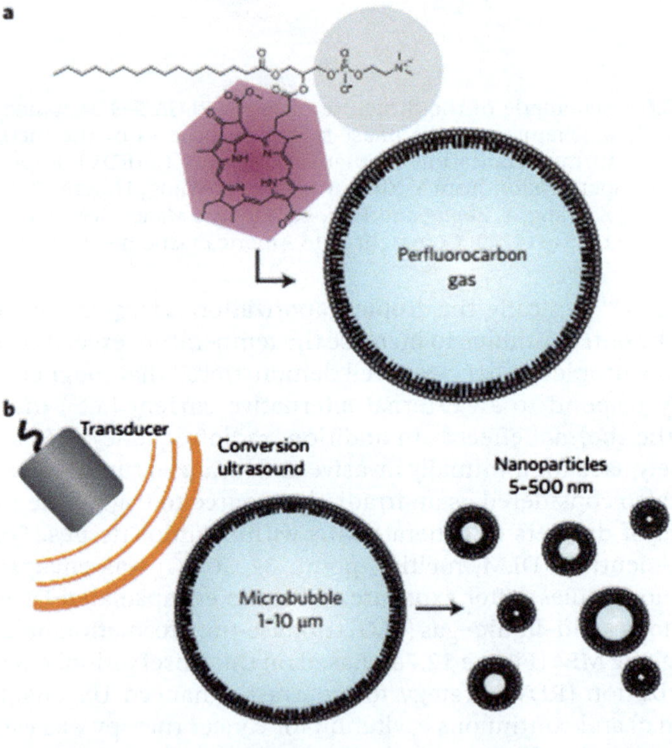

Figure 12.8 "Big-to-small" strategy for tumor imaging. Schematic of porphyrin microbubbles (pMBs) and their micro-to-nano conversion. (a) pMBs consist of a BChl-lipid shell encapsulating perfluorocarbon gas. (b) Conversion of pMBs to porphyrin nanoparticles (pNPs) *via* sonication with low-frequency, high-duty-cycle ultrasound (conversion ultrasound). Reprinted by permission from Springer Nature, *Nature Nanotechnology* (*In situ* conversion of porphyrin microbubbles to nanoparticles for multimodality imaging, E. Huynh, B. Y. C. Leung, B. L. Helfield, M. Shakiba, J. Gandier *et al.*,) Copyright 2015.

potential for use as a US-triggered intra-tumor drug delivery system and for photoacoustic imaging (PAI), as the intravascularly generated nanobubbles could enter the tumor tissue and be retained within over a long duration.

Compressible objects contract and expand as they experience the compression and rarefaction cycles of passing acoustic waves. These volumetric oscillations can facilitate drug release, increase drug uptake, and elicit strong backscattered echoes that can be used for imaging. The cavitation threshold can be easily achieved when low ultrasound frequencies (in the kHz range) are used. The bubbles are formed through acoustic droplet vaporization and are subjected to cavitation, thus promoting cellular uptake and/or release of the entrapped drugs in the tumor site. This has resulted in significant therapeutic efficacy and suppression of metastatic dissemination. To improve the tumor targeting specificity of these PFC nanodroplets, Gianneschi *et al.* functionalized them with aptamers.[64] Low-frequency ultrasounds have also been used to promote the delivery of drugs through the skin. For example, high penetration of siRNA-loaded liposomes has led to significant inhibition of the progression of melanocytic lesions.[65] Bubble liposomes increased the transfection efficiency of pDNA-loaded liposomes, probably also by enhancing ultrasound-mediated endosomal escape. Compared with other approaches, ultrasound-triggered drug delivery allows the drug to be released into the cytosol as a result of pore formation in the cell membrane, thus bypassing the degradative endocytotic pathway. This is especially useful in DNA/RNA delivery.

Ultrasonic energy is dissipated to thermal energy as the acoustic waves propagate through tissue and generate friction.[66] Focused ultrasound can lead to localized tissue heating, which may be monitored and guided by magnetic resonance imaging (MRI) and other thermometric imaging techniques. Tissue hyperthermia can be used to release drugs from thermo-sensitive carriers designed to destabilize above specific temperatures. Combination thermosensitive drug carriers and high-intensity focused ultrasound (HIFU) enables triggered drug release, with only a mild increase in temperature.[67] Another application of focused US (FUS) is to open the blood–brain barrier (BBB) non-invasively, by creating transient pores in BBB that will close on their own. This therapeutic modality can be applied to the treatment of brain-specific diseases. In these applications, MBs are employed to facilitate the opening of BBB by providing the acoustic emission to the endothelium and subsequently triggering the temporal cell deformation to generate temporal vascular-parenchyma-enhanced permeability of the central nervous system. However, the violent cavitation effect of FUS could potentially damage brain endothelium or parenchyma. Comparatively, bubbles with sub-micrometer or nanometer sizes are difficult to induce enough acoustic cavitation effect to disrupt the BBB because of the low US scattering efficiency. To overcome these limitations, Huang *et al.* have designed a new magnetically guidable system comprised of submicrometer bubbles or nanobubbles (MNBs).[68] By using polystyrene (PS) NPs as the hard template, hollow silica NPs with embedded magnetic Fe_3O_4 NPs within the shell were synthesized with the assistance of octyltriethoxysilane (OTES), tetraethyl orthosilicate (TEOS) and (3-aminopropyl)triethoxysilane (APTES), using an established surface-engineering method (Figure 12.14d). These elaborately designed perfluoropentane-encapsulated MNBs have been shown to efficiently induce the disruption of the BBB under magnetic guidance to guarantee sufficiently high acoustic emissions. This research provides a new strategy to achieve BBB disruption by adopting sub-micrometer/nanometer particles under magnetic guidance. When applied concurrently with administration of nanotherapeutics, such as FUS, the approach could potentially enhance nanodelivery to central nervous systems through transient opening of the BBB.

12.3.2 Thermoresponsive Theranostic Systems

Thermoresponsive theranostics are among the most investigated stimuli-responsive strategies, and have been widely explored in oncology. Thermoresponsiveness is usually governed by a nonlinear sharp change in the properties of at least one component of the nanocarrier material with temperature. Such a sharp response triggers the release of the drug following a variation in the surrounding temperature. Ideally, thermosensitive nanocarriers should retain their load at body temperature (~37 °C), and rapidly deliver the drug within a locally heated tumor (~40–42 °C) to counteract rapid blood-passage time and washout from the tumor.

Thermoresponsive systems are generally liposomes, polymer micelles or nanoparticles (usually poly(*N*-isopropyl acrylamide) (PNIPAM)) exhibiting a lower critical solution temperature. For liposomes, thermoresponsiveness usually arises from a phase transition of the constituent lipids and the associated conformational variations in the lipid bilayers. *In vivo*, heat is generally applied by using temperature-controlled water sacks, radiofrequency oscillators or miniature annular-phased array microwave applicators.[69–73] In the past few years, the focus has been on rapid and quantitative drug-release performance. Thermosensitive liposomes (TSLs) are perhaps the most advanced thermoresponsive nanosystems, as shown by their use in several clinical trials. Doxorubicin-loaded TSLs (TermoDox, Celsion Corporation), in association with hyperthermia or radiofrequency ablation, are at present being investigated in phase II trials for the treatment of breast cancer and colorectal cancer liver metastasis, and have reached phase III trials for the treatment of hepatocellular carcinoma. More recently, improved liposomal formulations have been shown to release their loads shortly after the onset of hyperthermia (~40–45 °C).[74] An alternative approach uses leucine zipper peptide–liposome hybrids, which combine the advantages of traditional TSLs with the dissociative, unfolding properties of a temperature-sensitive peptide (Figure 12.9a).[75] Also promising are thermoresponsive bubble generating liposomal systems. These rely on the creation of permeable defects in the lipid bilayer by means of the generation of carbon dioxide bubbles through decomposition of ammonium bicarbonate at mild hyperthermia (~42 °C). The hyperechogenic carbon dioxide bubbles generated by such

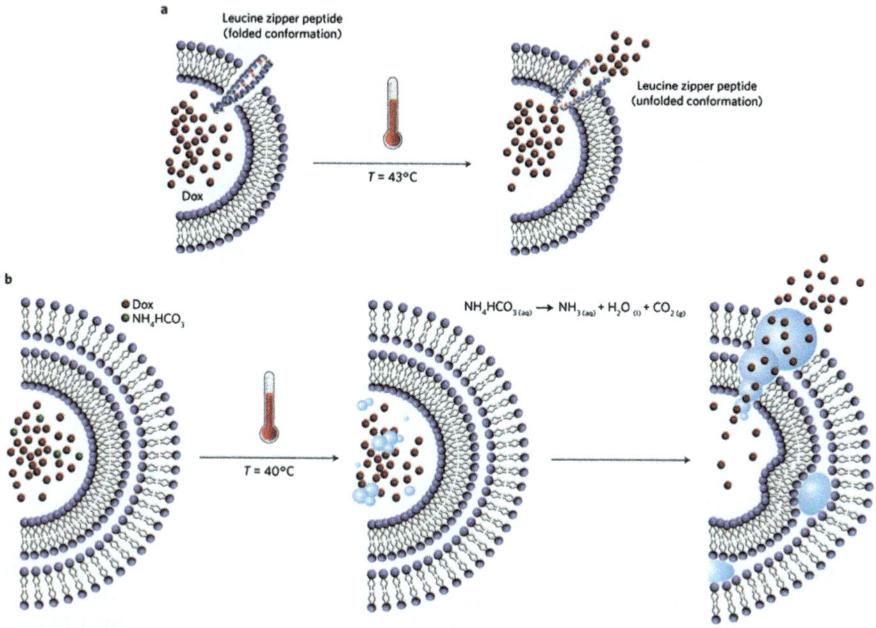

Figure 12.9 Temperature-based actuation mechanisms for liposomal drug delivery. (a) The temperature-triggered unfolding of a leucine zipper peptide inserted in the membrane of a doxorubicin (Dox)-carrying liposome opens a channel through which the drug is released. (b) Drug-permeable pores can also be created by the temperature-triggered generation of bubbles from the decomposition of encapsulated ammonium bicarbonate. Reprinted by permission of Springer Nature, *Nature Materials*, (Stimuli-responsive nanocarriers for drug delivery, S. Mura, J. Nicolas and P. Couvreur), Copyright 2013.

liposomal vesicles, not only facilitate rapid release of doxorubicin (Figure 12.9b)[76] but also improved ultrasound imaging of tissues. Thermosensitive liposomes can also be functionalized with ligands for specific targeting, such as the human epidermal growth factor receptor 2 affibody against breast-cancer.

Local hyperthermia has also been used as a stimulus for the on-off control of the activity of cell-penetrating peptides (CPP). In this case, the temperature-triggered assembly of diblock-copolymer elastin-like polypeptides allowed arginine residues to be displayed at the periphery of the resulting micelles. This resulted in a greater than eight-fold increase in HeLa-cell uptake (Figure 12.10).[77]

In general, the challenge in the design of thermoresponsive nanodevices lies in the use of materials that are both safe and sensitive enough to respond to slight temperature changes around the physiological temperature of 37 °C. Liposomal systems are at present the most advanced and thus have the highest potential for clinical applications.

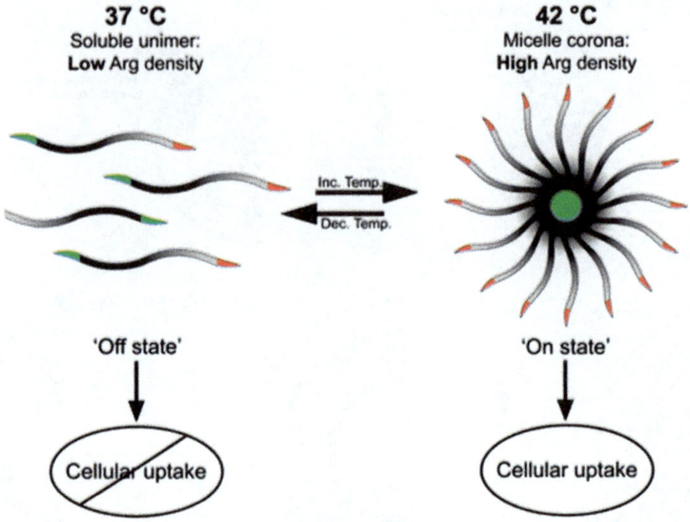

Figure 12.10 'Off state': at 37 °C, below the CMT, Arg5-ELPBCs exist as soluble unimers, whose limited number of Arg residues is below the threshold necessary for efficient cellular uptake. 'On state': above the CMT at 42 °C (a temperature that can be achieved with mild clinical hyperthermia) Arg5-ELPBCs assemble into spherical micelles. The high density of Arg decorating the micelle corona exceeds the threshold of Arg residues necessary for internalization, leading to cellular uptake of the micelles. Reproduced with permission from S. R. Macewan, A. Chilkoti, *Nano Lett*, 2012, **12**, 3322–3328. Copyright 2012 American Chemical Society.

12.3.3 Magnetic-responsive Theranostic Systems

Magnetic theranostic guidance is typically obtained by focusing an extracorporeal magnetic field on the biological target during the injection of a magnetically responsive nanocarrier. The advantage of using a magnetic field relies on the different nature that the magnetic response can take, which can be a magnetic guidance under a permanent magnetic field, a temperature increase when an alternating magnetic field is applied, or both when alternately used. Therefore, magnetically responsive theranostics systems allow for diversity in the imaging and drug delivery pathway. For example, Xie *et al.* successfully developed an effective magnetic nanocrystal (MNCs)-mediated theranostics strategy by using MRI and AMF in hyperthermic therapy. The PEGylated Mn–Zn ferrite nanocrystals showed excellent performance in the magnetism, relaxivity coefficient, and specific absorption rate (SAR) values (Figure 12.11).[78] Real time MRI monitoring by AMF indicated effective suppression of tumor growth. Moreover, the MRI monitoring guided smart magnetic DDSs can realize accurate long-lasting tumor hyperthermia. Furthermore, there is the possibility of performing magnetic resonance imaging, and hence to associate diagnostics and therapy within a single system[79,80] (the so-called theranostic approach).

This concept has demonstrated great potential in experimental cancer therapy because of improved drug accumulation inside solid-tumor models.

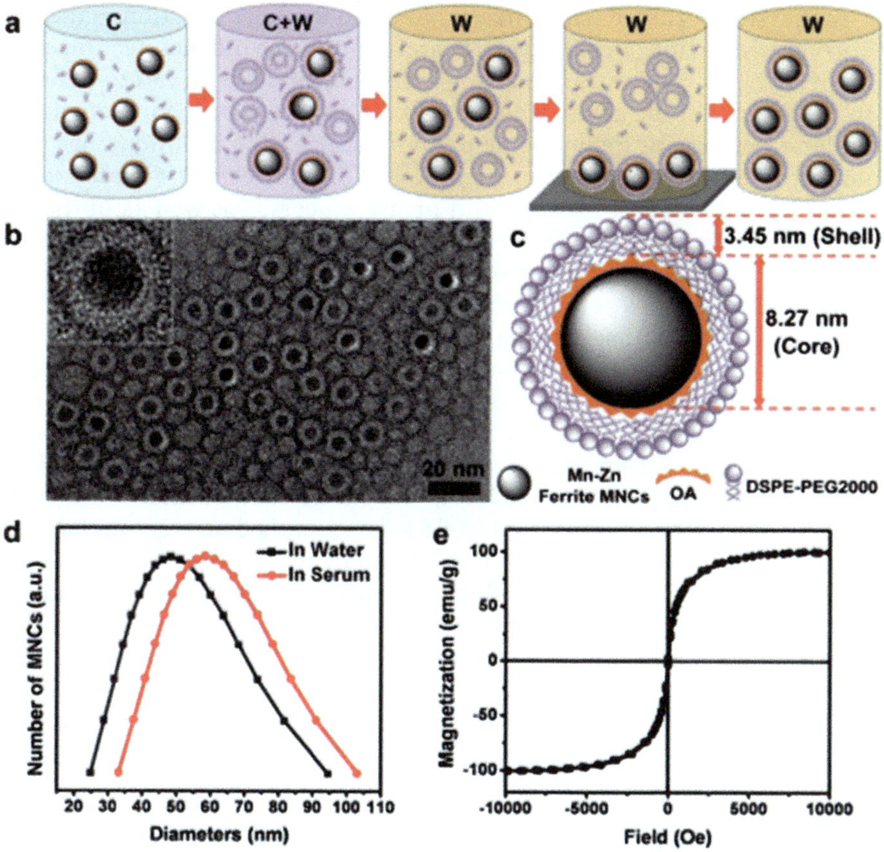

Figure 12.11 (a) A schematic diagram of PEGylated MneZn ferrite MNCs synthesis (C: chloroform; W: water). (b) TEM and high-resolution TEM (HRTEM) images of PEGylated MNCs with core shell structure negatively stained with 2% phosphotungstic acid to give a white layer surrounding the magnetic cores indicating the DSPE-PEG2000 coating layer. (c) A schematic diagram of PEGylated MNCs with 8.27 nm magnetic core and 3.45 nm lipid coating. (d) DLS of PEGylated MNCs in water and serum. (e) Hysteresis loops of PEGylated MNCs. Reproduced from *Biomaterials*, **35**, J. Xie, Y. Zhang, C. Yan, L. Song, S. Wen, F. Zang, G. Chen, Q. Ding, C. Yan and N. Gu, High-performance PEGylated Mn–Zn ferrite nanocrystals as a passive-targeted agent for magnetically induced cancer theranostics, 9126–9136, Copyright 2014, with permission from Elsevier.

Candidate nanosystems for such a therapeutic approach are core–shell nanoparticles (a magnetic core made of magnetite (Fe_3O_4) coated with silica or polymer),[81–85] magnetoliposomes (Fe_3O_4 or maghemite (Fe_2O_3) nanocrystals encapsulated in liposomes) and porous metallic nanocapsules. Most core–shell nanoparticles have shown promising results *in vitro*, yet only some of them have demonstrated improved tumor accumulation and anticancer pharmacological efficacy in various *in vivo* models.

Magnetically guided nanocarriers have also found application in the delivery of nucleic acids, including siRNA and genes. Such magnetofection experiments are generally performed using nanoassemblies with cationic coatings to condense nucleic acids, which results in higher transfection efficiencies under a permanent magnetic field. For instance, this technique has led to improved effectiveness in the transfection of siRNA *in vitro* and/or *in vivo* when directed against prostate and breast cancers,[86] as well as in the gene transfer to oligodendrocyte precursors for neural repair. For DNA vaccine delivery, two different strategies have been successful: the complexation of plasmid DNA (pDNA) at the surface of PEI coated Fe_3O_4 nanoparticles,[87] and the use of bacterial magnetic particles as carriers of a recombinant DNA.[88,89] Magnetic vaccines have also been demonstrated in the form of magnetic-nanoparticle formulations associated to a replication-defective adenovirus. Cell based gene therapies employing human monocytes-cells that naturally migrate from the bloodstream into tumors that were transfected with therapeutic genes, loaded with magnetic nanoparticles and placed under the influence of a magnetic field led to a marked increase in the extravasation of monocytes into the tumor. Magnetic nanoparticles have also been used for the delivery of antioxidant enzymes (catalase and superoxide dismutase, for example) with the aim of increasing resistance to oxidative stress *in vitro* (Figure 12.12).[90,91]

The use of magnetically responsive theranostics nanoparticles, either for magnetic guidance or local hyperthermia, is generally limited to accessible tumor nodules, but not metastasis or disseminated tumors. Even if most of these tumors are accessible for surgical resection, some are not surgically resectable because they are too hemorrhagic or localized in tissues with a high risk of healthy-tissue injury (as is the case for some brain cancers). In these situations, magnetically responsive nanoparticles represent a promising therapeutic option. However, magnetic guidance is hampered by the complexity involved in the setup of external magnetic fields, which need adequate focusing and deep penetration into the tissues to reach the diseased area with sufficient strength. In this respect, efforts to identify the best magnetic and irradiation technologies are needed.

12.3.4 Light-responsive Theranostic Systems

Owing to their non-invasiveness and the possibility of remote spatiotemporal control, a large variety of photoresponsive theranostics systems have been engineered in the past few years to achieve on-demand imaging and drug release in response to illumination with a specific wavelength (in the ultraviolet, visible or NIR regions).[92–94] For instance, the ultraviolet-visible reversible photoisomerization of the azobenzene group (and its derivatives), from *trans* to *cis* on irradiation at 300–380 nm and from *cis* to *trans* by shining light in the visible region, enables photo-regulated control of the drug release. Ultraviolet light can also activate reversible shrinkage in spiropyran-PEGylated lipid nanoparticles, thus enabling deeper tissue penetration. The

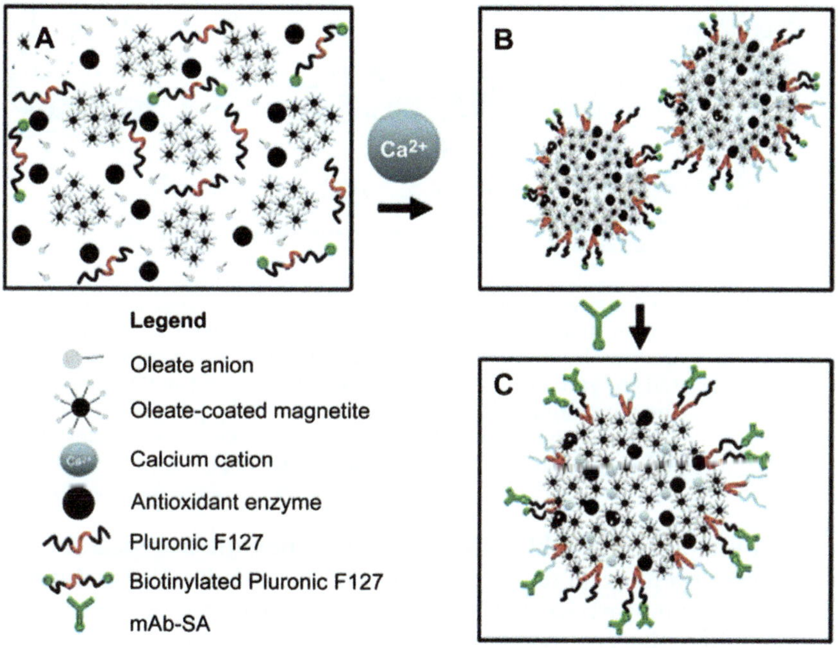

Figure 12.12 Endothelial targeted antioxidant nanoparticles formation scheme by controlled precipitation. Legend symbols signify as follows: grey dotted line represents oleate anion, and beneath that it surrounds a small black sphere indicating the oleate coated magnetite. Calcium cations that bind to free and magnetite-bound oleate anions are shown as blue spheres, and antioxidant enzymes (AOE; either catalase or SOD) are large black spheres. The amphiphilic block copolymer Pluronic F127 is indicated by a blue (for hydrophilic polyoxyethylene (POE) moieties) and red (for hydrophobic polyoxypropylene (PPE)) line and then below that the dual biotinylation of the molecule is indicated with green spheres on terminal POE groups. The green symbol labeled mAb-SA indicates monoclonal antibody functionalized with streptavidin. Throughout the text "Ab" represents either human or mouse anti-PECAM; Ab62 or Mec13.3 respectively. (A) Combination of aqueous suspension of oleate coated magnetite, AOE and Pluronic F127þ/biotinylation with CaCl2 drives the controlled precipitation and forms 300e400 nm PACs (B) with surface exposed biotinylated PEO chains. Streptavidin modified Ab/IgG are added, binding to biotinylated PACs (C). Unbound fractions are separated by magnetic decantation. (For interpretation of the references to color in this figure legend, the reader is referred to the web version of this article.) Reproduced from *Biomaterials*, **35**, E. D. Hood, M. Chorny, C. F. Greineder, I. S. Alferiev, R. J. Levy, V. R. Muzykantov, Endothelial targeting of nanocarriers loaded with antioxidant enzymes for protection against vascular oxidative stress and inflammation, 3708–3715, Copyright 2014, with permission from Elsevier.

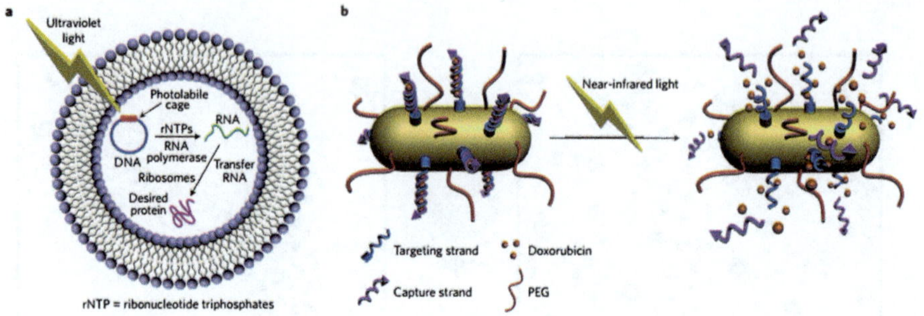

Figure 12.13 (a) Schematic representation of an encapsulated *in vitro* transcription–translation liposomal system triggered by irradiating caged DNA with light. (b) Delivery of doxorubicin through the near-infrared-triggered induction of dehybridization of the DNA conjugated at the surface of gold nanorods. Reprinted by permission of Springer Nature, *Nature Materials*, (Stimuli-responsive nanocarriers for drug delivery, S. Mura, J. Nicolas and P. Couvreur), Copyright 2013.

different strategies available rely on either a one-time or repeatable on-off drug release event triggered by photosensitiveness-induced structural alterations of the nanocarriers.[95] Photolabile caging groups have been exploited for the light-activated production of proteins. Lipid vesicles loaded with caged DNA and with the reacting and enzymatic components needed for transcription and translation enabled, upon local ultraviolet irradiation-mediated DNA uncaging, the activation of green fluorescent protein and the synthesis of enzymatically active luciferase (Figure 12.13a).[96]

The major drawback of the light-triggered theranostics system is the low tissue penetration depth (~10 mm) due to the strong scattering properties of soft tissues in the ultraviolet-visible region of the spectrum (less than 700 nm). A conventional light-induced theranostics system is thus only applicable to regions of the body that can be directly illuminated (such as the eye or the skin). However, by using photosensitive groups that respond to higher wavelengths or exploiting two-photon technology, it is possible to replace the UV-visible light source by a NIR laser (700–1000 nm range) with deeper tissue penetration, lower scattering properties and minimal harm to tissues.[97–99] This makes NIR responsive systems more promising for clinical applications. The capacity of NIR-absorbing materials to convert the photon energy adsorbed during irradiation into heat has been used to trigger the heat imaging and chemotherapeutic molecules release from NIR responsive nanodevices.[100] Light-to-heat transduction mediated by NIR irradiation of gold nanorods caused a rapid rise in the local temperature, which was exploited to induce dehybridization of DNA helices conjugated at the gold surface, allowing the release of doxorubicin molecules bound to consecutive cytosine-guanine base pairs (Figure 12.13b).[101] Alternatively, DNA duplexes can be used to cap the pores of a mesoporous silica shell, with dehybridization leading to the release of the host molecules. The increase in temperature resulting from the surface plasmon resonance of gold on NIR irradiation may also cause a phase transition in polymers or lipids, allowing the leakage of a preloaded drug.[100] Accordingly, PLGA nanoparticles (with a glass transition temperature of 45 °C) entirely or half coated with gold layer gold-shell-coated lipidic nanomicelles, polymer-coated gold nanocages or gold nanorods combined with thermosensitive liposomes have been designed, and some have shown significant anti-tumor activity in experimental models of breast cancer and glioma. Also, overheating gold nanoparticles with a short laser pulse in the NIR range can generate plasmonic nanobubbles. Although promising from a conceptual point of view, the safety and/or biodegradability of the typical materials used in light-responsive nanoparticles for theranostics applications (Au–Ag, gold nanorods, azobenzene and *o*-nitro benzyl derivatives) is questionable. Finding biocompatible photosensitive materials will therefore play a critical part in the potential clinical translation of these systems.

12.3.5 Electroresponsive Theranostic Systems

Weak electric fields (typically about 1 V) can be used to achieve pulsed or sustained drug and imaging agent release through a variety of actuation mechanisms. For instance, nanoparticles based on polypyrrole, a conductive polymer, exhibited tailored drug and imaging agent release profiles as a result of a synergistic process of electrochemical reduction-oxidation and electric-field driven movement of charged molecules (Figure 12.14).[102] Multi-walled carbon nanotubes could be used as a conductive additive to increase the electrical sensitivity of theranostics systems. Montmorillonite, when formulated in a chitosan nanohydrogel, finely tuned the release of drug and imaging agents upon electrostimulation, and preserved

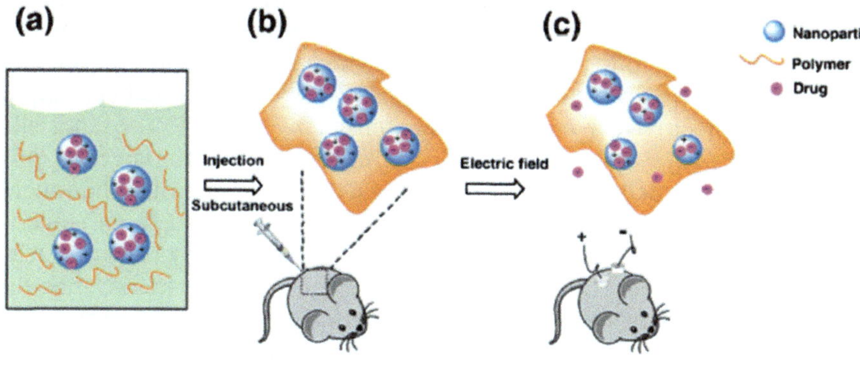

Figure 12.14 General scheme for the application of this system. (a) The nanoparticlepolymer solution is (b) subcutaneously injected into a mouse, followed by (c) application of a DC electric field to induce release of the drug cargo inside the nanoparticles. Reprinted with permission from E. N. Jun Ge, T. J. Cahill III, R. E. Beygui and R. N. Zare, *ACS Nano*, 2012 **6**, 227–233. Copyright 2012 American Chemical Society.

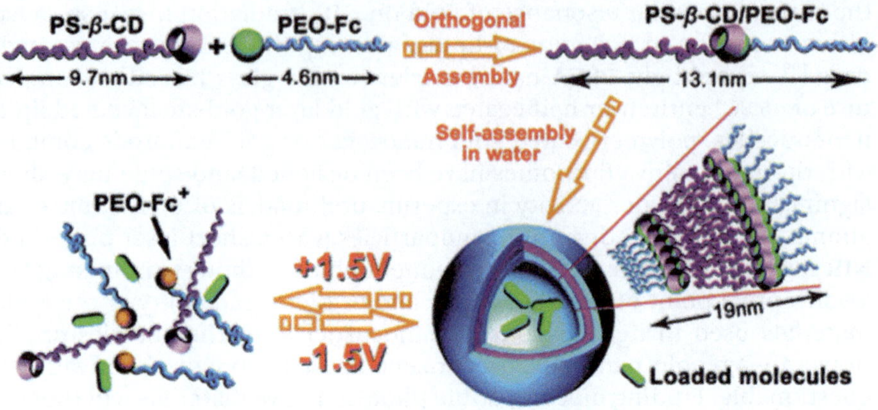

Figure 12.15 Structure of PS-β-CD and PEO-Fc and Schematic of the voltage-responsive controlled assembly and disassembly of PS-β-CD/PEO-Fc supramolecular vesicles. Reprinted with permission from J. Y. Q. Yan, Z. Cai, Y. Xin, Y. Kang and Y. Yin, *J. Am. Chem. Soc.*, 2010 **132**, 9268–9270. Copyright 2010 American Chemical Society.[103]

responsiveness and reversibility after consecutive on-off switching operations. An electric field also activated the reversible scission of supramolecular polymersomes that formed through host–guest complexation between the end groups (β-cyclodextrin and ferrocene) of a pair of homopolymers (one hydrophilic and one hydrophobic) (Figure 12.15).[103] Similarly, an oxidizing voltage activated the splitting of a vesicle membrane (composed of a redox-responsive self-assembled amphiphilic rod-coil tetraaniline-PEG) into smaller puck-like micelles, which could reassemble upon application of a reductive voltage.[104]

Electroporation, the application of a (typically high) transmembrane voltage to cause the formation of pores in cell membranes and thus increase their permeability to drugs, has been shown to be an efficient pathway for electro-responsive drug delivery. In the past few years, it has been applied to nucleic acid delivery against cancer, either by using PEG-coated silica nanoparticles with opposite polarities to enhance gene transfection or by using transferrin-decorated liposomes loaded with exogenous oligonucleotides. Similarly, iontophoresis, which uses an electric field to enhance the transdermal delivery of charged compounds is particularly versatile. It has been recently applied to various types of nano-scaled systems, including organic nanocarriers such as PLGA nanoparticles loaded with estradiol and liposome containing insulin, as well as inorganic gold nanoparticles against traumatic tendinitis.[105] Iontophoresis has also been applied to ocular delivery, for instance in the transport of dexamethasone across the human sclera with egg lecithin–taurocholate micelles as nanocarriers.[106] Overall, electro-responsive theranostic systems offer significant freedom in the design of the device that controls the electrical signal. However, similar to other external stimuli, the low tissue penetration depth and the need to avoid undesired tissue damage may limit their application in therapy.

12.4 Summary and Future Outlook

We have discussed here various aspects of stimuli-responsive materials in theranostics. In recent years, it has been demonstrated that stimuli-responsive materials, when applied to the design of theranostic systems, can provide advanced therapy and better ways to diagnose different diseases. Stimuli-responsive theranostic systems are often decorated with some specific functional groups for disease diagnosis and therapy. However, in general, less than 5% of the injected dose is able to reach the disease tissues by the stimuli-responsive systems. Meanwhile, controversies exist in the clinical translations of nano-theranostic systems utilizing the EPR effect and active targeting. Hence, although *in vitro* proofs have been provided for the application of stimuli-responsive theranostic systems, only a limited number of them have been verified in *in vivo* preclinical models, and even less have reached the clinical trial stage.

As we have shown in this chapter, immense progress in DDS for theranostics has led to the design of smart stimuli-responsive theranostic systems. A variety of responsive modalities have been developed to build these systems, using different architecture, chemistry and material. However, focus should now be shifted towards clinically acceptable systems that are non-toxic, efficacious, and more sensitive to discrete variations in specific stimuli, in order to benefit cancer patients.

Acknowledgements

The authors thank the financial support from NIH/NCI (R01CA199668 & 3R01CA115483), NIH/NIBIB (5R01EB012569), NIH/NICHD (1R01HD086195) and the DoD PRMRP Award (W81XWH-13-1-0490).

References

1. Y. P. Li, W. W. Xiao, K. Xiao, L. Berti, J. T. Luo, H. P. Tseng, G. Fung and K. S. Lam, Well-Defined, Reversible Boronate Crosslinked Nanocarriers for Targeted Drug Delivery in Response to Acidic pH Values and cis-Diols, *Angew. Chem., Int. Ed.*, 2012, **51**, 2864–2869.
2. Y. P. Li, K. Xiao, W. Zhu, W. B. Deng and K. S. Lam, Stimuli-responsive cross-linked micelles for on-demand drug delivery against cancers, *Adv. Drug Delivery Rev.*, 2014, **66**, 58–73.
3. S. G. Wang, X. Li, Y. Chen, X. J. Cai, H. L. Yao, W. Gao, Y. Y. Zheng, X. An, J. L. Shi and H. R. Chen, A Facile One-Pot Synthesis of a Two-Dimensional MoS2/Bi2S3 Composite Theranostic Nanosystem for Multi-Modality Tumor Imaging and Therapy, *Adv. Mater.*, 2015, **27**, 2775–2782.
4. B. Q. Spring, R. B. Sears, L. Z. Zheng, Z. M. Mai, R. Watanabe, M. E. Sherwood, D. A. Schoenfeld, B. W. Pogue, S. P. Pereira, E. Villa and T. Hasan, A photoactivable multi-inhibitor nanoliposome for tumour control and simultaneous inhibition of treatment escape pathways, *Nat. Nanotechnol.*, 2016, **11**, 378–387.

5. B. Jang, H. Kwon, P. Katila, S. J. Lee and H. Lee, Dual delivery of biological therapeutics for multimodal and synergistic cancer therapies, *Adv. Drug Delivery Rev.*, 2016, **98**, 113–133.
6. J. Lin, M. Wang, H. Hu, X. Y. Yang, B. Wen, Z. T. Wang, O. Jacobson, J. B. Song, G. F. Zhang, G. Niu, P. Huang and X. Y. Chen, Multimodal-Imaging-Guided Cancer Phototherapy by Versatile Biomimetic Theranostics with UV and gamma-Irradiation Protection, *Adv. Mater.*, 2016, **28**, 3273–3279.
7. B. R. Smith and S. S. Gambhir, Nanomaterials for *in vivo* Imaging, *Chem. Rev.*, 2017, **117**, 901–986.
8. Y. C. Wang, M. S. Shim, N. S. Levinson, H. W. Sung and Y. N. Xia, Stimuli-Responsive Materials for Controlled Release of Theranostic Agents, *Adv. Funct. Mater.*, 2014, **24**, 4206–4220.
9. M. Molina, M. Asadian-Birjand, J. Balach, J. Bergueiro, E. Miceli and M. Calderon, Stimuli-responsive nanogel composites and their application in nanomedicine, *Chem. Soc. Rev.*, 2015, **44**, 6161–6186.
10. L. L. Tan, H. W. Li, Y. C. Qiu, D. X. Chen, X. Wang, R. Y. Pan, Y. Wang, S. X. A. Zhang, B. Wang and Y. W. Yang, Stimuli-responsive metal-organic frameworks gated by pillar 5 arene supramolecular switches, *Chem. Sci.*, 2015, **6**, 1640–1644.
11. L. G. Xu, S. Zhao, W. Ma, X. L. Wu, S. Li, H. Kuang, L. B. Wang and C. L. Xu, Multigaps Embedded Nanoassemblies Enhance *In situ* Raman Spectroscopy for Intracellular Telomerase Activity Sensing, *Adv. Funct. Mater.*, 2016, **26**, 1602–1608.
12. G. X. Y. Zheng, B. T. Lau, M. Schnall-Levin, M. Jarosz, J. M. Bell, C. M. Hindson, S. Kyriazopoulou-Panagiotopoulou, D. A. Masquelier, L. Merrill, J. M. Terry, P. A. Mudivarti, P. W. Wyatt, R. Bharadwaj, A. J. Makarewicz, Y. Li, P. Belgrader, A. D. Price, A. J. Lowe, P. Marks, G. M. Vurens, P. Hardenbol, L. Montesclaros, M. Luo, L. Greenfield, A. Wong, D. E. Birch, S. W. Short, K. P. Bjornson, P. Patel, E. S. Hopmans, C. Wood, S. Kaur, G. K. Lockwood, D. Stafford, J. P. Delaney, I. Wu, H. S. Ordonez, S. M. Grimes, S. Greer, J. Y. Lee, K. Belhocine, K. M. Giorda, W. H. Heaton, G. P. McDermott, Z. W. Bent, F. Meschi, N. O. Kondov, R. Wilson, J. A. Bernate, S. Gauby, A. Kindwall, C. Bermejo, A. N. Fehr, A. Chan, S. Saxonov, K. D. Ness, B. J. Hindson and H. P. Ji, Haplotyping germline and cancer genomes with high-throughput linked-read sequencing, *Nat. Biotechnol.*, 2016, **34**, 303–311.
13. W. L. Ye, Y. P. Zhao, R. Na, F. Li, Q. B. Mei, M. G. Zhao and S. Y. Zhou, Actively Targeted Delivery of Doxorubicin to Bone Metastases by a pH-Sensitive Conjugation, *J. Pharm. Sci.*, 2015, **104**, 2293–2303.
14. S. Biswas, P. Kumari, P. M. Lakhani and B. Ghosh, Recent advances in polymeric micelles for anti-cancer drug delivery, *Eur. J. Pharm. Sci.*, 2016, **83**, 184–202.
15. D. Neri and C. T. Supuran, Interfering with pH regulation in tumours as a therapeutic strategy, *Nat. Rev. Drug Discovery*, 2011, **10**, 767–777.
16. A. M. Jhaveri and V. P. Torchilin, Multifunctional polymeric micelles for delivery of drugs and siRNA, *Front. Pharmacol.*, 2014, **5**, 1–26.

17. C. F. Anderson and H. G. Cui, Protease-Sensitive Nanomaterials for Cancer Therapeutics and Imaging, *Ind. Eng. Chem. Res.*, 2017, **56**, 5761–5777.
18. S. Bazban-Shotorbani, M. M. Hasani-Sadrabadi, A. Karkhaneh, V. Serpooshan, K. I. Jacob, A. Moshaverinia and M. Mahmoudi, Revisiting structure-property relationship of pH-responsive polymers for drug delivery applications, *J. Controlled Release*, 2017, **253**, 46–63.
19. R. J. Stubbs, D. A. Hughes, A. M. Johnstone, E. Rowley, C. Reid, M. Elia, R. Stratton, H. Delargy, N. King and J. E. Blundell, The use of visual analogue scales to assess motivation to eat in human subjects: a review of their reliability and validity with an evaluation of new hand-held computerized systems for temporal tracking of appetite ratings, *Br. J. Nutr.*, 2000, **84**, 405–415.
20. Y. M. Bae, Y. I. Park, S. H. Nam, J. H. Kim, K. Lee, H. M. Kim, B. Yoo, J. S. Choi, K. T. Lee, T. Hyeon and Y. D. Suh, Endocytosis, intracellular transport, and exocytosis of lanthanide-doped upconverting nanoparticles in single living cells, *Biomaterials*, 2012, **33**, 9080–9086.
21. T. G. Iversen, T. Skotland and K. Sandvig, Endocytosis and intracellular transport of nanoparticles: Present knowledge and need for future studies, *Nano Today*, 2011, **6**, 176–185.
22. Z. S. Ge and S. Y. Liu, Functional block copolymer assemblies responsive to tumor and intracellular microenvironments for site-specific drug delivery and enhanced imaging performance, *Chem. Soc. Rev.*, 2013, **42**, 7289–7325.
23. F. Bai, C. Wang, Q. Lu, M. Zhao, F.-Q. Ban, D.-H. Yu, Y.-Y. Guan, X. Luan, Y.-R. Liu and H.-Z. Chen, Nanoparticle-mediated drug delivery to tumor neovasculature to combat P-gp expressing multidrug resistant cancer, *Biomaterials*, 2013, **34**, 6163–6174.
24. E. Fleige, M. A. Quadir and R. Haag, Stimuli-responsive polymeric nanocarriers for the controlled transport of active compounds: Concepts and applications, *Adv. Drug Delivery Rev.*, 2012, **64**, 866–884.
25. P. C. Balaure and A. M. Grumezescu, Smart Synthetic Polymer Nanocarriers for Controlled and Site-Specific Drug Delivery, *Curr. Top. Med. Chem.*, 2015, **15**, 1424–1490.
26. S. Mura, J. Nicolas and P. Couvreur, Stimuli-responsive nanocarriers for drug delivery, *Nat. Mater.*, 2013, **12**, 991–1003.
27. Y. G. Wang, K. J. Zhou, G. Huang, C. Hensley, X. N. Huang, X. P. Ma, T. Zhao, B. D. Sumer, R. J. DeBerardinis and J. M. Gao, A nanoparticle-based strategy for the imaging of a broad range of tumours by nonlinear amplification of microenvironment signals, *Nat. Mater.*, 2014, **13**, 204–212.
28. X. N. Huang, G. Huang, S. R. Zhang, K. Sagiyama, O. Togao, X. P. Ma, Y. G. Wang, Y. Li, T. C. Soesbe, B. D. Sumer, M. Takahashi, A. D. Sherry and J. M. Gao, Multi-Chromatic pH-Activatable F-19-MRI Nanoprobes with Binary ON/OFF pH Transitions and Chemical-Shift Barcodes, *Angew. Chem., Int. Ed.*, 2013, **52**, 8074–8078.

29. T. T. Wang, D. G. Wang, H. J. Yu, M. W. Wang, J. P. Liu, B. Feng, F. Y. Zhou, Q. Yin, Z. W. Zhang, Y. Z. Huang and Y. P. Li, Intracellularly Acid-Switchable Multifunctional Micelles for Combinational Photo/Chemotherapy of the Drug-Resistant Tumor, *ACS Nano*, 2016, **10**, 3496–3508.
30. H. J. Yu, C. Y. Guo, B. Feng, J. P. Liu, X. Z. Chen, D. G. Wang, L. S. Teng, Y. X. Li, Q. Yin, Z. W. Zhang and Y. P. Li, Triple-Layered pH-Responsive Micelleplexes Loaded with siRNA and Cisplatin Prodrug for NF-Kappa B Targeted Treatment of Metastatic Breast Cancer, *Theranostics*, 2016, **6**, 14–27.
31. M. Kanamala, W. R. Wilson, M. M. Yang, B. D. Palmer and Z. M. Wu, Mechanisms and biomaterials in pH-responsive tumour targeted drug delivery: A review, *Biomaterials*, 2016, **85**, 152–167.
32. H. Z. Deng, J. J. Liu, X. F. Zhao, Y. M. Zhang, J. F. Liu, S. X. Xu, L. D. Deng, A. J. Dong and J. H. Zhang, PEG-b-PCL Copolymer Micelles with the Ability of pH-Controlled Negative-to-Positive Charge Reversal for Intracellular Delivery of Doxorubicin, *Biomacromolecules*, 2014, **15**, 4281–4292.
33. J. D. Dai, T. Nagai, X. Q. Wang, T. Zhang, M. Meng and Q. Zhang, pH-sensitive nanoparticles for improving the oral bioavailability of cyclosporine A, *Int. J. Pharm.*, 2004, **280**, 229–240.
34. J. Liu, Y. R. Huang, A. Kumar, A. Tan, S. B. Jin, A. Mozhi and X. J. Liang, pH-Sensitive nano-systems for drug delivery in cancer therapy, *Biotechnol. Adv.*, 2014, **32**, 693–710.
35. Y. N. Zhong, K. Goltsche, L. Cheng, F. Xie, F. H. Meng, C. Deng, Z. Y. Zhong and R. Haag, Hyaluronic acid-shelled acid-activatable paclitaxel prodrug micelles effectively target and treat CD44-overexpressing human breast tumor xenografts *in vivo*, *Biomaterials*, 2016, **84**, 250–261.
36. S. J. Sonawane, R. S. Kalhapure and T. Govender, Hydrazone linkages in pH responsive drug delivery systems, *Eur. J. Pharm. Sci.*, 2017, **99**, 45–65.
37. W. X. Hou, X. Zhao, X. Q. Qian, F. Pan, C. L. Zhang, Y. M. Yang, J. M. de la Fuente and D. X. Cui, pH-Sensitive self-assembling nanoparticles for tumor near-infrared fluorescence imaging and chemo-photodynamic combination therapy, *Nanoscale*, 2016, **8**, 104–116.
38. H. Kim, S. Kim, C. Park, H. Lee, H. J. Park and C. Kim, Glutathione-Induced Intracellular Release of Guests from Mesoporous Silica Nanocontainers with Cyclodextrin Gatekeepers, *Adv. Mater.*, 2010, **22**, 4280–4283.
39. Y. E. Kurtoglu, R. S. Navath, B. Wang, S. Kannan, R. Romero and R. M. Kannan, Poly(amidoamine) dendrimer-drug conjugates with disulfide linkages for intracellular drug delivery, *Biomaterials*, 2009, **30**, 2112–2121.
40. W. Ong, Y. M. Yang, A. C. Cruciano and R. L. McCarley, Redox-Triggered Contents Release from Liposomes, *J. Am. Chem. Soc.*, 2008, **130**, 14739–14744.
41. C. Luo, J. Sun, D. Liu, B. J. Sun, L. Miao, S. Musetti, J. Li, X. P. Han, Y. Q. Du, L. Li, L. Huang and Z. G. He, Self-Assembled Redox Dual-Responsive Prodrug-Nanosystem Formed by Single Thioether-Bridged Paclitaxel-Fatty Acid Conjugate for Cancer Chemotherapy, *Nano Lett.*, 2016, **16**, 5401–5408.

42. Y. P. Li, T. Y. Lin, Y. Luo, Q. Q. Liu, W. W. Xiao, W. C. Guo, D. Lac, H. Y. Zhang, C. H. Feng, S. Wachsmann-Hogiu, J. H. Walton, S. R. Cherry, D. J. Rowland, D. Kukis, C. X. Pan and K. S. Lam, A smart and versatile theranostic nanomedicine platform based on nanoporphyrin, *Nat. Commun.*, 2014, **5**, 4712.
43. S. H. van Rijt, D. A. Bolukbas, C. Argyo, S. Datz, M. Lindner, O. Eickelberg, M. Konigshoff, T. Bein and S. Meiners, Protease-Mediated Release of Chemotherapeutics from Mesoporous Silica Nanoparticles to *ex Vivo* Human and Mouse Lung Tumors, *ACS Nano*, 2015, **9**, 2377–2389.
44. Q. Y. Hu, W. J. Sun, Y. Lu, H. N. Bomba, Y. Q. Ye, T. Y. Jiang, A. J. Isaacson and Z. Gu, Tumor Microenvironment-Mediated Construction and Deconstruction of Extracellular Drug-Delivery Depots, *Nano Lett.*, 2016, **16**, 1118–1126.
45. S. Tang, Q. S. Meng, H. P. Sun, J. H. Su, Q. Yin, Z. W. Zhang, H. J. Yu, L. L. Chen, Y. Chen, W. W. Gu and Y. P. Li, Tumor-Microenvironment-Adaptive Nanoparticles Codeliver Paclitaxel and siRNA to Inhibit Growth and Lung Metastasis of Breast Cancer, *Adv. Funct. Mater.*, 2016, **26**, 6033–6046.
46. H. Tang, Y. Zheng and Y. Chen, Materials Chemistry of Nanoultrasonic Biomedicine, *Adv. Mater.*, 2016, **29**, 1604105.
47. L. X. Xing, Q. S. Shi, K. L. Zheng, M. Shen, J. Ma, F. Li, Y. Liu, L. Z. Lin, W. Z. Tu, Y. R. Duan and L. F. Du, Ultrasound-Mediated Microbubble Destruction (UMMD) Facilitates the Delivery of CA19-9 Targeted and Paclitaxel Loaded mPEG-PLGA-PLL Nanoparticles in Pancreatic Cancer, *Theranostics*, 2016, **6**, 1573–1587.
48. S. D. Jo, S. H. Ku, Y. Y. Won, S. H. Kim and I. C. Kwon, Targeted Nanotheranostics for Future Personalized Medicine: Recent Progress in Cancer Therapy, *Theranostics*, 2016, **6**, 1362–1377.
49. H. S. Min, D. G. You, S. Son, S. Jeon, J. H. Park, S. Lee, I. C. Kwon and K. Kim, Echogenic Glycol Chitosan Nanoparticles for Ultrasound-Triggered Cancer Theranostics, *Theranostics*, 2015, **5**, 1402–1418.
50. Y. J. Ho, Y. C. Chang and C. K. Yeh, Improving Nanoparticle Penetration in Tumors by Vascular Disruption with Acoustic Droplet Vaporization, *Theranostics*, 2016, **6**, 392–403.
51. J. J. Li, F. Y. Liu, S. Gupta and C. Li, Interventional Nanotheranostics of Pancreatic Ductal Adenocarcinoma, *Theranostics*, 2016, **6**, 1393–1402.
52. C. Kang, W. Cho, M. Park, J. Kim, S. Park, D. Shin, C. Song and D. Lee, H2O2-triggered bubble generating antioxidant polymeric nanoparticles as ischemia/reperfusion targeted nanotheranostics, *Biomaterials*, 2016, **85**, 195–203.
53. T. Si, G. B. Li, Q. Wu, Z. Q. Zhu, X. S. Luo and R. X. Xu, Optical droplet vaporization of nanoparticle-loaded stimuli-responsive microbubbles, *Appl. Phys. Lett.*, 2016, **108**, 111109.
54. E. Huynh, M. A. Rajora and G. Zheng, Multimodal micro, nano, and size conversion ultrasound agents for imaging and therapy, *Wiley Interdiscip. Rev.: Nanomed. Nanobiotechnol.*, 2016, **8**, 796–813.

55. H. Yang, W. Cai, L. Xu, X. Lv, Y. Qiao, P. Li, H. Wu, Y. Yang, L. Zhang and Y. Duan, Nanobubble-Affibody: Novel ultrasound contrast agents for targeted molecular ultrasound imaging of tumor, *Biomaterials*, 2015, **37**, 279–288.
56. G. P. Luke, A. S. Hannah and S. Y. Emelianov, Super-resolution ultrasound imaging *in vivo* with transient laser-activated nanodroplets, *Nano Lett.*, 2016, **16**, 2556–2559.
57. E. Huynh, B. Y. Leung, B. L. Helfield, M. Shakiba, J.-A. Gandier, C. S. Jin, E. R. Master, B. C. Wilson, D. E. Goertz and G. Zheng, *In situ* conversion of porphyrin microbubbles to nanoparticles for multimodality imaging, *Nat. Nanotechnol.*, 2015, **10**, 325–332.
58. K. Wilson, K. Homan and S. Emelianov, Biomedical photoacoustics beyond thermal expansion using triggered nanodroplet vaporization for contrast-enhanced imaging, *Nat. Commun.*, 2012, **3**, 618–628.
59. D. Kagan, M. J. Benchimol, J. C. Claussen, E. Chuluun-Erdene, S. Esener and J. Wang, Acoustic droplet vaporization and propulsion of perfluorocarbon-loaded microbullets for targeted tissue penetration and deformation, *Angew. Chem., Int. Ed. Engl.*, 2012, **51**, 7519–7522.
60. J. S. Xu, Y. Chen, L. M. Deng, J. X. Liu, Y. Cao, P. Li, H. T. Ran, Y. Y. Zheng and Z. G. Wang, Microwave-activated nanodroplet vaporization for highly efficient tumor ablation with real-time monitoring performance, *Biomaterials*, 2016, **106**, 264–275.
61. Y. J. Zhao, W. X. Song, D. Wang, H. T. Ran, R. H. Wang, Y. Z. Yao, Z. G. Wang, Y. Y. Zheng and P. Li, Phase-shifted PFH@PLGA/Fe3O4 nanocapsules for MRI/US imaging and photothermal therapy with near-infrared irradiation, *ACS Appl. Mater. Interfaces*, 2015, **7**, 14231–14242.
62. W. P. Li, C. H. Su, Y. C. Chang, Y. J. Lin and C. S. Yeh, Ultrasound-Induced Reactive Oxygen Species Mediated Therapy and Imaging Using a Fenton Reaction Activable Polymersome, *ACS Nano*, 2016, **10**, 2017–2027.
63. K. Zhang, P. Li, H. R. Chen, X. W. Bo, X. L. Li and H. X. Xu, Continuous Cavitation Designed for Enhancing Radiofrequency Ablation *via* a Special Radiofrequency Solidoid Vaporization Process, *ACS Nano*, 2016, **10**, 2549–2558.
64. Y. Huang, A. M. Vezeridis, J. Wang, Z. Wang, M. Thompson, R. F. Mattrey and N. C. Gianneschi, Polymer-Stabilized Perfluorobutane Nanodroplets for Ultrasound Imaging Agents, *J. Am. Chem. Soc.*, 2016, **139**, 15–19.
65. H. Xing, K. Hwang and Y. Lu, Recent Developments of Liposomes as Nanocarriers for Theranostic Applications, *Theranostics*, 2016, **6**, 1336–1352.
66. R. J. Paproski, A. Forbrich, E. Huynh, J. Chen, J. D. Lewis, G. Zheng and R. J. Zemp, Porphyrin Nanodroplets: Sub-micrometer Ultrasound and Photoacoustic Contrast Imaging Agents, *Small*, 2016, **12**, 371–380.
67. Y. F. You, Z. G. Wang, H. T. Ran, Y. Y. Zheng, D. Wang, J. S. Xu, Z. B. Wang, Y. Chen and P. Li, Nanoparticle-enhanced synergistic HIFU ablation and transarterial chemoembolization for efficient cancer therapy, *Nanoscale*, 2016, **8**, 4324–4339.

68. H. Y. Huang, H. L. Liu, P. H. Hsu, C. S. Chiang, C. H. Tsai, H. S. Chi, S. Y. Chen and Y. Y. Chen, A Multitheragnostic Nanobubble System to Induce Blood-Brain Barrier Disruption with Magnetically Guided Focused Ultrasound, *Adv. Mater.*, 2015, **27**, 655–661.
69. J. M. Cha, D. G. You, E. J. Choi, S. J. Park, W. Um, J. Jeon, K. Kim, I. C. Kwon, J. C. Park, H. R. Kim and J. H. Park, Improvement of Antitumor Efficacy by Combination of Thermosensitive Liposome with High-Intensity Focused Ultrasound, *J. Biomed. Nanotechnol.*, 2016, **12**, 1724–1733.
70. Y. Shi, C. Ma, Y. Du and G. Yu, Microwave-responsive polymeric core–shell microcarriers for high-efficiency controlled drug release, *J. Mater. Chem. B*, 2017, **5**, 3541–3549.
71. C. Cheng, B. Huang, X. Mao, Z. Zhang, Z. Zhang, Z. Geng, P. Xue and H. Chen, A graphene based frequency quadrupler, *Sci. Rep.*, 2017, **7**, 46605.
72. P. Zhao, M. Zheng, Z. Luo, P. Gong, G. Gao, Z. Sheng, C. Zheng, Y. Ma and L. Cai, NIR-driven Smart Theranostic Nanomedicine for On-demand Drug Release and Synergistic Antitumour Therapy, *Sci. Rep.*, 2015, **5**, 14258.
73. B. Taghizadeh, S. Taranejoo, S. A. Monemian, Z. Salehi Moghaddam, K. Daliri, H. Derakhshankhah and Z. Derakhshani, Classification of stimuli-responsive polymers as anticancer drug delivery systems, *Drug Delivery*, 2015, **22**, 145–155.
74. T. Tagami, W. D. Foltz, M. J. Ernsting, C. M. Lee, I. F. Tannock, J. P. May and S.-D. Li, MRI monitoring of intratumoral drug delivery and prediction of the therapeutic effect with a multifunctional thermosensitive liposome, *Biomaterials*, 2011, **32**, 6570–6578.
75. Z. S. Al-Ahmady, W. T. Al-Jamal, J. V. Bossche, T. T. Bui, A. F. Drake, A. J. Mason and K. Kostarelos, Lipid–peptide vesicle nanoscale hybrids for triggered drug release by mild hyperthermia *in vitro* and *in vivo*, *ACS Nano*, 2012, **6**, 9335–9346.
76. K.-J. Chen, H. F. Liang, H.-L. Chen, Y. Wang, P.-Y. Cheng, H.-L. Liu, Y. Xia and H.-W. Sung, A Thermoresponsive Bubble-Generating Liposomal System for Triggering Localized Extracellular Drug Delivery, *ACS Nano*, 2013, **7**, 438–446.
77. S. R. Macewan and A. Chilkoti, Digital switching of local arginine density in a genetically encoded self-assembled polypeptide nanoparticle controls cellular uptake, *Nano Lett.*, 2012, **12**, 3322–3328.
78. J. Xie, Y. Zhang, C. Yan, L. Song, S. Wen, F. Zang, G. Chen, Q. Ding, C. Yan and N. Gu, High-performance PEGylated Mn–Zn ferrite nanocrystals as a passive-targeted agent for magnetically induced cancer theranostics, *Biomaterials*, 2014, **35**, 9126–9136.
79. H. Y. Huang, H. L. Liu, P. H. Hsu, C. S. Chiang, C. H. Tsai, H. S. Chi, S. Y. Chen and Y. Y. Chen, A multitheragnostic nanobubble system to induce blood-brain barrier disruption with magnetically guided focused ultrasound, *Adv. Mater.*, 2015, **27**, 655–661.

80. Y. Zhao, W. Song, D. Wang, H. Ran, R. Wang, Y. Yao, Z. Wang, Y. Zheng and P. Li, Phase-Shifted PFH@PLGA/Fe3O4 Nanocapsules for MRI/US Imaging and Photothermal Therapy with near-Infrared Irradiation, *ACS Appl. Mater. Interfaces*, 2015, **7**, 14231–14242.
81. L. Cheng, Y. Liu, B. Zou, Y. Yu, W. Ruan and Y. Wang, Template-etching route to construct uniform rattle-type Fe3O4@SiO2 hollow microspheres as drug carrier, *Mater. Sci. Eng., C*, 2017, **75**, 829–835.
82. H. Ding, Y. Zhang, S. Xu and G. Li, A wrinkle to sub-100 nm yolk/shell Fe3O4@SiO2 nanoparticles, *Nano Res.*, 2016, **9**, 3632–3643.
83. K. W. Jianan Zhang, Q. Xu, Y. Zhou, F. Cheng and S. Guo, Beyond Yolk Shell Nanoparticles: Fe3O4@Fe3C Core@Shell Nanoparticles as Yolks and Carbon Nanospindles as Shells for Efficient Lithium Ion Storage, *ACS Nano*, 2015, **9**, 3369–3376.
84. C. G. S. d. Souza, J. B. Souza, W. Beck and L. C. Varanda, Luminomagnetic Silica-Coated Heterodimers of Core/Shell FePt/Fe3O4 and CdSe Quantum Dots as Potential Biomedical Sensor, *J. Nanomater.*, 2017, **2017**, 1–9.
85. Y. K. Huang, C. H. Su, J. J. Chen, C. T. Chang, Y. H. Tsai, S. F. Syu, T. T. Tseng and C. S. Yeh, Fabrication of Silica-Coated Hollow Carbon Nanospheres Encapsulating Fe3O4 Cluster for Magnetical and MR Imaging Guided NIR Light Triggering Hyperthermia and Ultrasound Imaging, *ACS Appl. Mater. Interfaces*, 2016, **8**, 14470–14480.
86. S. Prijic, L. Prosen, M. Cemazar, J. Scancar, R. Romih, J. Lavrencak, V. B. Bregar, A. Coer, M. Krzan, A. Znidarsic and G. Sersa, Surface modified magnetic nanoparticles for immuno-gene therapy of murine mammary adenocarcinoma, *Biomaterials*, 2012, **33**, 4379–4391.
87. D. Zhang, J. Wang, Z. Wang, R. Wang, L. Song, T. Zhang, X. Lin, P. Shi, H. Xin and X. Pang, Polyethyleneimine-Coated Fe3O4 Nanoparticles for Efficient siRNA Delivery to Human Mesenchymal Stem Cells Derived from Different Tissues, *Sci. Adv. Mater.*, 2015, **7**, 1058–1064.
88. D. L. Xia, Y. P. Chen, C. Chen, Y. F. Wang, X. D. Li, H. He and H. Y. Gu, Comparative Study of Biosafety, DNA, and Chromosome Damage of Different-Materials-Modified Fe3O4 in Rats, *Appl. Biochem. Biotechnol.*, 2015, **177**, 1069–1082.
89. S. Xiao, R. Castro, J. Rodrigues, X. Shi and H. Tomás, PAMAM Dendrimer/pDNA Functionalized-Magnetic Iron Oxide Nanoparticles for Gene Delivery, *J. Biomed. Nanotechnol.*, 2015, **11**, 1370–1384.
90. E. D. Hood, M. Chorny, C. F. Greineder, I. S. Alferiev, R. J. Levy and V. R. Muzykantov, Endothelial targeting of nanocarriers loaded with antioxidant enzymes for protection against vascular oxidative stress and inflammation, *Biomaterials*, 2014, **35**, 3708–3715.
91. H. W. Li, X. L. Jiang, Y. M. Liu, J. J. Hu, J. Liang and X. Liao, Screening of lipase inhibitors from Scutellaria baicalensis extract using lipase immobilized on magnetic nanoparticles and study on the inhibitory mechanism, *Anal. Bioanal. Chem.*, 2016, **408**, 2275–2283.

92. C. Chandola, S. Kalme, M. G. Casteleijn, A. Urtti and M. Neerathilingam, Application of aptamers in diagnostics, drug-delivery and imaging, *J. Biosci.*, 2016, **41**, 535–561.
93. R. Ni, J. L. Zhou, N. Hossain and Y. Chau, Virus-inspired nucleic acid delivery system: Linking virus and viral mimicry, *Adv. Drug Delivery Rev.*, 2016, **106**, 3–26.
94. C. Yue, P. Liu, M. Zheng, P. Zhao, Y. Wang, Y. Ma and L. Cai, IR-780 dye loaded tumor targeting theranostic nanoparticles for NIR imaging and photothermal therapy, *Biomaterials*, 2013, **34**, 6853–6861.
95. C. Yao, P. Wang, X. Li, X. Hu, J. Hou, L. Wang and F. Zhang, Near-Infrared-Triggered Azobenzene-Liposome/Upconversion Nanoparticle Hybrid Vesicles for Remotely Controlled Drug Delivery to Overcome Cancer Multidrug Resistance, *Adv. Mater.*, 2016, **28**, 9341–9348.
96. A. Schroeder, M. S. Goldberg, C. Kastrup, Y. Wang, S. Jiang, B. J. Joseph, C. G. Levins, S. T. Kannan, R. Langer and D. G. Anderson, Remotely activated protein-producing nanoparticles, *Nano Lett.*, 2012, **12**, 2685–2689.
97. H. Huang, L. Liu and M. O. Ngadi, Assessment of intramuscular fat content of pork using NIR hyperspectral images of rib end, *J. Food Eng.*, 2017, **193**, 29–41.
98. C. T. Zeng, W. T. Shang, X. Y. Liang, X. Liang, Q. S. Chen, C. W. Chi, Y. Du, C. H. Fang and J. Tian, Cancer Diagnosis and Imaging-Guided Photothermal Therapy Using a Dual-Modality Nanoparticle, *ACS Appl. Mater. Interfaces*, 2016, **8**, 29232–29241.
99. Y. Shimizu, T. Temma, I. Hara, R. Yamahara, E. Ozeki, M. Ono and H. Saji, Development of novel nanocarrier-based near-infrared optical probes for *in vivo* tumor imaging, *J. Fluoresc.*, 2012, **22**, 719–727.
100. J. You, R. Zhang, C. Xiong, M. Zhong, M. Melancon, S. Gupta, A. M. Nick, A. K. Sood and C. Li, Effective photothermal chemotherapy using doxorubicin-loaded gold nanospheres that target EphB4 receptors in tumors, *Cancer Res.*, 2012, **72**, 4777–4786.
101. Z. Xiao, C. Ji, J. Shi, E. M. Pridgen, J. Frieder, J. Wu and O. C. Farokhzad, DNA self-assembly of targeted near-infrared-responsive gold nanoparticles for cancer thermo-chemotherapy, *Angew. Chem., Int. Ed. Engl.*, 2012, **51**, 11853–11857.
102. J. Ge, E. Neogytou, T. J. Cahill III, R. E. Beygui and R. N. Zare, Drug Release from ElectricField-Responsive Nanoparticles, *ACS Nano*, 2012, **6**, 227–233.
103. J. Y. Qiang Yan, Z. Cai, Y. Xin, Y. Kang and Y. Yin, Voltage-Responsive Vesicles Based on Orthogonal Assembly of Two Homopolymers, *J. Am. Chem. Soc.*, 2010, **132**, 9268–9270.
104. H. Kim, S. M. Jeong and J. W. Park, Electrical switching between vesicles and micelles *via* redox-responsive self-assembly of amphiphilic rod-coils, *J. Am. Chem. Soc.*, 2011, **133**, 5206–5209.

105. M. B. Dohnert, M. Venancio, J. C. Possato, R. C. Zeferino, L. H. Dohnert, A. I. Zugno, C. T. De Souza, M. M. Paula and T. F. Luciano, Gold nanoparticles and diclofenac diethylammonium administered by iontophoresis reduce inflammatory cytokines expression in Achilles tendinitis, *Int. J. Nanomed.*, 2012, 7, 1651–1657.
106. P. Chopra, J. Hao and S. K. Li, Sustained release micellar carrier systems for iontophoretic transport of dexamethasone across human sclera, *J. Controlled Release*, 2012, **160**, 96–104.

CHAPTER 13

Stimuli-responsive Material Inspired Drug Delivery Systems and Devices

SHIVAPRASAD MANCHINEELLA AND THIMMAIAH GOVINDARAJU*

Bioorganic Chemistry Laboratory, New Chemistry Unit, Jawaharlal Nehru Centre for Advanced Scientific Research (JNCASR), Jakkur, Bengaluru, 560064, Karnataka, India
*E-mail: tgraju@jncasr.ac.in

13.1 Introduction

Drugs have been used for centuries to treat various pathological conditions, restore the altered physiological functions and improve health, rescue and extend the lives of humans. Delivering the drug to the target site is as important as the drug discovery itself and plays a decisive role in the development of novel therapeutic agents and methods. Drug delivery systems or devices provide a platform for the loading and delivery of drugs, therapeutic agents, and diverse biologically active molecules.[1–4] The controlled loading and retaining of the (bio)activity of the cargo plays a vital role in designing a potential drug delivery system (DDS). Apart from providing a platform for the loading and retaining the activity, an ideal DDS must be capable of releasing the drug temporally on demand at the targeted site. On-demand

Biomaterials Science Series No. 1
Stimuli-responsive Drug Delivery Systems
Edited by Amit Singh and Mansoor M. Amiji

Published by the Royal Society of Chemistry, www.rsc.org

temporal delivery systems help to downplay the unwanted release of active molecules, and also helps to achieve the targeted delivery exclusively at the site of requirement thus maximising the utility and minimising the drug side-effects.[5] Stimuli responsiveness, the property of materials to respond to specific physical, chemical or biological signals, render them responsive and allows on-demand delivery from such smart materials. Owing to the precise control over the triggered release of the payload, stimuli-responsive materials have been actively researched to develop systems and devices for various drug delivery applications.[4,6–8]

Recent advancements in pharmaceutical and engineering technologies have facilitated the successful employment of various stimuli in designing smart DDS that would release payloads temporally in response to specific stimuli. Moreover, the recent progress in the field of microfabrication has extended the development of miniaturized devices for drug storage and triggered release spatiotemporally. The plethora of stimuli used for the development of stimuli-responsive DDS and devices include, but are not limited to, pH, temperature, redox potentials, light, magnetic, electric, biomolecule, and enzyme action.[9–18] The synthetic details of such materials are discussed in preceding chapters and hence discussion on synthetic aspects is avoided in this chapter. We introduce the concept of stimuli-responsive material inspired drug delivery systems for each selected stimuli and discuss the potential application of such devices in drug delivery and therapeutics. In particular, this chapter introduces the reader to selected and representative examples of stimuli-responsive on-demand drug delivery systems based on pH, glucose, protease, glutathione, magnetic field, light and ultrasound triggers.

13.2 pH-Responsive Drug Delivery Systems

pH plays a vital role in various biological functions and is regulated differentially under normal physiological and disease conditions. pH change represents one of the most promising internal stimuli for the development of on-demand DDS with precise control over delivery owing to the established differential pH under normal and disease conditions. A variety of pH-responsive materials have been successfully used for drug delivery applications.[19–21] Iatrou and co-workers have developed pH-responsive hybrid organic/inorganic nanoparticles as drug delivery vehicles for cancer treatment (Figure 13.1a).[22] The inorganic mesoporous silica nanoparticles (MSNs) were used because of their unparalleled properties such as large specific pore volume for drug encapsulation, ease of surface functionalisation, along with the established biocompatibility.[23] A uniform layer of poly (L-histidine) (PHis) polypeptide was employed as a surface shell to overcome the unwanted diffusion controlled release and impart pH-responsive release property to the MSNs. PHis undergoes a solubility transition among physiological and tumour pH owing to the presence of an imidazole moiety in the structure, and functions as a pH-responsive gatekeeper for the drug delivery from MSNs.[22]

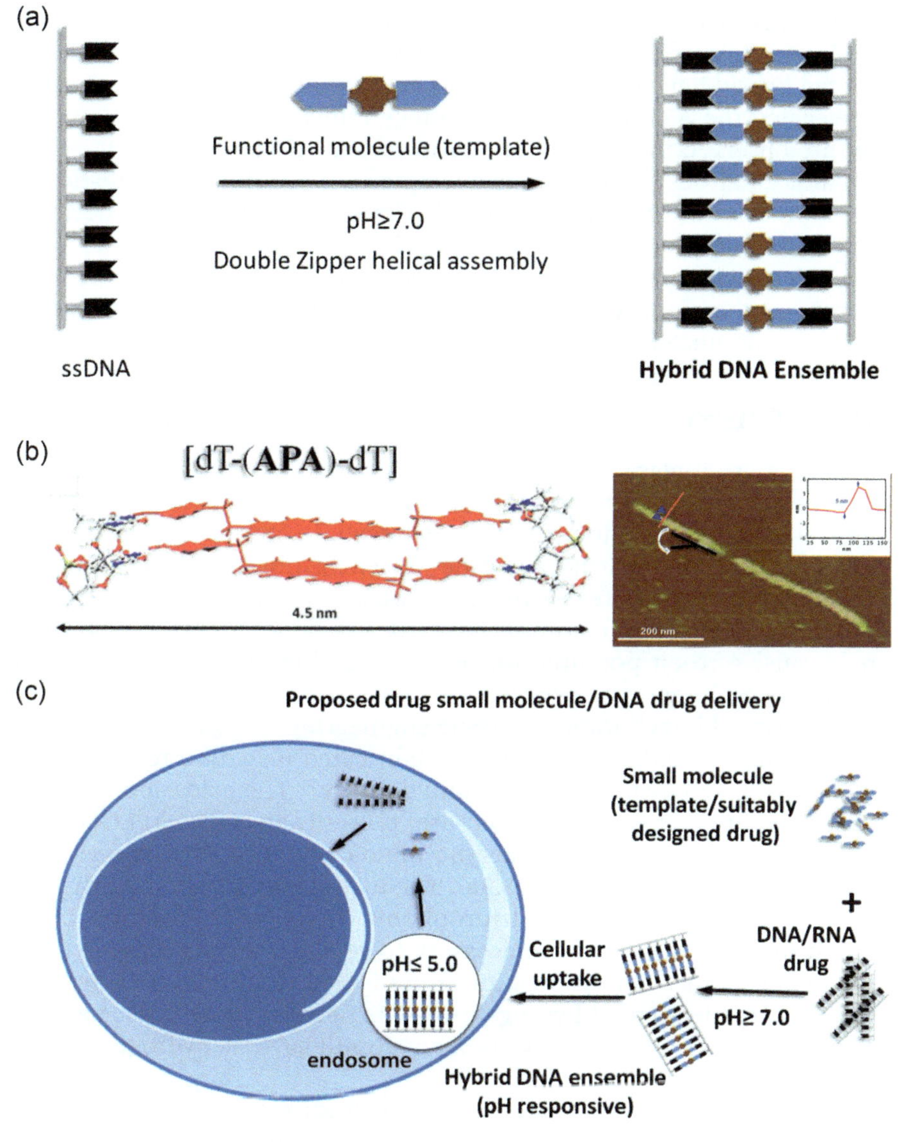

Figure 13.2 pH-responsive hybrid DNA ensembles. (a) Schematic illustration of the formation of hybrid DNA ensembles of APA with deoxy oligonucleotides through molecular zipper assembly. (b) Molecular model of APA hybrid ensemble with dT and AFM micrograph showing left-handed DNA ensemble of [dT_{20}-APA-dT_{20}]. Reproduced from ref. 30 with the permission of The Royal Society of Chemistry. (c) Schematic representation of proposed drug delivery mechanism using APA-DNA ensembles.

Figure 13.2b. Thus, the clever molecular design of APA has enabled the double zipper molecular templating of random coil deoxyoligonucleotides into ordered hybrid helical DNA ensembles at physiological pH, and are capable of releasing the APA and oligonucleotides under the acidic pH conditions that are observed in endosomes and in tumour tissues (Figure 13.2c). The APA molecular design represents a simple yet powerful minimalistic rational approach for the fabrication of stimuli (pH) responsive devices in a bottom-up approach (Figure 13.2c). The hybrid DNA ensembles of APA with oligonucleotides might be extended for the pH-triggered molecular delivery systems and devices for biologically important molecules, small molecule and nucleic acid-based drugs.

13.3 Glucose-triggered Drug Delivery Systems

Hydrogels are a class of soft biomaterials with potential utility in drug delivery and therapeutic applications, apart from other materials related applications. The resemblance to the extracellular matrix (ECM) conditions and ability to load the drugs and other bioactive molecules make the hydrogels promising candidates for the development of smart DDS.[33,34] Hydrogels that respond to specific stimuli have attracted the attention of researchers because of their potential for temporal and on-demand delivery. Such responsive hydrogels have successfully been used for applications ranging from sensing, biotechnology to tissue engineering.[33,34] The research groups of Fu-Sheng Du and Zi-Chen Li have fabricated oxidation-responsive and glucose triggered hydrogels for insulin delivery.[35] The oxidation-responsive hydrogels were prepared using four-arm polyethylene glycol (PEG) containing reactive oxygen cleavable acrylic bonds supported by phenylboronic ester pendant groups (Figure 13.3a). The hydrogels contain majorly hydrophilic PEG and a fraction of hydrophobic linker molecules and possess a good mechanical strength necessary to retain a macroscopic gel structure under culture conditions, and were found to be non-toxic with LO2 cells *in vitro*. Cleavage of hydrogels and the release of payloads is achieved only in the presence of reactive oxygen stimuli such as hydrogen peroxide (H_2O_2).[36,37]

Glucose oxidase (GOx), the enzyme that oxidizes glucose to gluconolactone and produces H_2O_2 is encapsulated into the hydrogels.[37] GOx incorporation has bestowed the hydrogels with a glucose responsive nature. Hydrogels with GOx and FITC-insulin were employed to demonstrate the glucose triggered release of insulin under *in vitro* conditions. The release of encapsulated insulin (FITC-insulin) in the presence of external H_2O_2 and glucose was monitored using fluorescence measurements and the loss of gel integrity was measured by rheology and visual disintegration of the hydrogels (Figure 13.3b). The release of encapsulated insulin was found to be dependent on the concentration of triggering molecules used (either H_2O_2 or glucose) and a faster release of insulin along with degradation of hydrogels was observed at high concentrations of stimuli molecules. Strikingly, the GOx

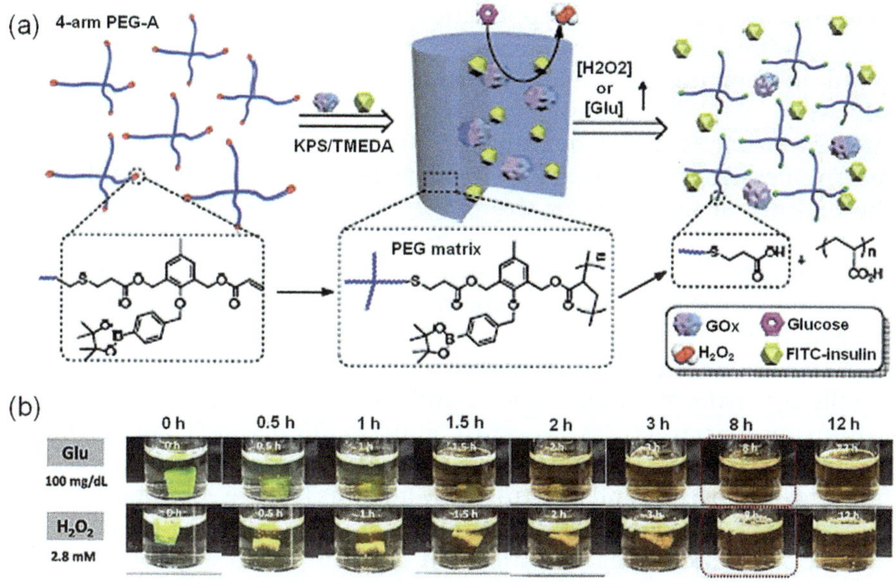

Figure 13.3 Glucose triggered insulin releasing hydrogels. (a) Schematic representation of synthesis, encapsulation of GOx, FITC-insulin into hydrogels and glucose/H_2O_2 triggered release through degradation. (b) Photographs showcasing the time-dependent release of FITC-insulin and degradation in the presence of H_2O_2 and glucose. Reprinted with permission from M. Zhang, C. Song, F. Du and Z. Li, *ACS Appl. Mater. Interfaces*, 2017, **9**, 25905, Copyright 2017 American Chemical Society.[35]

encapsulated hydrogels were found to be responsive to the pathological glucose concentration (20 μM) and were triggered to release the insulin under *in vitro* conditions.

Collectively, GOx encapsulated oxidation sensitive PEG hydrogels with high selectivity, sensitivity and cytocompatibility offered highly sensitive hydrogels for the temporal on-demand release of insulin with the possibility to tune the release of insulin as required. The glucose triggered insulin releasing hydrogels can be used for insulin delivery, and especially for the wound dressings of diabetic foot ulcers. The ability of hydrogels to release payloads in the presence of H_2O_2 stimuli further extends the suitability in the treatment of ROS associated diseases and tissue engineering.

13.4 Enzyme-responsive Drug Delivery Devices

Self-assembling short peptides and their derivatives have been successfully used for various biomedical applications including drug delivery, tissue engineering, and regenerative medicine.[38,39] The ease of synthesis, incorporation of diverse structural and functional moieties and ability to tune the self-assembling properties makes peptides promising molecular platforms for drug delivery applications. Peptide substrates incorporated with specific enzyme cleavable groups enables the design of enzyme triggered on-demand drug delivery systems under physiological conditions. Ulijn and co-workers have exploited the differential concentration of specific enzymes under healthy and disease conditions, and developed matrix metalloproteinase-9 (MMP-9) cleavable peptide amphiphiles for the delivery of hydrophobic drugs.[40] MMPs are the class of zinc-dependent endopeptidases involved in organogenesis and ECM remodelling, and overexpressed in various cancers. The selective overexpression of MMPs in cancer tissues makes them ideal triggers of choice for targeted drug delivery.[41]

Phenylacetyl functionalised hexapeptide (1a: PhAc-FFAGLDD) with hydrophobic and hydrophilic amino acids and a MMP cleavable site, was designed to form micellar structures owing to its amphiphilic nature (Figure 13.4a). The un-cleaved peptide 1a formed anticipated micellar aggregates as evidenced from the microscopy and spectroscopic analysis (Figure 13.4b), whereas the formation of hydrogels with nanofibrous morphology was observed with peptide 1b (PhAc-FFAG). The triggered transformation of peptide micellar aggregates into self-assembling fibres (1c: PhAc-FFAGL) was observed in the presence of MMP-9 (Figure 13.4c). The hydrophobic cancer drug DOX was studied for the MMP triggered release from encapsulated peptide micelles and self-assembled fibre assemblies. DOX encapsulated peptide micelles triggered by MMP-9 transformed into drug incorporated fibre bundles (Figure 13.4d) and released the DOX in a spatiotemporally controlled fashion. Overall, the MMP cleavable peptide amphiphile provided an interesting molecular platform to load and trigger release of the hydrophobic drugs, which are otherwise difficult due to solubility issues. Most importantly, this peptide amphiphile design enables a two-stage on-demand drug delivery for therapeutic applications.

Fakhrullin and co-workers have developed clay-based halloysite nanotube (HNT) delivery vehicles for intracellular enzyme triggered drug release for cancer treatment.[42] HNTs were judiciously chosen for the development of enzyme triggered drug delivery devices because of their proven relatively high internalisation into the cancer cells, as compared to normal cells.[43,44] HNTs with a lumen provide room for the loading of diverse therapeutic molecules through simple physical entrapment and release of the loaded drugs through diffusion. However, the diffusion-mediated passive drug release from pristine HNTs suffers from non-selective and uncontrolled drug release. Stimuli responsiveness and triggered active release properties were imparted through surface functionalisation of HNTs with intracellular glycosyl hydrolase degradable polysaccharide dextrin (DX-HNT) (Figure 13.5a).[45] A dextrin coating acts as an end stopper and provides enzyme responsiveness to the HNTs loaded with the desired therapeutic molecules. The *in vitro* triggered release studies of the model anticancer drug triazole brilliant green (BG) loaded DX-HNTs exhibited cell-specific internalisation, dextrin end stopper hydrolysis triggered by intracellular glycosyl hydrolases and resulted in

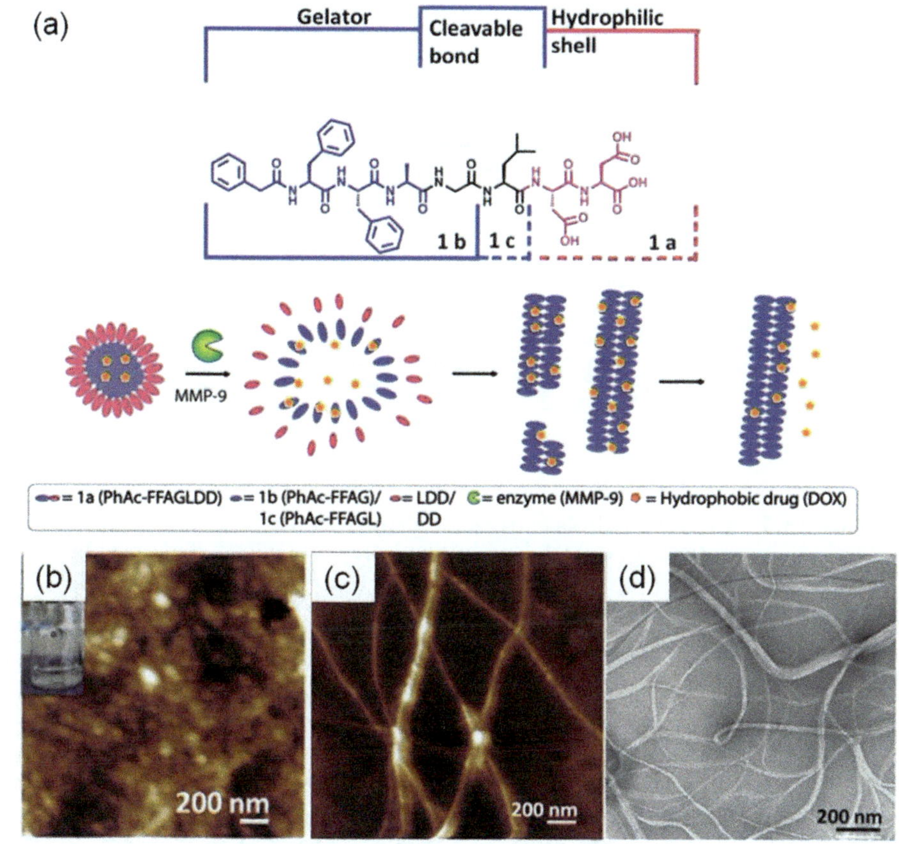

Figure 13.4 MMP responsive on-demand cancer drug delivery. (a) Molecular structure and schematic representation of MMP triggered micelle to fibre transformation of peptide assembly and on-demand cancer drug delivery. AFM micrographs of (b) micellar aggregates of peptide 1a and (c) fibres of peptide 1c. (d) TEM images showing the fibre formation after MMP cleavage of the doxorubicin-loaded peptide 1a micelles. Reproduced from ref. 40 with the permission of The Royal Society of Chemistry.

cancer cell death (Figure 13.5b). Overall, the DX-HNTs represent easily accessible smart drug delivery devices for encapsulation and active on-demand intracellular release.

13.5 Redox-sensitive Stomatocyte Nanomotors

Self-reliant synthetic nanovesicles with stimuli-responsive triggered payload release at the site of action are promising nanomedical systems for drug delivery applications.[46,47] Wilson and co-workers have demonstrated redox-responsive stomatocyte nanomotor systems developed from the

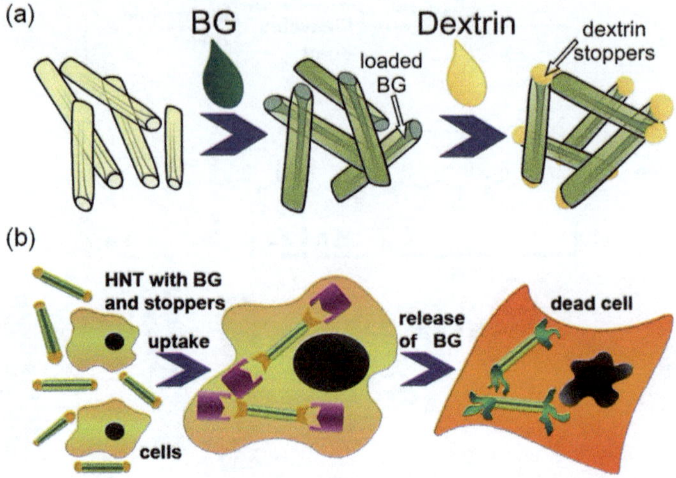

Figure 13.5 Glycosyl hydrolase triggered drug delivery devices. (a) Fabrication of anticancer model drug brilliant green (BG) loaded glycosyl hydrolase responsive halloysite nanotubes (HNTs). (b) Schematic representation of enzyme triggered intracellular drug release from HNTs leading to cancer cell death. Adapted from ref. 42, http://dx.doi.org/10.1038/srep10560. Copyright 2015 The Authors. Published under the terms of the CC BY 4.0 licence, https://creativecommons.org/licenses/by/4.0/.

self-assembly of amphiphilic block copolymers.[48] Stomatocytes are bowl-shaped vesicular structures with a potential to encapsulate drugs and therapeutic molecules. Hydrophilic poly(ethylene glycol) (PEG) and the hydrophobic polystyrene (PS) were judiciously chosen for amphiphilic self-assembly into nanovesicles and subsequently stomatocyte structures. Nanovesicles were bestowed with stimuli-responsive self-destruction and payload release in the presence of biological oxidising agent glutathione (GSH) by incorporating disulphide bridges in co-block polymer between PEG and PS (PEG–SS–PS). Further platinum nanoparticles (PtNP) were loaded into the nanovesicles as a model catalyst to drive the stomatocyte nanomotors by generating H_2O_2 fuel.

Self-assembly of DOX-loaded PEG–SS–PS into nanovesicles and the formation of PtNP loaded stomatocyte nanomotors was achieved through sequential assembly and solvent exchange. The formation of stomatocyte structures of redox-responsive nanomotors was evaluated using scanning electron microscopy analysis. Directionally propelled Brownian motion of nanomotors was observed in response to the concentration of H_2O_2 under simulated physiological conditions. The redox-responsive self-destruction of stomatocyte nanomotors into aggregates and release of the encapsulated drug DOX was observed in the presence of GSH, whose concentration is similar to that found in pathological conditions. Aggregation of self-destructed stomatocyte nanomotors after the disulphide cleavage is evidenced from the SEM analysis, which confirmed the proposed mechanism of GSH triggered drug release. Furthermore, the DOX-loaded stomatocyte nanomotors exhibited promising cellular internalisation and GSH triggered release under *in vitro* conditions. The redox-sensitive stomatocyte nanomotor exploits the differential concentration of biomolecules across the healthy and pathological tissues and utilises the dual stimuli responses for on-demand drug delivery. The self-propelled nature of nanomotors further exemplifies the potential of design strategy for the development of nanomedical devices.

13.6 Magnetically Triggered Drug Delivery Devices

Magnetically responsive materials are a class of smart materials employed for designing DDS owing to non-invasive, controlled and sustained on-demand drug delivery.[49] Cai and co-workers have developed a multi-reservoir device that delivers the payloads exclusively upon applying a magnetic trigger.[50] Biodegradable polymer poly(D,L-lactic acid) (PDLLA) was employed as a substrate to overcome the surgical intervention to remove the device post delivery from the body. The device, with near precise dimensions, was fabricated using a soft lithography technique. The device contains 9 x 9 microarray of reservoirs with the precise reservoir dimensions of 500 μm x 500 μm x 500 μm (L x W x D) and facilitates the loading of multiple drugs and therapeutic molecules into a single device. Porous polycarbonate, another biodegradable polymer, was utilised to seal the reservoirs and Fe_3O_4 particles, that match the pore size of the polycarbonate membrane, were used to impart magnetic response and reversible drug delivery to the device (Figure 13.6a,b).

The magnetically-triggered release of payloads from the multi-reservoir device was evaluated using model drugs, DNA and vitamin B12. The payloads were judiciously chosen as representative examples of biomolecules, hydrophobic drugs and other bioactive molecules for the delivery. Drug release studies demonstrated the controlled delivery of consignment DNA and vitamin B12 in response to the external magnetic field and with good control over the delivery under *in vitro* conditions. Altogether, the biodegradable multi-reservoir device with magnetically triggered active drug delivery represents a smart DDS with an external control for controlled drug delivery. Even though the device is not tested under *in vivo* and clinical settings, the ability to load multiple drugs and tune their release patterns independently may inspire the development of advanced devices for real-life applications.

Kohane and co-workers have fabricated thermoresponsive nanogels containing nanocomposite membrane delivery devices for magnetically triggered drug release.[51] The nanomembranes were designed incorporating multiple responsive materials that respond in cascade for on-demand drug delivery. Superparamagnetic iron nanoparticles in the ethylcellulose membranes provide the magnetic response and also act as a local heat source upon applying an external magnetic field. Poly(*N*-isopropylacrylamide)

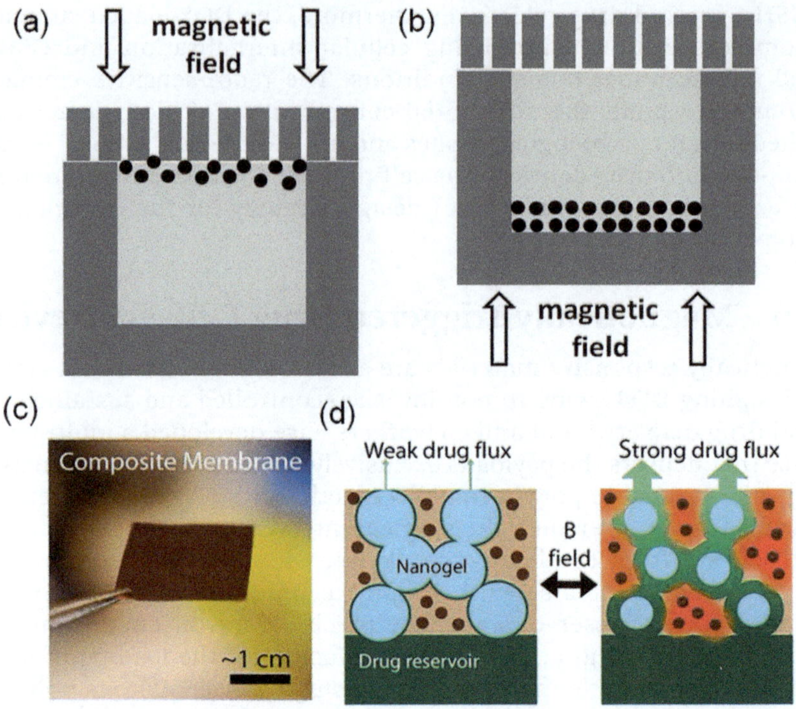

Figure 13.6 Magnetically triggered drug delivery devices. (a), (b) Schematic representations of magnetically triggered control of drug release in response to the external magnetic field. Reproduced with permission from ref. 50, Copyright © 2009 WILEY-VCH Verlag GmbH & Co.KGaA, Weinheim. (c) Photograph of magnetically responsive ethylcellulose membranes containing ferromagnetic nanoparticles and nanogels. (d) Schematic representation of composition and magnetically triggered drug release from nanocomposite membranes. Reprinted with permission from ref. 51. T. Hoare, B. P. Timko, J. Santamaria, G. F. Goya, S. Irusta, S. Lau, C. F. Stefanescu, D. Lin, R. Langer and D. S. Kohane, *Nano Lett.*, 2011, **11**, 1395. Copyright 2011 American Chemical Society.

(pNIPAm) thermoresponsive nanogels, that change volume reversibly with response to the local temperature, are incorporated into the nanocomposite membrane to assist the transport of encapsulated drugs across the DDS (Figure 13.6c). The nanocomposite membranes successfully showed on-demand delivery of diverse model drugs such as bupivacaine, a small molecule-based amphiphilic drug, and high molecular weight dextran (~40 kDa) (Figure 13.6d). Furthermore, the delivery rate and dose can be modulated either by tuning the composition of the membrane device or simply by controlling the duration of the external magnetic field. The ability to load diverse bioactive molecules and control their sustained release, both at the design stage and clinical stage, makes the nanocomposite membranes smart devices for a plethora of drug delivery applications.

13.7 Light Triggered Drug Delivery Devices

Photoresponsive materials are ideal candidates for the development of next generation drug delivery devices owing to the ability to control the non-invasive and precise temporal drug delivery, remotely.[52] Yavuz and co-workers have developed near-infrared (NIR) light triggered drug releasing gold nanocage platform for spatiotemporal drug delivery (Figure 13.7a,b).[53] The use of NIR light as a stimuli for the triggered drug release addresses the issue of tissue attenuation associated with the use of high energy ultraviolet (UV) light. The design involves the use of gold nanocages surface decorated with

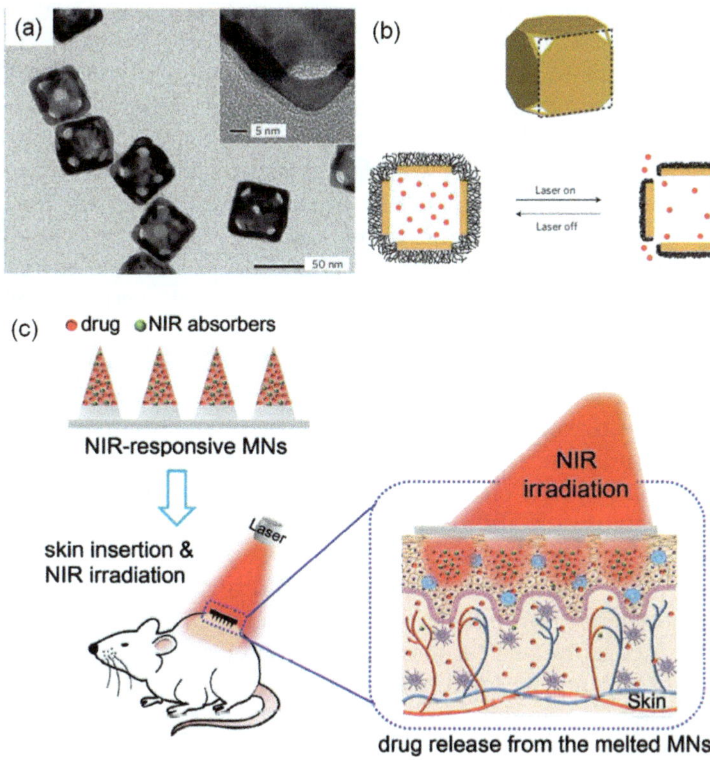

Figure 13.7 NIR-triggered drug delivery devices. (a) TEM micrographs of gold nanocages surface coated with the thermoresponsive pNIPA-Am-co-pAAm copolymer. The inset shows the representative corner of the gold nanocages. (b) Schematic representation of NIR-triggered (laser) reversible opening and drug release from the gold nanocages. Adapted from ref. 53 with the permission of Nature Publishing Group. (c) Schematic illustration of NIR-triggered on-demand transdermal drug delivery from polycaprolactone microneedle patches consisting of LaB6@SiO2 nanostructures (NIR absorbers). Reprinted with permission from ref. 54 M. Chen, M. Ling, K. Wang, Z. Lin, B. Lai and D. Chen, *Biomacromolecules*, 2015, **16**, 598. Copyright 2012 American Chemical Society.

thermoresponsive polymers for the preparation of NIR triggered drug delivery devices. Gold cubes were chosen because of their proven inert nature, biocompatibility, absorption and local heating in the NIR region. pNIPAm was employed to surface functionalise the gold nanocubes and fabricate the thermoresponsive gold nanocages, because of its phase transition behaviour resulting in the collapse of the structure and release of payloads at its lower critical solution temperature (LCST, 32 °C) (Figure 13.7a). The LCST of the polymers was further tuned by incorporating acrylamide side chains into the polymer (pNIPAm-*co*-pAAm) to match the physiological conditions and to avoid the deleterious hyperthermia effects on cells and tissues.

The utility of gold nanocages was successfully demonstrated for the triggered release of diverse model drugs such as alizarin-PEG, DOX, and lysozyme under *in vitro* conditions. Alizarin-PEG represents a medium-sized payload for therapeutic applications. Gold nanocages with encapsulated chemotherapeutic drug DOX exhibited NIR triggered drug release and cancer cell death when studied under *in vitro* laboratory conditions (Figure 13.7b). Interestingly, the bacterial cell wall damaging enzyme lysozyme was loaded into the nanocages without losing its bioactivity and showed NIR triggered bacterial death against *Micrococcus lysodeikticus*. Thus, the ease of fabrication, surface functionalisation, ability to load a plethora of drugs and use of tissue friendly NIR light trigger highlights the design strategy of the gold nanocage device for drug delivery.

Chen and co-workers have developed a light responsive microneedle (MN) device for NIR-triggered transdermal drug delivery (Figure 13.7c).[54] The pyramidal microneedle patches were fabricated employing a polycaprolactone (PCL) polyester as the matrix material using polydimethylsiloxane (PDMS) molds. PCL was chosen for MN fabrication owing to its proven biocompatibility, bioresorbability, low melting point and extensive use in delivery devices.[55] NIR responsive nanostructures of silica-coated lanthanum hexaboride ($LaB_6@SiO_2$) were incorporated into the microneedle device to impart photothermal properties for the triggered drug delivery.[56] Thus, the $LaB_6@SiO_2$ incorporated PCL microneedle patches melt in response to the NIR irradiation from the local heating of nanostructures and acts as smart device platform for on-demand drug delivery (Figure 13.7c). DOX-loaded NIR-responsive microneedles were inserted into the rat dorsal skin to study the *in vivo* triggered drug release. The study exemplified the potential of NIR-responsive microneedle patch devices for the on-demand drug release with a precise control over the release by tuning the laser irradiation time and light intensity. The NIR-responsive microneedle devices can be extended for the use of transdermal drug delivery and therapeutic applications across disease types with minimal invasion and precise control.

13.8 Ultrasound Responsive Drug Delivery Vesicles

Smart materials that respond to ultrasound stimuli have attracted the attention of researchers owing to the non-invasive nature of DDS developed based on these materials. Due to the unique properties, such as easy accessibility,

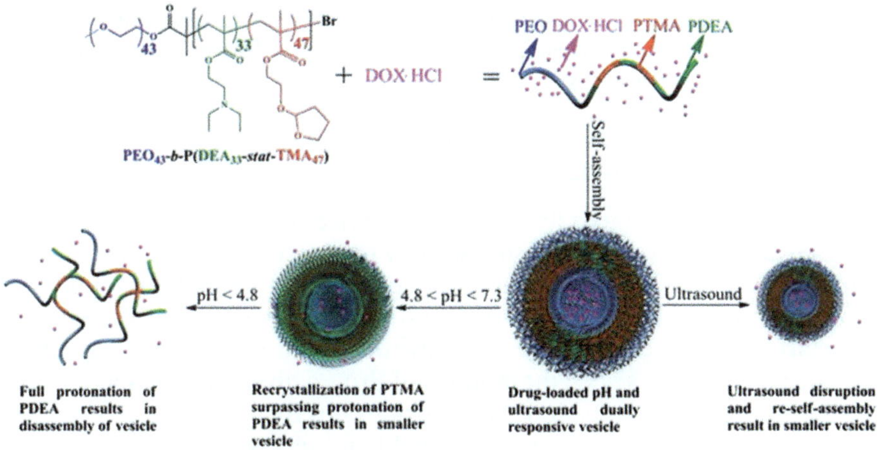

Figure 13.8 Schematic illustrations of formation and drug release from ultrasound and pH-responsive self-assembled polymer vesicles. Adapted from ref. 58 with the permission of Nature Publishing Group.

deep tissue penetration along with the non-invasive nature, ultrasound has been successfully used for diagnosis and therapeutic applications.[57] Du and co-workers have prepared ultrasound responsive non-toxic self-assembled polymer vesicles for cancer drug delivery (Figure 13.8).[58] Interestingly, the polymer vesicles were incorporated with a pH-responsive nature alongside the ultrasound to devise a smart DDS with dual responsive nature for advanced chemotherapeutic applications. The dual stimuli responsive self-assembling polymer vesicles were prepared from the block copolymer from hydrophilic poly(ethylene oxide) (PEO), pH-sensitive 2-(diethylamino) ethyl methacrylate (DEA) and ultrasound sensitive (2-tetrahydrofuranyloxy) ethyl methacrylate (TMA) and PEO$_{43}$-*b*-P(DEA$_{33}$-*stat*-TMA$_{47}$) through atom transfer radical polymerization (ATRP).

The haemolysis experiments demonstrated the non-toxic nature of the polymer vesicles at concentrations less than 100 μg mL^{-1}. Furthermore, the polymer vesicles showed good cytocompatibility with human liver cancer (HCCLM3) and normal liver (L02) cells under *in vitro* conditions. A model cancer drug doxorubicin hydrochloride (DOX.HCl) was used to demonstrate the successful encapsulation and stimuli-responsive controlled drug delivery properties of the polymer vesicles. The *in vitro* drug release studies indicated the controlled release of DOX from the polymer vesicle in the presence of ultrasound, as well as a change in solution pH. Polymer vesicles disrupted and underwent re-arrangement to form smaller vesicles in the presence of ultrasound and resulted in the release of vesicle contents (Figure 13.8). Similarly, recrystallisation of PTMA (4.8 < pH < 7.3) and protonation induced collapse of polymer vesicles (pH < 4.8) under acidic conditions enabled the drug release in response to the pH change (Figure 13.8). Interestingly, the combination of pH and ultrasound accelerated the drug release and can be employed as a means of tuning the release kinetics. Overall, the self-assembled polymer vesicles with external ultrasound and internal pH-responsive nature present a smart dual responsive design strategy for the development of next generation drug delivery vehicles.

13.9 Conclusions

Stimuli-responsive materials are a smart choice for the development of triggered drug delivery and have successfully been used for the construction of various drug delivery systems and devices. Access to a plethora of physical, chemical and biological stimuli have resulted in the design and development of smart drug delivery systems with controlled spatiotemporal drug delivery. We discussed a few selected examples of drug delivery systems and devices with stimuli-responsive on-demand drug delivery based on pH, glucose, enzyme, glutathione, light, magnetic field and ultrasound triggers under *in vitro* conditions, and an example of NIR-triggered *in vivo* transdermal drug delivery. Despite the tremendous progress of smart drug delivery systems for various diseases employing diverse stimuli triggers, there is still a need for next generation stimuli-responsive drug delivery systems and devices with more precise and on-demand delivery. Importantly, most of the reported stimuli-responsive drug delivery systems and devices are tested under either *in vitro* or *in cellulo* settings, but not under *in vivo* conditions involving real pathological situations. In spite of the recent progress in stimuli-responsive smart delivery devices, there is a plenty of room for the development of next-generation multifunctional smart drug delivery systems and devices that are biocompatible and biodegradable, in addition to being spatiaotemporal, target specific, for on-demand delivery.

Acknowledgements

We thank Prof. C. N. R. Rao FRS for constant support and encouragement, JNCASR, SwarnaJayanti Fellowship, the Department of Science and Technology (DST), Government of India (Grant: DST/SJF/CSA-02/2015-2016) and Sheikh Saqr Laboratory (SSL), ICMS-JNCASR, for financial support and a research fellowship to SM from CSIR, India.

References

1. A. N. Zelikin, C. Ehrhardt and A. M. Healy, *Nat. Chem.*, 2016, **8**, 997.
2. M. J. Webber and R. Langer, *Chem. Soc. Rev.*, 2017, **46**, 6600.
3. Y. Wang and D. S. Kohane, *Nat. Rev. Mater.*, 2017, **2**, 17020.
4. J. R. Lopes, G. Santos, P. Barata, R. Oliveira and C. M. Lopes, *Curr. Pharm. Des.*, 2013, **19**, 7169.
5. R. Langer, *Science*, 2001, **293**, 58.
6. A. Nelson, *Nat. Mater.*, 2008, **7**, 523.
7. F. Liu and M. W. Urban, *Prog. Polym. Sci.*, 2010, **35**, 3.

8. D. Roy, J. N. Cambre and B. S. Sumerlin, *Prog. Polym. Sci.*, 2010, **35**, 278.
9. S. Ganta, H. Devalapally, A. Shahiwala and M. Amiji, *J. Controlled Release*, 2008, **126**, 187.
10. C. Alvarez-Lorenzo, L. Bromberg and A. Concheiro, *Photochem. Photobiol.*, 2009, **85**, 848.
11. S. C. Abeylath, S. Ganta, A. K. Iyer and M. Amiji, *Acc. Chem. Res.*, 2011, **44**, 1009.
12. G. F. Luo, X. D. Xu, J. Zhang, J. Yang, Y. H. Gong, Q. Lei, H. Z. Jia, C. Li, R. X. Zhuo and X. Z. Zhang, *ACS Appl. Mater. Interfaces*, 2012, **4**, 5317.
13. S. Mura, J. Nicolas and P. Couvreur, *Nat. Mater.*, 2013, **12**, 991.
14. V. P. Torchilin, *Nat. Rev. Drug Discovery*, 2014, **13**, 813.
15. A. P. Blum, J. K. Kammeyer, A. M. Rush, C. E. Callmann, M. E. Hahn and N. C. Gianneschi, *J. Am. Chem. Soc.*, 2015, **137**, 2140.
16. S. Gadde, *Med. Chem. Commun.*, 2015, **6**, 1916.
17. C. Ding, L. Tong, J. Feng and J. Fu, *Molecules*, 2016, **21**, 1715.
18. B. Shen, Y. Ma, S. Yu and C. Ji, *ACS Appl. Mater. Interfaces*, 2016, **8**, 24502.
19. L. E. Gerweck and K. Seetharaman, *Cancer Res.*, 1996, **56**, 1194.
20. Q. Duan, Y. Cao, Y. Li, X. Hu, T. Xiao, C. Lin, Y. Pan and L. Wang, *J. Am. Chem. Soc.*, 2013, **135**, 10542.
21. Y. J. Zhu and F. Chen, *Chem.–Asian J.*, 2015, **10**, 284.
22. P. Bilalis, L.-A. Tziveleka, S. Varlas and H. Iatrou, *Polym. Chem.*, 2016, **7**, 1475.
23. B. G. Trewyn, I. I. Slowing, S. Giri, H.-T. Chen and V. S.-Y. Lin, *Acc. Chem. Res.*, 2007, **40**, 846.
24. L. Song, V. H. B. Ho, C. Chen, Z. Yang, D. Liu, R. Chen and D. Zhou, *Adv. Healthcare Mater.*, 2013, **2**, 275.
25. C. M. Alxander, M. M. Maye and J. C. Dabrowiak, *Chem. Commun.*, 2011, **47**, 3418.
26. N. C. Seeman, *Nature*, 2003, **421**, 427.
27. P. W. K. Rothemund, *Nature*, 2006, **440**, 297.
28. Y. Krishnan and F. C. Simmel, *Angew. Chem., Int. Ed.*, 2011, **50**, 3124.
29. J. Sharma, R. Chhabra, H. Yana and Y. Liu, *Chem. Commun.*, 2007, 477.
30. N. Narayanaswamy, G. Suresh, U. D. Priyakumar and T. Govindaraju, *Chem. Commun.*, 2015, **51**, 5493.
31. N. Soh and T. Ueda, *Talanata*, 2011, **85**, 1233.
32. M. B. Avinash and T. Govindaraju, *Adv. Mater.*, 2012, **24**, 3905.
33. A. S. Hoffman, *Adv. Drug Delivery Rev.*, 2012, **64**, 18.
34. H. R. Culver, J. R. Clegg and N. A. Peppas, *Acc. Chem. Res.*, 2017, **50**, 170.
35. M. Zhang, C. Song, F. Du and Z. Li, *ACS Appl. Mater. Interfaces*, 2017, **9**, 25905.
36. L. C. Lo and C. Y. Chu, *Chem. Commun.*, 2003, 2728.
37. N. Narayanaswamy, S. Narra, R. Nair, D. K. Saini, P. Kondaiah and T. Govindaraju, *Chem. Sci.*, 2016, **7**, 2832.
38. S. Zhang, *Nat. Biotechnol.*, 2003, **21**, 1171.
39. A. N. Moore and J. D. Hartgerink, *Acc. Chem. Res.*, 2017, **50**, 714.

40. D. Kalafatovic, M. Nobis, N. Javid, P. W. J. M. Frederix, K. I. Anderson, B. R. Saunders and R. V. Ulijn, *Biomater. Sci.*, 2015, **3**, 246.
41. M. Egeblad and Z. Werb, *Nat. Rev. Cancer*, 2002, **2**, 161.
42. M. R. Dzamukova, E. A. Naumenko, Y. M. Lvov and R. F. Fakhrullin, *Sci. Rep.*, 2015, **5**, 10560.
43. T. G. Shutava, R. F. Fakhrullin and Y. M. Lvov, *Curr. Opin. Pharmacol.*, 2014, **18**, 141.
44. Y. Lee, G.-E. Jung, S. J. Cho, K. E. Geckeler and H. Fuchs, *Nanoscale*, 2013, **5**, 8577.
45. A. Bernardos, L. Mondragon, E. Aznar, M. D. Marcos, R. Martinez-Mañez, F. Sancenon, J. Soto, J. M. Barat, E. Perez-Paya, C. Guillem and P. Amoros, *ACS Nano*, 2010, **4**, 6353.
46. W. Gao and J. Wang, *Nanoscale*, 2014, **6**, 10486.
47. W. Duan, W. Wang, S. Das, V. Yadav, T. E. Mallouk and A. Sen, *Annu. Rev. Anal. Chem.*, 2015, **8**, 311.
48. Y. Tu, F. Peng, P. B. White and D. A. Wilson, *Angew. Chem., Int. Ed.*, 2017, **56**, 7620.
49. X. Zhao, J. Kim, C. A. Cezar, N. Huebsch, K. Lee, K. Bouhadir and D. J. Mooney, *Proc. Natl. Acad. Sci. U. S. A.*, 2011, **108**, 67.
50. K. Cai, Z. Luo, Y. Hu, X. Chen, Y. Liao, L. Yang and L. Deng, *Adv. Mater.*, 2009, **21**, 4045.
51. T. Hoare, B. P. Timko, J. Santamaria, G. F. Goya, S. Irusta, S. Lau, C. F. Stefanescu, D. Lin, R. Langer and D. S. Kohane, *Nano Lett.*, 2011, **11**, 1395.
52. M. Karimi, P. S. Zangabad, S. Baghaee-Ravari, M. Ghazadeh, H. Mirshekari and M. R. Hamblin, *J. Am. Chem. Soc.*, 2017, **139**, 4584.
53. M. S. Yavuz, Y. Cheng, J. Chen, C. M. Cobley, Q. Zhang, M. Rycenga, J. Xie, C. Kim, K. H. Song, A. G. Schwartz, L. V. Wang and Y. Xia, *Nat. Mater.*, 2009, **8**, 935.
54. M. Chen, M. Ling, K. Wang, Z. Lin, B. Lai and D. Chen, *Biomacromolecules*, 2015, **16**, 598.
55. M. A. Woodruff and D. W. Hutmacher, *Prog. Polym. Sci.*, 2010, **35**, 1217.
56. C. J. Chen and D. H. Chen, *Chem. Eng. J.*, 2012, **180**, 337.
57. W. G. Pitt, G. A. Husseini and B. J. Staples, *Expert Opin. Drug Delivery*, 2004, **1**, 37.
58. W. Chen and J. Du, *Sci. Rep.*, 2013, **3**, 2162.

CHAPTER 14

Regulatory and Commercialization Challenges with Stimuli-responsive Nanomedicines

AMIT SINGH[a] AND MANSOOR M. AMIJI*[b]

[a]Allexcel Inc., 135 Wood St., West Haven, Connecticut, USA 06516; [b]Department of Pharmaceutical Sciences, School of Pharmacy, Bouve College of Health Sciences, Northeastern University, 360 Huntington Avenue – 140 TF, Boston, Massachusetts, USA 02115
*E-mail: m.amiji@neu.edu

14.1 Introduction

Nanotechnology has revolutionized the healthcare industry and several innovative drug products under the general umbrella of "nanomedicines" are being investigated for multiple indications. However, since the approval of Doxil® by the United States Food and Drug Administration (FDA) in 1995 as the first liposomal nanoparticle formulation of doxorubicin, the approval of novel nanotherapeutic products has been slow and challenging with only a few products managing to make successful transition from bench-side to the bed-side.[1] Nanoparticle-based drug delivery systems can be viewed as a complex three-dimensional (3D) construct where each component confers a

Biomaterials Science Series No. 1
Stimuli-responsive Drug Delivery Systems
Edited by Amit Singh and Mansoor M. Amiji

Published by the Royal Society of Chemistry, www.rsc.org

unique property such as targeting, evading immune recognition or providing a hospitable environment to labile drugs in the body until they reach their site of action. Such complex constructs require precision in design, extensive physico-chemical and biological characterization, clear understanding of the spatial arrangement of the components and acquiring manufacturing adequacies for large batches. The properties of a nanomedicine construct are a delicate interplay of the individual components and a subtle change in the property of a component can adversely affect the characteristics of the end product. A comprehensive understanding of the properties of the individual components and their impact on the overall property of the construct is therefore key to designing a commercially viable nanotherapeutics product. Therefore, developing a nano-based therapeutic product has fundamental differences compared to conventional drug development, that warrants a revised approach to drug product design, chemistry, manufacturing and controls (CMC) and quality assurance criteria. The following sections will discuss the unique challenges in development of nanomedicines and efforts from regulatory agencies to develop a harmonized platform to support their development.

14.2 Challenges with Developing Nanomedicines

There are multiple challenges to developing nanomedicines for clinical use because of added levels of complexity associated with these three-dimensional constructs. Like conventional drugs, manufacturing quality control, batch-to-batch product consistency, detailed characterization, appropriate analytical assays to assess safety and efficacy have to be established with proper standards to be able to get the product approved for human use. Some of the considerations for nanomedicine development should include:

- Through understanding of the individual components that are to be used for product development and their interactions;
- Identification of critical attributes that govern the physical and biological properties of the product and impact of variability;
- Identification of in-process parameters that would aid in manufacturing of product quality batches;
- Validated analytical assays for detailed characterization of the drug product and their degraded by-products;
- Ability to produce sterile drug products;
- Ability to target the drug product to the desired site of action and mitigate side-effects;
- Clinically acceptable stability, efficacy, safety and product half-life.

Nanotechnology exploits size-dependent properties of materials that allows designing of materials at the atomic and molecular level to develop and design functionally superior drug products. Typical nanomaterial-based medicines contain multiple components that collectively impart unique characteristics to the drug product and therefore are popularly referred to as "*Non-biological complex drugs (NBCDs)*". The inherent complexity of nanoparticles as drugs or drug carriers warrants a precision in product design to achieve a consistent product quality and therefore "Quality by Design (QbD)" is a highly recommended approach. QbD is a systematic approach to product development with a predefined objective and emphasis on understanding the product, process and CQAs based on sound scientific rationale and quality risk. It is therefore important to identify the CQAs that define the process, as well as the product, and relate to the performance and quality. This becomes critical especially for drug products containing nanomaterials because a subtle change in physical properties such as size, shape, surface charge, surface functionality, drug release profile or targeting ligand concentration can change the stability and biological properties of the drug product. Size, for example, is a critical parameter for nanomaterials because the surface-to-volume, and thus the surface properties of the nanoparticles, change dramatically with a change in size and the manner in which they interact with the biological systems.[2]

The "*nano*" component of a drug product might be an active pharmaceutical ingredient (API) by itself such as drug nanocrystals or it could merely be a nanoparticle drug carrier where it may serve as an excipient with no pharmacological significance. However, even as a drug delivery system, the physicochemical properties and biological performance of nanotherapeutics are highly dependent on properties such as size, surface charge, shape, surface coating or drug release kinetics. There are a wide variety of nanoparticle delivery systems such as polymers, liposomes, proteins, inorganic materials like gold, silver, iron oxide, quantum dots, micelles and each of these nanosystem are unique in their physiochemical properties and synthetic procedures. A "*one-fit-all*" approach to define properties and critical quality attributes (CQAs) for nanomaterials is therefore not viable; each nanomaterial has to be addressed independently to understand and anticipate the challenges associated with its development.

14.2.1 Controlling Physicochemical Properties

Nanotechnology by definition implies a length scale-based identity to a material; size range between macroscopic and atomic scale where the material properties are dramatically different from either extreme. Size and its distribution in a population therefore becomes the most critical parameter in the development of nanomedicines. Prior evidence clearly suggests that the size of the nanomaterial governs its pharmacokinetic behavior, efficacy and safety. Smaller particles with size less than 20 nm are rapidly removed from circulation by renal clearance, while particles larger than 200 nm are efficiently taken up by cells comprising the mononuclear phagocytic system (MPS). Similarly, larger sized particles tend to accumulate more in the spleen and liver and are most likely to undergo hepatic clearance from the body. It is impossible to synthesize nanoparticles without any size distribution in a population and therefore the size distribution characteristics of a nanoparticle system should be an important consideration when anticipating the biodistribution. It is also crucial to acknowledge that the biodistribution of the nanoparticles in the body does not solely depend on the size of the particle, but shape and surface charge play a major role. Smaller sized spherical nanoparticles prefer to accumulate in the liver and spleen, but the larger sized nanoparticles accumulate in the kidney. Similarly, negatively charged particles tend to remain in systemic circulation for longer, but positively charged nanoparticles are highly prone to uptake by phagocytic cells in the lungs, liver and spleen. Figure 14.1 shows a schematic representation of nanoparticle biodistribution as a function of size, shape and surface charge.[3] Upon entering the circulation, nanoparticles are rapidly coated with blood components, a process called opsonization; a natural defense mechanism that is meant to flag the foreign particles and activate the complement system to aid in their clearance from the body.[4] Therefore the nanoparticle surface properties themselves change upon contact with a biological system, a hydrophilic surface can become more hydrophobic due to protein coating, while a hydrophobic nanoparticle would be taken up by macrophages without the need for opsonization.[5] Polyethylene glycol (PEG) and other polymers and biomolecules such as CD47 have been shown to help in evading recognition by MPS and prolonging residence time of nanoparticles in the body.[6] Similarly, attempts have been made to engineer the nanoparticle shape, size, surface charge, surface characteristics and elasticity to improve circulation time, cell penetration, mitigate non-specific uptake and control their drug release kinetics and achieve a better safety and efficacy profile.[7]

14.2.2 Challenges with Analytical Characterization

Development of analytical methods for characterizing nanoparticles as a stand-alone entity and in formulation is another challenging task that limits our ability to understand the nanosystems and predict their clinical performance. An analytical method should be able to analyze an analyte in a complex matrix and predict its concentration, variability from batch to batch, impact of such variability on the overall property of a drug product and define selection criteria for acceptance or rejection of a product batch. The well evolved conventional characterization tools or assays may not be amenable to accommodate nanosystems as an analyte and depending on where they are, data analysis and interpretation could be tricky.[8] A nanoparticle has a complex 3D geometry with multiple components that have functional roles in the end product and so it would be of utmost importance to understand qualitatively and analyze quantitatively the spatial arrangement and stoichiometry of these components and define the product selection criteria. Abraxane® could be taken as a case example where the manufacturing processes, source of albumin protein and nanoparticle preparation methods with paclitaxel posed issues with consistency in the product quality due to the variability in physical characteristics.[5] It is therefore extremely important to understand the effect of a process variable on the product quality and its

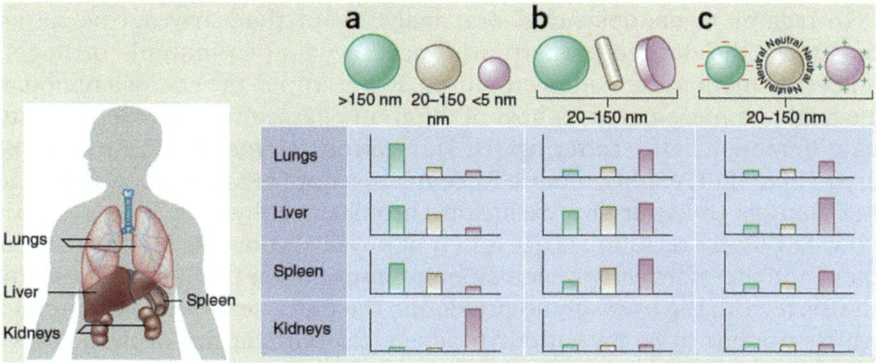

Figure 14.1 Nanoparticle size, shape and surface charge dictate biodistribution among the different organs including the lungs, liver, spleen and kidneys. (**a**) Spherical particles, including gold nanoparticles, liposomes and polymeric micelles/nanoparticles can vary in size and display disparate *in vivo* fates. Large rigid particles with diameters >2000 nm accumulate readily within the spleen and liver, as well as in the capillaries of the lungs. Nanoparticles in the range of 100–200 nm have been shown to extravasate through vascular fenestrations of tumors (the EPR effect) and escape filtration by the liver and spleen. As size increases beyond 150 nm, more and more nanoparticles are entrapped within the liver and spleen. Small-sized nanoparticles (<5 nm) are filtered out by the kidneys. (**b**) Novel 'top-down' and 'bottom up' fabrication techniques have enabled the exploration of different geometries of nanoparticles, including cylindrical and discoidal shapes, which have been shown to exhibit pronounced effects on pharmacokinetics and biodistribution. Different nanoparticle shapes exhibit unique flow characteristics that substantially alter circulating lifetimes, cell membrane interactions and macrophage uptake, which in turn affect biodistribution among the different organs. (**c**) Charge of nanoparticles stemming from distinct surface chemistries influences opsonization, circulation times and interaction with resident macrophages of organs comprising the MPS, with positively charged particles more prone to sequestration by macrophages in the lungs, liver and spleen. Neutral and slightly negatively charged nanoparticles have longer circulation lifetimes and less accumulation in the aforementioned organs of the MPS. In both **b** and **c**, the size of the nanoparticles is assumed to range from 20–150 nm. Individual panels represent *in vivo* fates of nanoparticles, taking into account singular design parameters of size, shape and surface charge independent of one another, and for this reason, respective scales vary from one panel to the next. It is important to note that *in vivo* biodistribution will undoubtedly vary based on the interplay of several of the above parameters. Reprinted with permission from Nature Publishing Group, ref. 3, Copyright © 2015, Rights Managed by Nature Publishing Group.

effect thereafter and be able to precisely measure the change in the product quality. Simultaneously, it becomes important to understand the instrumental limitations, parameters that can increase or decrease variability, sampling size, possible contaminations and any other factor that can affect precise measurement of the desired property.

In regards to nanoparticles, one analytical method may not be suitable to discern the desired property with accuracy and so multiple orthogonal analytical methods are often employed. For example, the size of a nanoparticle could be measured by a host of different characterization methods such as differential light scattering (DLS), electron microscopy (TEM, HRTEM, cryo-TEM, SEM, eSEM), atomic force microscopy (AFM), nanoparticle tracking analysis (NTA) or size exclusion chromatography with light scattering (SEC-LS). These techniques use different physical concepts for size measurement and therefore will provide different information from the same sample. In this regard, the user should pre-define the purpose for the measurement: which size has to be measured? Independent particle size or particles with agglomeration; what is the selection criteria for the sample (acceptance criteria)? Is hydrodynamic size is needed or actual particle size is desired? Is the technique to be used suitable for measuring the particle size of the sample with statistical accuracy (system suitability)? Does shape play a role in the particle size determination? The analytical characterization of nanoparticles therefore becomes an arduous task because of the complexity of the system. In addition, the analytical technique may not be able to differentiate between a functionally active nanoparticle system from an inactive or less active form and therefore the physical characterization should be correlated with the biological activity to corroborate the analysis. A lack of suitable standards to validate an analytical method becomes another challenge, which will be discussed in more detail in the subsequent section on the regulatory challenges in nanoparticle characterization.

14.2.3 Manufacturing Challenges with Nanomaterials

Manufacturing a product in large scale repeatedly is a prerequisite for product development and becomes overly difficult for complex multicomponent nanoparticles. In this regard, identification of key parameters that can be suitably monitored as in-process controls during the nanoparticle manufacturing is critical. The earlier these parameters are identified in the drug development process, the more useful they will be in devising a robust, reproducible and scalable process for nanoparticle synthesis. Understanding of components of a nanoparticle and their interaction in a 3D nanoconstruct is therefore of paramount importance to determine the CQAs for a process. Simultaneously, a foresight has to be developed while adopting a process at the bench or small scale so that the process can be seamlessly integrated into a large batch process during manufacturing. The scaling up issue could be as simple as removal of a by-product that might have an adverse side effect clinically, which can be easily achieved by a process at the bench scale but may not be amenable to integration at the larger scale. Typical drug products usually do not have complex components that could be sensitive to physical parameters such as temperature, pH, viscosity, ionic strengths or solvents. Similarly, parameters such as concentration, drug-carrier ratio, stochiometry of the targeting moiety or ratios of the other excipients should be in perfect balance to obtain a finished product of high quality. Many a time, nanoparticle systems have labile molecules such as proteins, peptides, antibody fragments, nucleic acids or aptamers which may not function as API, but could be critical as targeting ligands or protective coatings to evade immune response, and thereby capable of altering the therapeutic outcome of a drug product. The manufacturing process therefore should be conducive to accommodate such molecules in their active form to perform the intended function.

Reproducibility is the major concern with the manufacturing process, where under defined parameters, the end result of the batch should be consistent with the previous batches produced under same conditions. Nanoparticle production is often a multistep and sequential process, and therefore with multiple components involved in the process, it is difficult to achieve a robust process to maintain reproducibility. It becomes critical to define the process parameters adequately, identify the key attributes that govern the process flow and define the acceptability criteria that are appropriate for regulatory purposes. Implementation of these principles at the large scale warrants extensive preliminary experimentation at the small scale to have a clear understanding of the factors that govern the process. This further cements the need for a well-defined and robust analytical characterization cascade that can be applied as a quick but reliable source of information to develop a reproducible process with adequate control. The intended end use of the finished product often governs the process development for any drug product, and so nanomaterial containing drugs are no exception.

A prior knowledge of the end use including the route of administration, physical form of the drug product and extent of its exposure to the biological elements is important in designing a product acceptable to the regulatory agencies. Injectable drugs, for example, have to undergo a very strict regulatory approval consideration because they are directly exposed in circulation as opposed to topical creams, which have limited systemic exposure. Sterility of the finished drug product becomes extremely important in this regard. It is clearly established that nanoparticles are 3D complex structure and their structural integrity is the key to their optimal clinical performance. The majority of the terminal sterilization techniques therefore cannot be applied to nanomaterials containing drug products because they compromise the architecture of the nanomaterial and hence the activity.[9] Filtration can be an option that could work well with softer nanoparticles, such as micelles or liposomes, but may not be equally easy to apply to hard nanoparticles based formulations. Filtration through a 220 nm filter is essential to ensure sterility, but may not be applied to nanoparticles in that size range, could lead to product loss, may be difficult to process viscous formulations or may lead to contamination due to loss in membrane integrity. Besides, filtration will be able to remove particulate microorganisms, but would not be able to eliminate their toxins or other pyrogens, which could be detrimental to the drug product as well as the humans receiving them. Aseptic manufacturing is another choice to maintain sterility during the course of such processes, but maintaining an aseptic environment throughout the manufacturing process for a multi-step synthetic scheme is highly cumbersome, expensive and prone to failure.

14.2.4 Challenges with Biological Characterization

Biological activity is the ultimate goal that has to be demonstrated conclusively for a drug product to move to the clinic. A typical drug product is initially tested for its *in vitro* activity in multiple cell culture models to establish key biological feature such as safety, efficacy and immunological concerns. Use of the standard assays suitable for small molecule drugs may not translate with similar efficiency for nanoproducts. Small molecule drugs tend to remain as a stable solution during the course of incubation with cells and diffuse with ease even under static conditions, but nanoparticles tend to settle down over time. Therefore, using a similar strategy and protocols for small molecule drug products and nanomaterials may bias the results in favor of one or the other. Drug release kinetics is a recommended study for drug nanocarriers because it provided information on the drug availability under physiological conditions and helps anticipate the PKPD profile. However, performing such experiments on nanocarriers is challenging due to their complex drug release characteristics and difficulty with maintaining a biologically relevant environment for such a study. Due to the lack of a robust method for drug release testing, the reliability of *in vitro* data is often confirmed by *in vivo* PKPD and bioequivalence data that could be time consuming, expensive and often equally challenging.

Determining the PK/PD profile, ADME, efficacy and safety of small molecule drugs can be studied easily because of well-established methods, but the same methods may not be applicable for nanoparticle drugs. In a dynamic system like blood circulation, small molecule drugs can diffuse easily and thus achieve an equilibrium state in blood or plasma and therefore their PK profile can be related to that of the target tissue to some extent. However, the same assumption may not provide an accurate PK profile of the nanomaterials because unlike small molecules, they are 3D colloidal suspensions with different mobility characteristics and diffusion capability. It might be more relevant to study the PK profile of the target tissue to estimate the achievable drug levels in such cases. In this regard, considering the "*nanoparticle concentration*" in the target tissue may also not prove to be an appropriate yardstick for predicting therapeutic efficiency and rather the actual API in the formulation should be measured for PK, efficacy and toxicity considerations. It would also prove important to confirm that a nanoformulation does not have free, unencapsulated drug or has not released the drug in formulation such that two drug entities (free and encapsulated) exist even before the drug product is injected into the blood of the animal being studied. In such cases, the PK profile could be a combined profile of the drug encapsulated within the nanomaterial formulation and "free drug" that has been released from the nanocarrier.

Readers are encouraged to read an in-depth commentary by Stern *et al.* on the impact of unencapsulated drug on the PK profile of a nanomedicine and challenges in discerning the information.[10]

The challenges with *in vitro* testing and correlating them to their *in vivo* performance has left a huge void in our understanding of biological behavior of drug product containing nanomaterials. *In vitro–in vivo correlation* (IVIVC) models have been developed to predict the relation between an *in vitro* property and it's *in vivo* performance. The FDA defines IVIVC as "*A predictive mathematical model describing the relationship between an in vitro property of an extended release dosage form (usually the rate or extent of drug in vitro release or release) and a relevant in vivo response, for example, plasma drug concentration or amount of drug absorbed*". The central idea of developing IVIVC models is to reduce the number of *in vivo* tests required to get a product approved for clinical use. However, establishing a validated and reliable predictive model for nanomaterials is extremely challenging due to the lack of a suitable knowledge bank. Drug release in an *in vitro* environment is commonly studied by dialysis or the flow through cell method, which have poor predictability of physiological drug release conditions. Therefore, great efforts need to be made in developing and validating better methods to study drug release *in vitro* and apply the knowledge to ensure product quality.[11]

14.2.5 Challenges with Toxicological Profiling

Patient safety is always the primary concern with any drug product and pre-clinical toxicology data from animal studies is used for regulatory approval of a drug product for clinical studies. In theory, the existing methods for toxicology analysis should be suitable for predicting any adverse biological effect but the general consensus is that it may not be applied universally to all drug products, especially those containing nanomaterials. An ideal nanomedicine should be biocompatible, should not interfere with physiological processes, should not cause any short-term or long-term adverse effects and should be eventually eliminated from the body. Use of "generally regarded as safe (GRAS)" materials can ensure biocompatibility to some extent, but some of these desired characteristics would depend on multiple factors. For example, a nanoparticle that is rapidly cleared from the body should not pose unprecedented toxicological threats, but those that accumulate in tissues and are retained in the body for longer time (due to a slow clearance profile) may have long term effects, particularly under a multiple dosing regimen. The majority of toxicological analyses are short-term studies that may not be suitable for prediction of long-term exposure effects. It is also important to appreciate that different nanomedicines may have a different mode of action, which would be closely related to their physicochemical properties and thus product-specific toxicological probabilities have to be considered while designing the safety and tolerability studies. Physicochemical properties such as size, shape, surface charge, hydrophobicity/hydrophilicity characteristics or surface modification may contribute

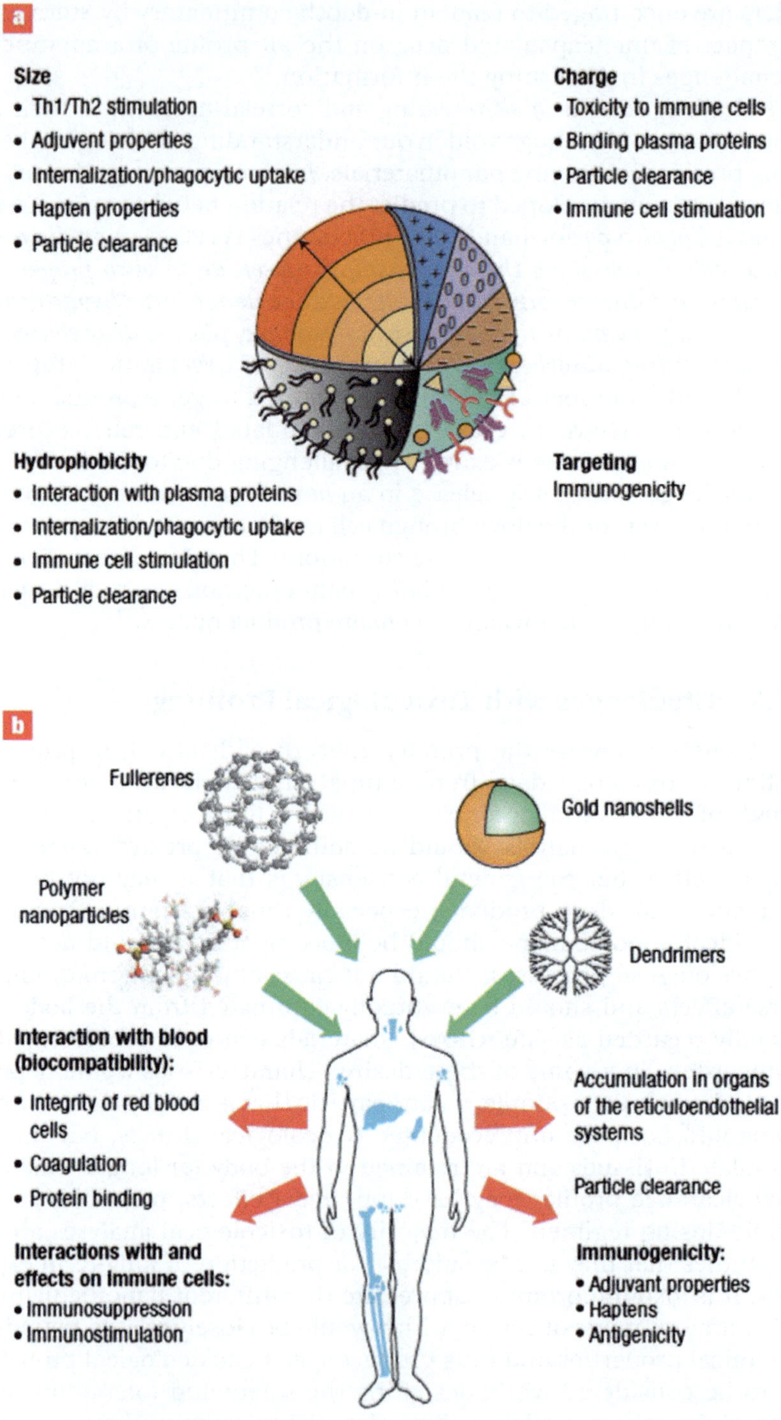

Figure 14.2 Nanoparticle properties determine their interaction with the immune system. (**a**) The effect of nanoparticle size, charge, hydrophobicity and targeting on immunotoxicity. (**b**) Some nanoparticles (shown

to the toxicological profile of a nanomaterial and thus a more thorough profiling may be required to identify the contributing factor.

14.2.6 Challenges with Immunological Profiling

Unlike toxicological profiling, the immunological profiling may not be as predictive in animal studies for several reasons. The immunological makeup of humans is fairly diverse compared to the animals used for tolerability and safety studies and thus the immunological profile from such studies may not represent the possible outcome in humans. More specifically, nanoparticles containing biological macromolecules such as proteins, peptides or nucleic acids may illicit an immune response in humans because these biomolecules can serve as antigens. In certain cases, the interaction of a drug-nanocarrier might change the conformation of the biomolecule to trigger the immune system, an example being albumin bound paclitaxel (Abraxane®) that shows immunological response in pigs, as opposed to albumin control where no effect was observed.[5] Nanoparticles themselves are capable of generating an immune response due to their physicochemical properties (Figure 14.2).[12] Scientific evidence confirms that the blood proteins bind to the nanoparticles in systemic circulation and the nature of bound protein tags the particles for uptake by immune cells. Nanoparticles that are readily opsonized can activate the complement pathway that can further lead to activation of the immunological cascade. Despite an abundance of effort to understand the interaction of nanoparticles with the blood components or toxicity associated with the route of administration, a distinct trend has not been established primarily due to lack of thorough evidence. While developing an immuno-toxicological profile of nanomaterials, the following aspects should be focused to obtain conclusive and indicative data:

- Physicochemical properties such as size, shape, surface charge, surface modifications, nanoparticle stability and their impact on the physiological interactions;
- The nature of surface modification of the nanoparticle in biological environment including the change in density, thickness and nature of surface coating;
- Correlation of the physicochemical property to the biological interaction of the nanoparticles;

---

schematically) can trigger certain immune responses as listed here. A characterization scheme for nanoparticles intended for biomedical applications must include testing for these responses. Such tests may exclude a potentially harmful drug candidate from the development pipeline and inform future studies relevant to the immunomodulatory properties of nanoparticles. Reprinted with permission from Nature Publishing Group, ref. 12, Copyright © 2007, Rights Managed by Nature Publishing Group.

- Variability in product quality during the manufacturing process and;
- A thorough immunological characterization.

Surface characteristics of the nanoparticle probably play the most critical role in determining how the nanoparticle will interact with blood components. The "*stealth effect*" offered by PEG modification of nanoparticle surface has been argued as a result of the decrease in protein adsorption in circulation.[13] On the other hand, positively charged nanoparticles have shown higher potential for erythrocyte damage, hemolysis and immunotoxicity compared to neutral or negatively charged nanoparticles.[14,15] In a study, it was demonstrated that smaller sized silica nanoparticles (30 and 70 nm) show higher cytokine levels *in vitro* and stronger inflammatory response *in vivo* than the large sized particles (300 and 1000 nm).[16] The mechanism of immune toxicity was associated to the generation of reactive oxygen species and activation of mitogen-activated protein kinases (MAPKs). Importantly, modification of the surface of the 70 nm silica particles with carboxyl groups reduced the MAPKs activation, resulting in the reduction of immune activation. Similarly, reduction of the cationic surface charge density has shown to reduce toxicity and immune response.[17] A general trend suggests that increasing the surface charge of nanoparticles leads to reduction in adsorption of blood protein, but a contrasting trend has also been reported.[12] Such discrepancy in the literature clearly indicates that a more thorough examination of factors contributing to immunotoxicity of nanoparticles is required to further our understanding and aid in the development of safer nanomedicines.

14.3 Regulatory Framework and Nanomaterials

The transitional path for any investigational drug into a drug product varies from country to country, but the general approval structure is consistent to maintain an overall uniformity. For the sake of clarity, the formal approval path for a drug by the FDA in the USA is discussed as a case example (Figure 14.3). A typical drug development path is initiated by a *preclinical phase* that includes discovery of a new drug, study of its pharmacokinetics/pharmacodynamics, understanding of its toxicity profile and dosage range and establishing its efficacy for the intended indication. The knowledge base developed from the preclinical phase is used to submit a formal application with the FDA as a investigational new drug (IND) to initiate a clinical testing for safety and efficacy in human subjects. The data obtained from the preclinical toxicology, efficacy and safety forms the key basis to support an IND application. Upon approval from the FDA, the "*clinical phase*" is initiated which can be divided into Phase I, II and III where the efficacy and safety of the drug is evaluated in human subjects. When a drug shows statistically significant improvement in efficacy compared to the existing approved drug without eliciting any adverse toxicity, the drug manufacturer can submit a new drug application (NDA), seeking license to sell the drug for clinical use

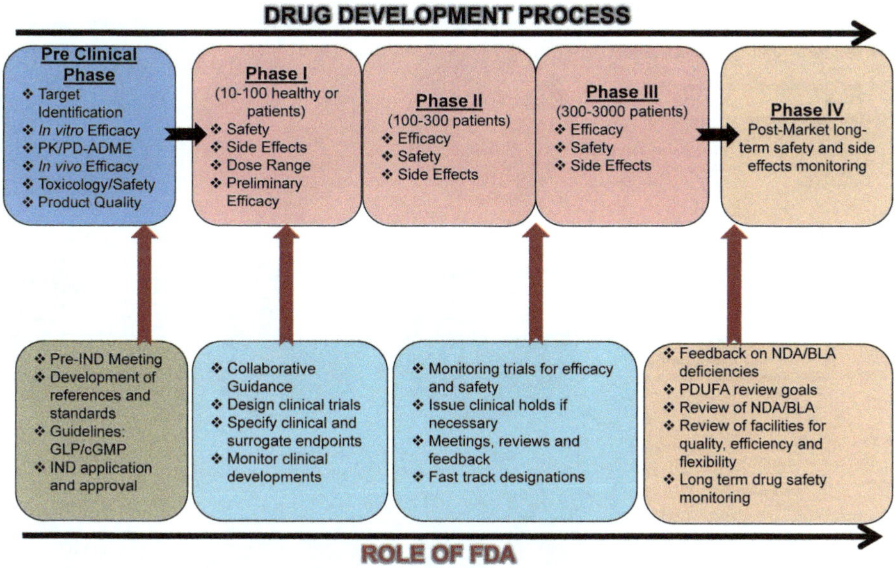

Figure 14.3 Schematic representation of various stages in drug development process and the role of the FDA in regulating the approval process.

with appropriate labels that guides a medical practitioner for proper use. Once the drug is in the market, the FDA continues to monitor the long term effect of the drug in the patients as *"Phase IV: Post-market safety monitoring"* and can request additional studies, changes in dosage, labeling, caution statements or in extreme adverse conditions, terminate the use of the drug. The FDA also continues to inspect the manufacturing facilities domestically or abroad to ensure that minimum quality standards are maintained in the manufacturing process.

The fierce pace at which the pharmaceutical companies are pursuing novel nano-based drug development has propelled endeavors from regulatory agencies to adopt new guidelines. A total for 359 applications for drug products containing nanomaterials have been submitted to the FDA between 1970 and 2015, of which 234 were IND applications (to determine if a product is safe for clinical application), 62 were NDA and 62 were abbreviated new drug application (ANDA).[18] Figure 14.4 shows the breakdown of nanomaterial containing drug product between 1970 and 2015 according to the class of nanomaterial, indication and the route of administration. With nanotherapeutics and nanopharmaceuticals being an emerging technology, the majority of the regulatory bodies are still trying to come to terms with a universally acceptable code for nomenclatures and definitions, CQAs to define the drug products and establishing their bioequivalence to an existing drug. The majority of the regulatory framework at the FDA was drafted before nanotechnology was officially applied for medical products, but they have adapted to the changing landscape. The FDA acknowledges the presence of

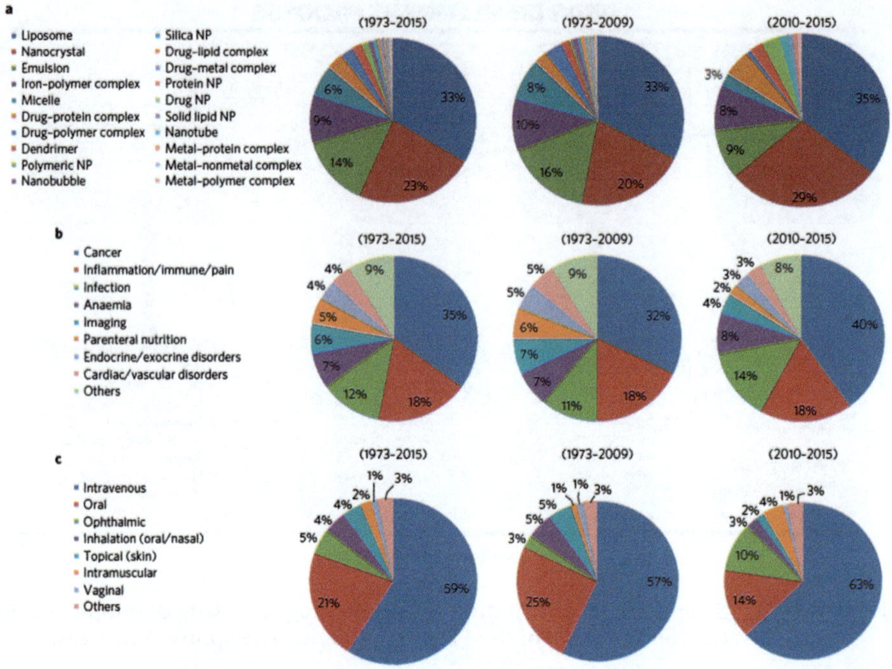

Figure 14.4 Distribution of nanomaterial use in drug products from 1973 to 2015. (**a**) Breakdown of the types of nanomaterials used in drug products. The nanotechnology terminologies do not represent any implication for CDER drug product labelling and were used only to describe/interpret the type of nanomaterials in identified drug products for the purpose of this analysis. (**b**) Breakdown of the types of indication for drug products containing nanomaterials. The uses (prophylactic/therapeutic/diagnostic) for each product were categorized into nine application areas on the basis of their intended or approved use. (**c**) Overview of routes of administration for drug products containing nanomaterials. NP, nanoparticle. Reprinted with permission from Nature Publishing Group, ref. 18, Copyright © 2017, Rights Managed by Nature Publishing Group.

nanomaterials with size-dependent properties in products, but takes a more inclusive approach to accommodate such applications instead of a case-to-case basis. The FDA enlists the following attributes to regulating drug products including nanomaterials:[19]

- The FDA maintains its product-focused, science-based regulatory policy;
- Respects the variations in legal standards for different product-classes;
- Where premarket review authority exists, attention to nanomaterials is incorporated into standing procedures;
- Where statuary authority does not exist for premarket review, consultation is encouraged to reduce the risk of unintended harm to human or animal health;

- The FDA will continue post-market monitoring;
- The industry is responsible for ensuring that its product meets all the legal standards including safety requirements;
- The FDA will collaborate with domestic and international counterparts on regulatory issues and;
- Both for products that are not subject to premarket review and those that are, the FDA will offer technical advice and guidance, as needed, to help industry meet its regulatory and statutory obligations.

The FDA issued a guidance document in 2014 titled, "*Considering whether an FDA-regulated product involves application of nanotechnology*", which reiterates that the FDA has not established a regulatory definition for "nanotechnology", "nanomaterial", "nanoscale" or other related terms. However, this guidance identifies two considerations that would be used to classify if a product involves application of nanotechnology: (1) whether a material or end product is engineered to have at least one external dimension, or an internal or surface structure, in the nanoscale range (approximately 1 nm to 100 nm) or (2) whether a material or end product is engineered to exhibit properties or phenomena, including physical or chemical properties or biological effects, that are attributable to its dimension(s), even if these dimensions fall outside the nanoscale range, up to one micrometer (1000 nm).[20] The conventional regulatory framework is robust but flexible, allowing the agency to assess potential risks involved with nanomaterials in drug products and exercise discretion in their decision making processes. As a result, an application that includes nanomaterials in a drug product is assessed through the same regulatory pathway as a conventional drug product. The Center for drug evaluation and research (CDER) is an organizational branch of the FDA that is responsible for evaluating the safety and efficacy of drugs that are going to be available for human use. Their responsibility includes regulation of over-the-counter (OTC) and prescription drugs including biological therapeutics and generic drugs. As is the case with conventional small molecule drugs, nanomaterials in drug products are also regulated by the Code of federal regulation (CFR), Chapter I, Subchapter D: Drug for Human Use, which includes information desired in submitted applications. As such, CDER seeks following information while reviewing any application for a drug product:[21]

- Description and composition of the drug product including active pharmaceutical ingredient, as well as excipients (inactive components);
- Critical quality attributes that define the drug product;
- Physicochemical characterization for the CQAs of the drug product;
- Details of CMC for the drug product;
- Proof of validated analytical methods to measure the CQAs;
- Stability criteria of the drug product for intended use including shelf life and expiration date.

The FDA has published guidance documents available in the public domain that provides recommendations on obtaining critical information about a drug product, suitable product-specific analytical considerations, CQAs and appropriate ways to report the findings.[22] An escalating number of submissions with nanomaterials in drug products has prompted the FDA to provide guidance documents specific to nanomaterial-based products. As a pioneering effort, the FDA published a guidance document titled, "*Liposome Drug Products*" in as early as August, 2002 and its first revised version in October, 2015,[23] which provides a yardstick for industry to gauge the FDA's expectations on a nanomaterial drug product.

The European medicines agency (EMA), the European counterpart of the FDA, has a strikingly similar approach to regulating nanomaterials containing drug products, they also process such submissions according to the conventional guidelines for a drug product. However, the European regulatory agencies have a slightly more elaborate and better defined structure to addressing drug products with nanomaterials. The European commission (EC) published a recommendation for the definition of nanomaterials with size as a critical factor (1–100 nm) but acknowledging that the upper limit on this size range may not be scientifically relevant for all nanomaterials.[24] The EMA's framework for nanomedicine highlights the following salient objectives that have been emphasized to aid in the development nanomedicines:

- Evaluation of nanomedicines based on benefit/risk analysis instead of solely relying on technology;
- Formation of academia and regulators expert group with multidisciplinary expertise;
- Close cooperation between scientific committees, networks and the EC;
- International cooperation with other regulatory agencies;
- Transparent dialogues through meetings and workshops.

The Committee for medicinal products for human use (CHMP) published a series of four reflection papers (similar to guidance documents from the FDA) on data requirements for intravenous iron-based nano-colloidal products, intravenous liposomal products, development of block-copolymer-micelle medicinal products and general issues for consideration regarding parenteral administration of coated nanomedicine products.[25] These reflection papers provide a developmental guideline for new nanomedicines as well as nanosimilars.[26] One such reflection paper published by the EMA in 2006 focused on "*Nanotechnology-based medicinal products for human use*" and states that "nanomedicinal products may exhibit a complex mechanism of action combining mechanical, chemical, pharmacological and immunological properties," and "additional specialized expertise may be required for the evaluation of the quality, safety, efficacy and risk management of such nanomedicinal products".[27] The regulatory agencies acknowledge that despite the long-standing presence of nanomaterials in drug product and their established safety and efficacy in previously approved drug products, there is a need for a better understanding of their functional aspect, more harmonized regulations for nomenclature and terminology and robust CQAs and analytical standards.

The United States government has made attempts to develop a common platform through the National Nanotechnology Initiative (NNI) launched in 2000 to increase dialogue and inter-agency collaboration among nearly 25 federal agencies involved in scientific and technical research and development of nanoscale materials. A total of nearly $19.4 billion has been invested in funding between the financial years 2001 to 2014 that is disbursed to different agencies including the National Institute of Health (NIH), National Science Foundation (NSF), Department of Defense (DoD), Department of Energy (DoE) and National Institute for Standards and Technology (NIST). The primary aim of the initiative has been directed towards the further understanding and advancement of nanomaterials. NNI has established four goals: (1) advance a world-class nanotechnology research program; (2) promote the transfer of new technologies into commercial products; (3) develop and sustain educational resources, a skilled workforce, infrastructure and toolset for advancing nanotechnology and (4) support responsible development of nanotechnology.

Our ability to characterize and understand the nanomaterial itself and its interaction with the other components of the drug product is central to nanomaterial containing drug development. The National Cancer Institute (NCI) in collaboration with the FDA and NIST founded the Nanotechnology Characterization Lab (NCL) in 2004 as a public–private partnership to accelerate the development of nanomedicines. The NCL works in close collaboration with academic, industry and other regulatory agencies locally and globally to understand the evolving trends in nanomedicine and develop characterization methods to further the research on nanomedicines. The NCL's mission is to perform and standardize preclinical characterization of nanomedicines for therapeutic and diagnostic application. The core objective of the NCL revolves around four main objectives:

(1) Characterize nanoparticles using standardized methods: Physicochemical, *in vitro* and *in vivo* characterization;
(2) Conduct structure activity relationship studies;
(3) Facilitate regulatory review of nanotech constructs by rigorous and thorough analysis of nanomaterials intended for clinical application and;
(4) Engage in educational and knowledge sharing efforts.

One of the initial aims of the NCL has been to develop an "*assay cascade*" of scientific tests to characterize nanomaterials for safety, efficacy and reproducibility. Since its inception, the NCL has tested more than 350 nanomaterials and developed a knowledge bank on the characterization of nanoformulations for their physicochemical and biological properties. Their experience with a variety of nanomaterials including polymers, liposomes, protein, micelles and metallic nanoparticles has led to the development of a battery of assays that is shared with the research community.[28] Their current pool of assays include tests for checking the sterility and endotoxin levels in nanoformulations, physicochemical characterization (size, size distribution, surface charge measurement and composition analysis), *in vitro* characterization for a host of biological properties (interaction of nanomaterials with blood components and immunological properties), toxicity analysis (cytotoxicity, oxidative stress studies, autophagy and apoptosis analysis), pharmacological analysis (drug release) and *in vivo* analysis for pharmacology and toxicity, efficacy and immunotoxicity.

The European Nano-Characterization Lab (EU-NCL) was established in July 2015 with funding support from the European Union's horizon 2020 research and innovation program. The EU-NCL has a mission and objectives similar to the NCL and is moving towards developing a work class nanomaterial characterization platform to support the development of nanomedicines in European nations. The EU-NCL works in partnership with 9 agencies globally including the FDA, NCL and EMA. The collaborative efforts from all these government agencies has led to significant advancement in the identification of avenues for improvement in nanomaterial-based product development. Industry as well as government agencies have realized a need for harmonized regulations to review drug products containing nanomaterials. The current challenges that need immediate attention include precise nomenclature and definitions for nanomaterials, the need for appropriate analytical standards for nanomaterial characterization and well-streamlined regulatory guidelines for the development of nanomedicines. The complex nature of the nanomaterials and the existing diversity in the type of nanomaterials being studied, suggests that a "one-size fit all" approach will not be suitable to tackle these challenges. The benefit/risk approach seems appropriate for the approval process, but more transparency in the "*desired*" characteristics of nanomaterials is necessary. International dialogues and knowledge sharing through conferences like the "Global Summit on Regulatory Science (GSRS)" are steps in the right direction to open up the discussion on nanomedicine globally and work collectively to address the challenges in nanomedicine development.

References

1. V. Weissig, T. K. Pettinger and N. Murdock, Nanopharmaceuticals (part 1): products on the market, *Int. J. Nanomed.*, 2014, **9**, 4357–4373, PubMed PMID: 25258527. Pubmed Central PMCID: 4172146.
2. A. M. Nystrom and B. Fadeel, Safety assessment of nanomaterials: implications for nanomedicine, *J. Controlled Release*, 2012, **161**(2), 403–408, PubMed PMID: 22306428.
3. E. Blanco, H. Shen and M. Ferrari, Principles of nanoparticle design for overcoming biological barriers to drug delivery, *Nat. Biotechnol.*, 2015, **33**(9), 941–951, PubMed PMID: 26348965. Pubmed Central PMCID: 4978509.

4. The complement system and innate immunity, in *Immunobiology: The Immune System in Health and Disease*, ed. C. Janeway, Garland Science, New York, 5th edn, 2001.
5. N. Desai, Challenges in development of nanoparticle-based therapeutics, *AAPS J.*, 2012, **14**(2), 282–295, PubMed PMID: 22407288. Pubmed Central PMCID: 3326161.
6. Y. Qie, H. Yuan, C. A. von Roemeling, Y. Chen, X. Liu and K. D. Shih, *et al.*, Surface modification of nanoparticles enables selective evasion of phagocytic clearance by distinct macrophage phenotypes, *Sci. Rep.*, 2016, **6**, 26269, PubMed PMID: 27197045. Pubmed Central PMCID: 4872535.
7. R. Toy and K. Roy, Engineering nanoparticles to overcome barriers to immunotherapy, *Bioeng. Transl. Med.*, 2016, **1**(1), 47–62.
8. M. A. Dobrovolskaia, B. W. Neun, J. D. Clogston, J. H. Grossman and S. E. McNeil, Choice of method for endotoxin detection depends on nanoformulation, *Nanomedicine*, 2014, **9**(12), 1847–1856, PubMed PMID: 24359551.
9. M. A. Vetten, C. S. Yah, T. Singh and M. Gulumian, Challenges facing sterilization and depyrogenation of nanoparticles: effects on structural stability and biomedical applications, *Nanomedicine*, 2014, **10**(7), 1391–1399, PubMed PMID: 24709329.
10. S. T. Stern, M. N. Martinez and D. M. Stevens, When Is It Important to Measure Unbound Drug in Evaluating Nanomedicine Pharmacokinetics?, *Drug Metab. Dispos.*, 2016, **44**(12), 1934–1939, PubMed PMID: 27670412. Pubmed Central PMCID: 5118636.
11. S. D'Souza, A review of *in vitro* drug release test methods for nano-sized dosage forms, *Adv. Pharm.*, 2014, **2014**, 12.
12. M. A. Dobrovolskaia and S. E. McNeil, Immunological properties of engineered nanomaterials, *Nat. Nanotechnol.*, 2007, **2**(8), 469–478, PubMed PMID: 18654343.
13. M. T. Peracchia, S. Harnisch, H. Pinto-Alphandary, A. Gulik, J. C. Dedieu and D. Desmaele, *et al.*, Visualization of *in vitro* protein-rejecting properties of PEGylated stealth polycyanoacrylate nanoparticles, *Biomaterials*, 1999, **20**(14), 1269–1275, PubMed PMID: 10403044.
14. M. Elsabahy, S. Zhang, F. Zhang, Z. J. Deng, Y. H. Lim and H. Wang, *et al.* Surface charges and shell crosslinks each play significant roles in mediating degradation, biofouling, cytotoxicity and immunotoxicity for polyphosphoester-based nanoparticles, *Sci. Rep.*, 2013, **3**, 3313, PubMed PMID: 24264796. Pubmed Central PMCID: 3837308.
15. M. Elsabahy and K. L. Wooley, Cytokines as biomarkers of nanoparticle immunotoxicity, *Chem. Soc. Rev.*, 2013, **42**(12), 5552–5576, PubMed PMID: 23549679. Pubmed Central PMCID: 3665642.
16. T. Morishige, Y. Yoshioka, H. Inakura, A. Tanabe, S. Narimatsu and X. Yao, *et al.*, Suppression of nanosilica particle-induced inflammation by surface modification of the particles, *Arch. Toxicol.*, 2012, **86**(8), 1297–1307, PubMed PMID: 22418595.

17. R. Shrestha, M. Elsabahy, S. Florez-Malaver, S. Samarajeewa and K. L. Wooley, Endosomal escape and siRNA delivery with cationic shell cross-linked knedel-like nanoparticles with tunable buffering capacities, *Biomaterials*, 2012, **33**(33), 8557–8568, PubMed PMID: 22901966. Pubmed Central PMCID: 3582407.
18. S. R. D'Mello, C. N. Cruz, M. L. Chen, M. Kapoor, S. L. Lee and K. M. Tyner, The evolving landscape of drug products containing nanomaterials in the United States, *Nat. Nanotechnol.*, 2017, **12**(6), 523–529, PubMed PMID: 28436961.
19. FDA, *FDA's Approach to Regulation of Nanotechnology Products*, [accessed 24 July 2017], available from: http://www.fda.gov/ScienceResearch/SpecialTopics/Nanotechnology/ucm301114.htm.
20. FDA, *Considering whether an FDA-regulated Product Involves the Application of Nanotechnology*, 2014, [cited 2017 23 July], available from: http://www.fda.gov/RegulatoryInformation/Guidances/ucm257698.htm.
21. K. M. Tyner, P. Zou, X. Yang, H. Zhang, C. N. Cruz and S. L. Lee, Product quality for nanomaterials: current U.S. experience and perspective, *Wiley Interdiscip. Rev.: Nanomed. Nanobiotechnol.*, 2015, **7**(5), 640–654, PubMed PMID: 25641690.
22. FDA, *Guidances (Drugs)*, [accessed 24 July 2017], available from: http://www.fda.gov/Drugs/GuidanceComplianceRegulatoryInformation/Guidances/default.htm.
23. FDA, *Liposomal Drug Products*, 2015, [cited 2017 24 July], available from: http://www.fda.gov/downloads/drugs/guidances/ucm070570.pdf.
24. EC, *Commission Recommendation of 18 October 2011 on the Definition of Nanomaterial 2011*, 2011, [accessed 24 July 2017], available from: eur-lex.europa.eu/LexUriServ/LexUriServ.do?uri=OJ:L:2011:275:0038:0040:EN:PDF.
25. EMA, *Multidisciplinary: Nanomedicines*, [accessed 24 July 2017], available from: http://www.ema.europa.eu/ema/index.jsp?curl=pages/regulation/general/general_content_000564.jsp&mid=WC0b01ac05806403e0.
26. F. Ehmann, K. Sakai-Kato, R. Duncan, D. Hernan Perez de la Ossa, R. Pita and J. M. Vidal, *et al.*, Next-generation nanomedicines and nanosimilars: EU regulators' initiatives relating to the development and evaluation of nanomedicines, *Nanomedicine*, 2013, **8**(5), 849–856, PubMed PMID: 23656268.
27. EMA, *Reflection Paper on Nanotechnology-based Medicinal Products for Human Use*, 2006, [cited 2017 24 July], available from: http://www.ema.europa.eu/docs/en_GB/document_library/Regulatory_and_procedural_guideline/2010/01/WC500069728.pdf.
28. NCL, [cited 2017 26 July], available from: ncl.cancer.gov/resources/assay-cascade-protocols#.

Subject Index